Selected Titles in This Series

14 Elliott H. Lieb and Michael Loss, Analysis, 1997
13 Paul C. Shields, The ergodic theory of discrete sample paths, 1996
12 N. V. Krylov, Lectures on elliptic and parabolic equations in Hölder spaces, 1996
11 Jacques Dixmier, Enveloping algebras, 1996 Printing
10 Barry Simon, Representations of finite and compact groups, 1996
9   Dino Lorenzini, An invitation to arithmetic geometry, 1996
8   Winfried Just and Martin Weese, Discovering modern set theory. I: The basics, 1996
7   Gerald J. Janusz, Algebraic number fields, second edition, 1996
6   Jens Carsten Jantzen, Lectures on quantum groups, 1996
5   Rick Miranda, Algebraic curves and Riemann surfaces, 1995
4   Russell A. Gordon, The integrals of Lebesgue, Denjoy, Perron, and Henstock, 1994
3   William W. Adams and Philippe Loustaunau, An introduction to Gröbner bases, 1994
2   Jack Graver, Brigitte Servatius, and Herman Servatius, Combinatorial rigidity, 1993
1   Ethan Akin, The general topology of dynamical systems, 1993

ANALYSIS

ANALYSIS

Elliott H. Lieb
Princeton University

Michael Loss
Georgia Institute of Technology

Graduate Studies
in Mathematics

Volume 14

American Mathematical Society

Editorial Board

Lance Small (Chair)
James E. Humphreys
Julius L. Shaneson
David Sattinger

1991 *Mathematics Subject Classification.* Primary 28-01, 42-01, 46-01, 49-01; Secondary 26D10, 26D15, 31B05, 31B15, 46E35, 46F05, 46F10, 49R05, 81Q05.

ABSTRACT. This book is a course in real analysis that begins with the usual measure theory yet brings the reader quickly to a level where a wider than usual range of topics can be appreciated, including some recent research. The reader is presumed to know only basic facts learned in a good course in calculus. Topics covered include L^p-spaces, rearrangement inequalities, sharp integral inequalities, distribution theory, Fourier analysis, potential theory and Sobolev spaces. To illustrate the topics, the book concludes with a chapter on the calculus of variations, with examples from mathematical physics.

The book will be of interest to beginning graduate students of mathematics, as well as to students of the natural sciences and engineering who want to learn some of the important tools of real analysis.

Library of Congress Cataloging-in-Publication Data
Lieb, Elliott H.
  Analysis / Elliott H. Lieb, Michael Loss.
    p. cm. — (Graduate studies in mathematics, ISSN 1065-7339; v. 14)
  Includes bibliographical references and index.
  ISBN 0-8218-0632-7 (alk. paper)
  1. Mathematical analysis. I. Loss, Michael, 1954- .
II. Title.   III. Series.
QA300.L54 1996
515–dc20   96-31605
                                                                         CIP

Copying and reprinting. Individual readers of this publication, and nonprofit libraries acting for them, are permitted to make fair use of the material, such as to copy a chapter for use in teaching or research. Permission is granted to quote brief passages from this publication in reviews, provided the customary acknowledgment of the source is given.

Republication, systematic copying, or multiple reproduction of any material in this publication (including abstracts) is permitted only under license from the American Mathematical Society. Requests for such permission should be addressed to the Assistant to the Publisher, American Mathematical Society, P. O. Box 6248, Providence, Rhode Island 02940-6248. Requests can also be made by e-mail to reprint-permission@ams.org.

© 1997 by the authors.
Printed in the United States of America.

∞ The paper used in this book is acid-free and falls within the guidelines established to ensure permanence and durability.

10 9 8 7 6 5 4 3 2 1   02 01 00 99 98 97

To

Christiane and Ute

Contents

Preface	xv
CHAPTER 1. Measure and Integration	1
1.1 Introduction	1
1.2 Basic notions of measure theory	4
1.3 Monotone class theorem	9
1.4 Uniqueness of measures	11
1.5 Definition of measurable functions and integrals	12
1.6 Monotone convergence	17
1.7 Fatou's lemma	18
1.8 Dominated convergence	19
1.9 Missing term in Fatou's lemma	21
1.10 Product measure	23
1.11 Commutativity and associativity of product measures	24
1.12 Fubini's theorem	25
1.13 Layer cake representation	26
1.14 Bathtub principle	28
1.15 Constructing a measure from an outer measure	29
Exercises	32
CHAPTER 2. L^p-Spaces	35
2.1 Definition of L^p-spaces	35

2.2	Jensen's inequality	38
2.3	Hölder's inequality	39
2.4	Minkowski's inequality	41
2.5	Hanner's inequality	43
2.6	Differentiability of norms	44
2.7	Completeness of L^p-spaces	45
2.8	Projection on convex sets	47
2.9	Continuous linear functionals and weak convergence	48
2.10	Linear functionals separate	50
2.11	Lower semicontinuity of norms	51
2.12	Uniform boundedness principle	52
2.13	Strongly convergent convex combinations	54
2.14	The dual of $L^p(\Omega)$	55
2.15	Convolution	57
2.16	Approximation by C^∞-functions	58
2.17	Separability of $L^p(\mathbb{R}^n)$	61
2.18	Bounded sequences have weak limits	62
2.19	Approximation by C_c^∞-functions	63
2.20	Convolutions of functions in dual $L^p(\mathbb{R}^n)$-spaces are continuous	64
2.21	Hilbert-spaces	64
Exercises		69

CHAPTER 3. Rearrangement Inequalities — 71

3.1	Introduction	71
3.2	Definition of functions vanishing at infinity	72
3.3	Rearrangements of sets and functions	72
3.4	The simplest rearrangement inequality	74
3.5	Nonexpansivity of rearrangement	75
3.6	Riesz's rearrangement inequality in one-dimension	76
3.7	Riesz's rearrangement inequality	79
3.8	General rearrangement inequality	85
3.9	Strict rearrangement inequality	85
Exercises		87

CHAPTER 4. Integral Inequalities — 89

    4.1 Introduction — 89

    4.2 Young's inequality — 90

    4.3 Hardy–Littlewood–Sobolev inequality — 98

    4.4 Conformal transformations and stereographic projection — 102

    4.5 Conformal invariance of the Hardy–Littlewood–Sobolev inequality — 106

    4.6 Competing symmetries — 109

    4.7 Proof of Theorem 4.3: Sharp version of the Hardy–Littlewood–Sobolev inequality — 111

    4.8 Action of the conformal group on optimizers — 112

Exercises — 113

CHAPTER 5. The Fourier Transform — 115

    5.1 Definition of the L^1 Fourier transform — 115

    5.2 Fourier transform of a Gaussian — 117

    5.3 Plancherel's theorem — 118

    5.4 Definition of the L^2 Fourier transform — 119

    5.5 Inversion formula — 120

    5.6 The Fourier transform in $L^p(\mathbb{R}^n)$ — 120

    5.7 The sharp Hausdorff–Young inequality — 121

    5.8 Convolutions — 122

    5.9 Fourier transform of $|x|^{\alpha-n}$ — 122

    5.10 Extension of 5.9 to $L^p(\mathbb{R}^n)$ — 123

Exercises — 125

CHAPTER 6. Distributions — 127

    6.1 Introduction — 127

    6.2 Test functions (The space $\mathcal{D}(\Omega)$) — 128

    6.3 Definition of distributions and their convergence — 128

    6.4 Locally summable functions, $L^p_{\mathrm{loc}}(\Omega)$ — 129

    6.5 Functions are uniquely determined by distributions — 130

    6.6 Derivatives of distributions — 131

    6.7 Definition of $W^{1,p}_{\mathrm{loc}}(\Omega)$ and $W^{1,p}(\Omega)$ — 132

6.8	Interchanging convolutions with distributions	134
6.9	Fundamental theorem of calculus for distributions	135
6.10	Equivalence of classical and distributional derivatives	136
6.11	Distributions with zero derivatives are constants	138
6.12	Multiplication and convolution of distributions by C^∞-functions	138
6.13	Approximation of distributions by C^∞-functions	139
6.14	Linear dependence of distributions	140
6.15	$C^\infty(\Omega)$ is 'dense' in $W^{1,p}_{\mathrm{loc}}(\Omega)$	141
6.16	Chain rule	142
6.17	Derivative of the absolute value	144
6.18	Min and Max of $W^{1,p}$-functions are in $W^{1,p}$	145
6.19	Gradients vanish on the inverse of small sets	146
6.20	Distributional Laplacian of Green's functions	148
6.21	Solution of Poisson's equation	149
6.22	Positive distributions are measures	151

Exercises   155

CHAPTER 7. The Sobolev Spaces H^1 and $H^{1/2}$ 157

7.1	Introduction	157
7.2	Definition of $H^1(\Omega)$	157
7.3	Completeness of $H^1(\Omega)$	158
7.4	Multiplication by functions in $C^\infty(\Omega)$	159
7.5	Remark about $H^1(\Omega)$ and $W^{1,2}(\Omega)$	160
7.6	Density of $C^\infty(\Omega)$ in $H^1(\Omega)$	160
7.7	Partial integration for functions in $H^1(\mathbb{R}^n)$	161
7.8	Convexity inequality for gradients	163
7.9	Fourier characterization of $H^1(\mathbb{R}^n)$	165
7.10	$-\Delta$ is the infinitesimal generator of the heat kernel	167
7.11	Definition of $H^{1/2}(\mathbb{R}^n)$	167
7.12	Integral formulas for $(f,|p|f)$ and $(f,\sqrt{p^2+m^2}\,f)$	170
7.13	Convexity inequality for the relativistic kinetic energy	171
7.14	Density of $C^\infty_c(\mathbb{R}^n)$ in $H^{1/2}(\mathbb{R}^n)$	171

7.15	Action of $\sqrt{-\Delta}$ and $\sqrt{-\Delta + m^2} - m$ on distributions	172
7.16	Multiplication of $H^{1/2}$ functions by C^∞-functions	173
7.17	Symmetric decreasing rearrangement decreases kinetic energy	174
7.18	Weak limits	176
7.19	Magnetic fields: The H_A^1-spaces	177
7.20	Definition of $H_A^1(\mathbb{R}^n)$	178
7.21	Diamagnetic inequality	179
7.22	$C_c^\infty(\mathbb{R}^n)$ is dense in $H_A^1(\mathbb{R}^n)$	180

Exercises 181

CHAPTER 8. Sobolev Inequalities 183

8.1	Introduction	183
8.2	Definition of $D^1(\mathbb{R}^n)$ and $D^{1/2}(\mathbb{R}^n)$	185
8.3	Sobolev's inequality for gradients	186
8.4	Sobolev's inequality for $|p|$	188
8.5	Sobolev inequalities in 1 and 2 dimensions	189
8.6	Weak convergence implies strong convergence on small sets	192
8.7	Weak convergence implies a.e. convergence	196
8.8	Sobolev inequalities for $W^{m,p}(\Omega)$	197
8.9	Rellich–Kondrashov theorem	198

Exercises 199

CHAPTER 9. Potential Theory and Coulomb Energies 201

9.1	Introduction	201
9.2	Definition of harmonic, subharmonic, and superharmonic functions	202
9.3	Properties of harmonic, subharmonic, and superharmonic functions	203
9.4	The strong maximum principle	207
9.5	Harnack's inequality	209
9.6	Subharmonic functions are potentials	210
9.7	Spherical charge distributions are 'equivalent' to point charges	212
9.8	Positivity properties of the Coulomb energy	214

	9.9 Mean value inequality for $\Delta - \mu^2$	215
	9.10 Lower bounds on Schrödinger 'wave' functions	217

Exercises 219

CHAPTER 10. Regularity of Solutions of Poisson's Equation 221

    10.1 Introduction 221
    10.2 Continuity and first differentiability of solutions of Poisson's equation 224
    10.3 Higher differentiability of solutions of Poisson's equation 226

CHAPTER 11. Introduction to the Calculus of Variations 231

    11.1 Introduction 231
    11.2 Schrödinger's equation 233
    11.3 Domination of the potential energy by the kinetic energy 234
    11.4 Weak continuity of the potential energy 238
    11.5 Existence of a minimizer for E_0 239
    11.6 Higher eigenvalues and eigenfunctions 242
    11.7 Regularity of solutions 243
    11.8 Uniqueness of minimizers 244
    11.9 Uniqueness of positive solutions 245
    11.10 The hydrogen atom 246
    11.11 The Thomas–Fermi problem 248
    11.12 Existence of an unconstrained Thomas–Fermi minimizer 249
    11.13 Thomas–Fermi equation 250
    11.14 The Thomas–Fermi minimizer 251
    11.15 The capacitor problem 253
    11.16 Solution of the capacitor problem 257
    11.17 Balls have smallest capacity 260

Exercises 261

List of Symbols 265

References 269

Index 273

Preface

A glance at the table of contents will reveal the somewhat unconventional nature of this introductory book on analysis, so perhaps we should explain our philosophy and motivation for writing a book that has elementary integration theory together with potential theory, rearrangements, regularity estimates for differential equations and the calculus of variations all sandwiched between the same covers.

Originally, we were motivated to present the essentials of modern analysis to physicists and other natural scientists, so that some modern developments in quantum mechanics, for example, would be understandable. From personal experience we realized that this task is little different from the task of explaining analysis to students of mathematics. At the present time there are many excellent texts available, but they mostly emphasize concepts in themselves rather than their useful relation to other parts of mathematics. It is a question of taste, but there are many students (and teachers) who, in the limited time available, prefer to go through a subject by doing something with the material, as it is learned, rather than wait for a full-fledged development of all basic principles.

The topics covered here are selected from those we have found useful in our own research and are among those that practicing analysts need in their kit-bag, such as basic facts about measure theory and integration, Fourier transforms, commonly used function spaces (including Sobolev spaces), distribution theory, etc. Our goal was to guide beginning students through these topics with a minimum of fuss and to lead them to the point where they can read current literature with some understanding. At the same time everything is done in a rigorous and, hopefully, pedagogical way.

Inequalities play a key role in our presentation and some of them are less standard, such as the Hardy–Littlewood–Sobolev inequality, Hanner's inequality and rearrangement inequalities. These and other unusual topics, such as $H^{1/2}$- and H^1_A-spaces, are included for a definite pedagogical reason: They introduce the student to some serious exercises in hard analysis (i.e., interesting theorems that take more than a few lines to prove), but ones that can be tackled with the elementary tools presented here. In this way we hope that relative beginners can get some of the flavor of research mathematics and the feeling that the subject is open-ended.

Throughout, our approach is 'hands on', meaning that we try to be as direct as possible and do not always strive for the most general formulation. Occasionally we have slick proofs, but we avoid unnecessary abstraction, such as the use of the Baire category theorem or the Hahn–Banach theorem, which are not needed for L^p-spaces. Our preference is to understand L^p-spaces and then have the reader go elsewhere to study Banach spaces generally (for which excellent texts abound), rather than the other way around. Another noteworthy point is that we try not to say, "there exists a constant such that ...". We usually give it, or at least an estimate of it. It is important for students of the natural sciences, *and* mathematics, to learn how to calculate. Nowadays, this is often overlooked in mathematics courses that usually emphasize pure existence theorems.

From some points of view, the topics included here are a curious mixture of the advanced-specialized together with the elementary but the reader will, we believe, see that there is a unity to it all. For example, most texts make a big distinction between 'real analysis' and 'functional analysis', but we regard this distinction as somewhat artificial. Analysis without functions doesn't go very far. On the other hand, Hilbert-space is hardly mentioned, which might seem strange in a book in which many of the examples are taken from quantum mechanics. This theory (beyond the linear algebra level) becomes truly interesting when combined with operator theory, and these topics are not treated here because they are covered in many excellent texts. Perhaps the severest rearrangement of the conventional order is in our treatment of Lebesgue integration. In Chapter 1 we introduce what is needed to understand and use integration, but we do not bother with the proof of the existence of Lebesgue measure; it suffices to *know* its existence. Finally, after the reader has acquired some sophistication, the proof is given in Exercise 6.5 as a corollary of Theorem 6.22 (positive distributions are measures).

Things the reader is expected to know: While we more or less start from 'scratch', we do expect the reader to know some elementary facts, all of which will have been learned in a good calculus course. These include:

vector spaces, limits, lim inf, lim sup, open, closed and compact sets in $\mathbb{R}^n$, continuity and differentiability of functions (especially in the multivariable case), convergence and uniform convergence (indeed, the notion of 'uniform', generally), the definition and basic properties of the Riemann integral, integration by parts (of which Gauss's theorem is a special case).

How to read this book: There is a great deal of material here but the following selection hits the main points. It is possible to cover them conveniently in a year's course of 25 weeks.

CHAPTER 1. The basic facts of integration can be gleaned from 1.1, 1.2, 1.5–1.8, 1.10, 1.12 (the statement only), 1.13.

CHAPTER 2. The essential facts about L^p-spaces are in 2.1–2.4, 2.7, 2.9, 2.10, 2.14–2.19.

CHAPTER 3. 3.3, 3.4, 3.7 are enough for a first reading about rearrangements. This serves as a useful exercise in manipulating integrals.

CHAPTER 4. Read the nonsharp proofs of Young's inequality, 4.2, and the HLS inequality, 4.3.

CHAPTER 5. Fourier transforms are basic in many applications. Read 5.1–5.8.

CHAPTER 6. 6.1–6.18, 6.20, 6.21, 6.22 (statement only).

CHAPTER 7. 7.1–7.10, 7.17, 7.18. $H^{1/2}$ spaces and H^1_A spaces are specialized examples, useful in quantum mechanics, and can be ignored at first.

CHAPTER 8. All except 8.4. Sobolev inequalities are essential for partial differential equations and it is necessary to be familiar with their statements, if not their proofs.

CHAPTER 9. Potential theory is classical and basic to physics and mathematics. 9.1–9.5, 9.7, 9.8 are the most important. 9.10 is a useful extension of Harnack's inequality and is worth studying.

CHAPTER 10. It is important to know how to go from weak to strong solutions of partial differential equations. 10.1 and the statements of 10.2, 10.3, if not the proofs, should be learned.

CHAPTER 11. The calculus of variations, especially as a key to solving some differential equations, is extremely useful and important. All the examples given here, 11.1–11.17 are worth learning, not only for their intrinsic value, but because they use many of the topics presented earlier in the book.

A word about notation. The book is organized around theorems, but frequently there are some pertinent remarks before and after the statement of a theorem. The symbol ● is used to denote the introduction of a new idea or discussion, while ■ is used for the end of a proof. Equations are numbered separately in each section. The notation 1.6(2), for example, means equation number (2) in Section 1.6. Exercise 1.15, for example means exercise number 15 in Chapter 1. To avoid unnecessary enumeration, (2) means equation number (2) of the section we are presently in; similarly, Exercise 15 refers to Exercise 15 of the present chapter. Bold-face is used whenever a bit of terminology appears for the first time.

According to Walter Thirring there are three things that are easy to start but very difficult to finish. The first is a war. The second is a love affair. The third is a trill. To this may be added a fourth: a book. Many students and colleagues helped over the years to put us on the right track on several topics and helped us eliminate some of the more egregious errors and turgidities. Our thanks go to Almut Burchard, Eric Carlen, E. Brian Davies, Evans Harrell, David Jerison, Richard Laugesen, Bruno Nachtergaele, Barry Simon, Avraham Soffer, Bernd Thaller, Kenji Yajima, our students at Georgia Tech and Princeton, several anonymous referees and to Lorraine Nelson for typing most of the manuscript.

Chapter 1

Measure and Integration

1.1 INTRODUCTION

The most important analytic tool used in this book is integration. The student of analysis meets this concept in a calculus course where an integral is defined as a Riemann integral. While this point of view of integration may be historically grounded and useful in many areas of mathematics, it is far from being adequate for the requirements of modern analysis. The difficulty with the Riemann integral is that it can be defined only for a special class of functions and this class is not closed under the process of taking pointwise limits of sequences (not even monotonic sequences) of functions in this class. Analysis, it has been said, is the art of taking limits, and the constraint of having to deal with an integration theory that does not allow taking limits is much like having to do mathematics only with rational numbers and excluding the irrational ones.

If we think of the graph of a real-valued function of n variables, the integral of the function is supposed to be the $(n+1)$-dimensional volume under the graph. The question is how to define this volume. The Riemann integral attempts to define it as 'base times height' for small, predetermined n-dimensional cubes as bases, with the height being some 'typical' value of the function as the variables range over that cube. The difficulty is that it may be impossible to define this height properly if the function is sufficiently discontinuous.

The useful and far-reaching idea of Lebesgue and others was to compute the $(n+1)$-dimensional volume 'in the other direction' by first computing the n-dimensional volume of the set where the function is greater than some number y. This volume is a well-behaved, monotone nonincreasing function of the number y, which then can be integrated in the manner of Riemann.

This method of integration not only works for a large class of functions (which is closed under taking pointwise limits), but it also greatly simplifies a problem that used to plague analysts: Is it permissible to exchange limits and integration?

In this chapter we shall first sketch in the briefest possible way the ideas about measure that are needed in order to define integrals. Then we shall prove the most important convergence theorems which permit us to interchange limits and integration. Many measure-theoretic details are not given here because the subject is lengthy and complicated and is presented in any number of texts, e.g. [Rudin, 1987]. The most important reason for omitting the measure theory is that the intricacies of its development are not needed for its exploitation. For instance, we all know the tremendously important fact that

$$\int (f+g) = \left(\int f\right) + \left(\int g\right),$$

and we can use it happily without remembering the proof (which actually does require some thought); the interested reader can carry out the proof, however, in Exercise 9. Nevertheless we want to emphasize that this theory is one of the great triumphs of twentieth century mathematics and it is the culmination of a long struggle to find the right perspective from which to view integration theory. We recommend its study to the reader because it is the foundation on which this book ultimately rests.

Before dealing with integration, let us review some elementary facts and notation that will be needed. The real numbers are denoted by $\mathbb{R}$, while the complex numbers are denoted by $\mathbb{C}$ and $\bar{z}$ is the complex conjugate of z. It will be assumed that the reader is equipped with a knowledge of the fundamentals of the calculus on **n-dimensional Euclidean space**

$$\mathbb{R}^n = \{(x_1, \ldots, x_n) : \text{each } x_i \text{ is in } \mathbb{R}\}.$$

The **Euclidean distance** between two points y and z in $\mathbb{R}^n$ is defined to be $|y - z|$ where, for $x \in \mathbb{R}^n$,

$$|x| := \left(\sum_{i=1}^{n} x_i^2\right)^{1/2}.$$

(The symbols $a := b$ and $b =: a$ mean that a is defined by b.) We expect the reader to know some elementary inequalities such as the **triangle inequality**,

$$|x| + |y| \geq |x - y|.$$

The definition of **open sets** (a set, each of whose points is at the center of some ball contained in the set), **closed sets** (the complement of an open set), **compact sets** (closed and bounded subsets of $\mathbb{R}^n$), limits, the Riemann integral and differentiable functions are among the concepts we assume known. $[a,b]$ denotes the **closed interval** in $\mathbb{R}$, $a \leq x \leq b$, while (a,b) denotes the **open interval** $a < x < b$. The notation $\{a : b\}$ means, of course, the set of all things of type a that satisfy condition b. We introduce here the useful notation

$$C^k(\Omega)$$

to describe the complex-valued functions on some open set $\Omega \in \mathbb{R}^n$ that are k times continuously differentiable (i.e., the partial derivatives $\partial^k f/\partial x_{i_1}, \ldots, \partial x_{i_k}$ exist at all points $x \in \Omega$ and are continuous functions on Ω). If a function f is in $C^k(\Omega)$ for all k, then we write $f \in C^\infty(\Omega)$.

In general, if f is a function from some set A (e.g., some subset of $\mathbb{R}^n$) with values in some set B (e.g., the real numbers), we denote this fact by $f : A \to B$. If $x \in A$, we write $x \mapsto f(x)$, the bar on the arrow serving to distinguish the image of a single point x from the image of the whole set A.

An important class of functions consists of the **characteristic functions** of sets. If A is a set we define

$$\chi_A(x) = \begin{cases} 1 & \text{if } x \in A, \\ 0 & \text{if } x \notin A. \end{cases} \tag{1}$$

These will serve as building blocks for more general functions (see Sect. 1.13, Layer cake representation). Note that $\chi_A \chi_B = \chi_{A \cap B}$.

Recall that the **closure of a set** $A \subset \mathbb{R}^n$ is the smallest closed set in $\mathbb{R}^n$ that contains A. We denote the closure by $\overline{A}$. Thus, $\overline{\overline{A}} = \overline{A}$. The **support of a continuous function** $f : \mathbb{R}^n \to \mathbb{C}$, denoted by $\text{supp}\{f\}$, is the *closure* of the set of points $x \in \mathbb{R}^n$ where $f(x)$ is nonzero, i.e.,

$$\text{supp}\{f\} = \overline{\{x \in \mathbb{R}^n : f(x) \neq 0\}}.$$

It is important to keep in mind that the above definition is a topological notion. Later, in Sect. 1.5, we shall give a definition of *essential* support for measurable functions. We denote the set of functions in $C^\infty(\Omega)$ whose support is bounded and contained in Ω by $C_c^\infty(\Omega)$. The subscript c stands for 'compact' since a set is closed and bounded if and only if it is compact.

Here is a classic example of a compactly supported, infinitely differentiable function on $\mathbb{R}^n$; its support is the unit ball $\{x \in \mathbb{R}^n : |x| \leq 1\}$:

$$j(x) = \begin{cases} \exp\left[-\frac{1}{1-|x|^2}\right] & \text{if } |x| < 1, \\ 0 & \text{if } |x| \geq 1. \end{cases} \tag{2}$$

The verification that j is actually in $C^\infty(\mathbb{R}^n)$ is left as an exercise.

This example can be used to prove a version of what is known as **Urysohn's lemma** in the $\mathbb{R}^n$ setting. Let $\Omega \subset \mathbb{R}^n$ be an open set and let $K \subset \Omega$ be a compact set. Then there exists a nonnegative function $\psi \in C_c^\infty(\Omega)$ with $\psi(x) = 1$ for $x \in K$. An outline of the proof is given in Exercise 15.

1.2 BASIC NOTIONS OF MEASURE THEORY

Before trying to define a measure of a set one must first study the structure of sets that are **measurable**, i.e., those sets for which it will prove to be possible to associate a numerical value in an unambiguous way. Not necessarily all sets will be measurable.

We begin, generally, with a set Ω whose elements are called **points**. For orientation one might think of Ω as a subset of $\mathbb{R}^n$, but it might be a much more general set than that, e.g., the set of paths in a path-space on which we are trying to define a 'functional integral'.

A distinguished collection, Σ, of subsets of Ω is called a **sigma-algebra** if the following axioms are satisfied:
 (i) If $A \in \Sigma$, then $A^c \in \Sigma$, where $A^c := \Omega \sim A$ is the **complement** of A in Ω. (Generally, $B \sim A := B \cap A^c$.)
 (ii) If $A_1, A_2, \ldots$ is a countable family of sets in Σ, then their union $\bigcup_{i=1}^\infty A_i$ is also in Σ.
 (iii) $\Omega \in \Sigma$.

Note that these assumptions imply that the empty set $\emptyset$ is in Σ and that Σ is also closed under countable intersections, i.e., if $A_1, A_2, \ldots \in \Sigma$, then $\bigcap_{i=1}^\infty A_i \in \Sigma$. Also, $A_1 \sim A_2$ is in Σ.

It is a trivial fact that *any* family $\mathcal{F}$ of subsets of Ω can be extended to a sigma-algebra (just take the sigma-algebra consisting of all subsets of Ω). Among all these extensions there is a special one. Consider all the sigma-algebras that contain $\mathcal{F}$ and take their intersection, which we call Σ, i.e., a subset $A \subset \Omega$ is in Σ if and only if A is in every sigma-algebra containing $\mathcal{F}$. It is easy to check that Σ is indeed a sigma-algebra. Indeed it is the **smallest sigma-algebra containing** $\mathcal{F}$; it is also called the **sigma-algebra generated by** $\mathcal{F}$. An important example is the sigma-algebra $\mathcal{B}$ of **Borel sets** of $\mathbb{R}^n$ which is generated by the open subsets of $\mathbb{R}^n$. Alternatively, it is generated by the **open balls** of $\mathbb{R}^n$, i.e., the family of sets of the form

$$B_{x,R} = \{y \in \mathbb{R}^n : |x - y| < R\}. \tag{1}$$

It is a fact that this Borel sigma-algebra contains the closed sets by (i) above. With the help of the axiom of choice one can prove that $\mathcal{B}$ does *not* contain *all* subsets of $\mathbb{R}^n$, but we emphasize that the reader does not need to know either this fact or the axiom of choice.

A **measure** (sometimes also called a **positive measure** for emphasis) μ, defined on a sigma-algebra Σ, is a function from Σ into the nonnegative real numbers (including infinity) such that $\mu(\emptyset) = 0$ and with the following crucial property of **countable additivity**. If $A_1, A_2, \ldots$ is a sequence of *disjoint sets* in Σ, then

$$\mu\left(\bigcup_{i=1}^{\infty} A_i\right) = \sum_{i=1}^{\infty} \mu(A_i). \tag{2}$$

The big breakthrough, historically, was the realization that countable additivity is an essential requirement. It is, and was, easy to construct finitely additive measures (i.e., where (2) holds with ∞ replaced by an arbitrary finite number), but a satisfactory theory of integration cannot be developed this way. Since $\mu(\emptyset) = 0$, equation (2) includes finite additivity as a special case. Three other important consequences of (2) are

$$\mu(A) \leq \mu(B) \qquad \text{if } A \subset B, \tag{3}$$

$$\lim_{j \to \infty} \mu(A_j) = \mu\left(\bigcup_{i=1}^{\infty} A_i\right) \qquad \text{if } A_1 \subset A_2 \subset A_3 \subset \cdots, \tag{4}$$

$$\lim_{j \to \infty} \mu(A_j) = \mu\left(\bigcap_{i=1}^{\infty} A_i\right) \qquad \text{if } A_1 \supset A_2 \supset \cdots \text{ and } \mu(A_1) < \infty. \tag{5}$$

The reader can easily prove (3)–(5) using the properties of a sigma-algebra.

A **measure space** thus has three parts: A set Ω, a sigma-algebra Σ and a measure μ. If $\Omega = \mathbb{R}^n$ (or, more generally, if Ω has open subsets, so that $\mathcal{B}$ can be defined) and if $\Sigma = \mathcal{B}$, then μ is said to be a **Borel measure**. We often refer to the elements of Σ as the **measurable sets**. Note that whenever Ω' is a measurable subset of Ω we can always define the measure subspace (Ω', Σ', μ), in which Σ' consists of the measurable subsets of Ω'. This is called the **restriction of μ to Ω'**.

A simple and important example in $\mathbb{R}^n$ is the **Dirac delta-measure**, δ_y, located at some arbitrary, but fixed, point $y \in \mathbb{R}^n$:

$$\delta_y(A) = \begin{cases} 1 & \text{if } y \in A, \\ 0 & \text{if } y \notin A. \end{cases} \tag{6}$$

In other words, using the definition of characteristic functions in 1.1(1),

$$\delta_y(A) = \chi_A(y). \tag{7}$$

Here, the sigma-algebra can be taken to be $\mathcal{B}$ or it can be taken to be *all* subsets of $\mathbb{R}^n$.

The second, and for us most important, example is **Lebesgue measure** on $\mathbb{R}^n$. Its construction is not easy, but it has the property of correctly giving the Euclidean volume of 'nice' sets. We do not give the construction because it can be found in many, many books, e.g., [Rudin, 1987]. However, the determined reader will be invited to construct Lebesgue measure as Exercise 5 in Chapter 6, with the aid of Theorem 6.22 (positive distributions are positive measures). Σ is taken to be $\mathcal{B}$ and the measure (or volume) of a set $A \in \mathcal{B}$ is denoted by $\mathcal{L}^n(A)$ or by the symbol

$$|A| := \mathcal{L}^n(A).$$

The Lebesgue measure of a ball is

$$\mathcal{L}^n(B_{x,r}) = |B_{0,1}|r^n = \frac{2\pi^{n/2} r^n}{n\Gamma(n/2)} = \frac{1}{n}|\mathbb{S}^{n-1}|r^n, \tag{8}$$

where

$$|\mathbb{S}^{n-1}| = 2\pi^{n/2}/\Gamma(n/2)$$

is the area of $\mathbb{S}^{n-1}$, which is the sphere of radius 1 in $\mathbb{R}^n$.

This measure is translation invariant—meaning that for every fixed $y \in \mathbb{R}^n$, $\mathcal{L}^n(A) = \mathcal{L}^n(\{x + y : x \in A\})$. Up to an over-all constant *it is the only translation invariant measure on* $\mathbb{R}^n$. The fact that the classical measure (8) can be extended in a *countably additive way* to a *sigma-algebra* containing all balls is a triumph which, having been achieved, makes integration theory relatively painless.

A small annoyance is connected with sets of measure zero, and is caused by the fact that a subset of a set of measure zero might not be measurable. An example is produced in the following fashion: Take a line ℓ in the plane $\mathbb{R}^2$. This set is a Borel set and $\mathcal{L}^2(\ell) = 0$. Now take any subset $\gamma \subset \ell$ that is not a Borel set in the one-dimensional sense. One can show that γ is also not a Borel set in the two-dimensional sense and therefore it is meaningless to say that $\mathcal{L}^2(\gamma) = 0$. One can get around this difficulty by declaring all subsets of sets of zero measure to be measurable and to have zero measure. But then, for consistency, these new sets have to be added to, and subtracted from, the Borel sets in $\mathcal{B}$. In this way Lebesgue measure can be extended to a larger class than $\mathcal{B}$, and it is easy to see that this class forms a sigma-algebra (Exercise 10). While this extension (called the **completion**) has its merits, we shall not use it in this book for it has no real value for us and causes problems, notably that the intersection of a measurable set in $\mathbb{R}^n$ with a hyperplane may not be measurable. For us, $\mathcal{L}^n$ is defined only on $\mathcal{B}$.

There is, however, one way in which subsets of sets of zero measure play a role. Given (Ω, Σ, μ) we say that some property holds **μ-almost everywhere** (or μ-a.e., or simply a.e. if μ is understood) whenever the subset of Ω for which the property fails to hold is a subset of a set of measure zero.

Lebesgue measure has two important properties called **inner regularity** and **outer regularity**. (See Theorem 6.22 and Exercise 6.5.) For every Borel set A

$$\mathcal{L}^n(A) = \inf\{\mathcal{L}^n(O) : A \subset O \text{ and } O \text{ is open}\} \quad \textit{outer regularity,} \quad (9)$$
$$\mathcal{L}^n(A) = \sup\{\mathcal{L}^n(C) : C \subset A \text{ and } C \text{ is compact}\} \quad \textit{inner regularity.} \quad (10)$$

Formula (9) will play an important role in establishing that measurable functions can be approximated by continuous functions (see Theorem 2.16 (approximation by C^∞ functions)).

Another important property of Lebesgue measure is its **sigma-finiteness**. A measure space (Ω, Σ, μ) is sigma-finite if there are countably many sets $A_1, A_2, \ldots$ such that $\mu(A_i) < \infty$ for all $i = 1, 2, \ldots$ and such that $\Omega = \bigcup_{i=1}^{\infty} A_i$. If sigma-finiteness holds it is easy to prove that the A_i's can be taken to be disjoint. In the case of $\mathcal{L}^n$ we can, for instance, take the A_i's to be cubes of unit side length.

As a final topic in this section we explain **product sigma-algebras** and **product measures**. Given two spaces Ω_1, Ω_2 with sigma-algebras Σ_1 and Σ_2 we can form the **product space**

$$\Omega = \Omega_1 \times \Omega_2 = \{(x_1, x_2) : x_1 \in \Omega_1, \ x_2 \in \Omega_2\}.$$

A good example is to think of Ω_1 as $\mathbb{R}^m$ and Ω_2 as $\mathbb{R}^n$ and $\Omega = \mathbb{R}^{m+n}$. The product sigma-algebra $\Sigma = \Sigma_1 \times \Sigma_2$ of sets in Ω is defined by first declaring all **rectangles** to be members of Σ. A rectangle is a set of the form

$$A_1 \times A_2 = \{(x_1, x_2) : x_1 \in A_1, \ x_2 \in A_2\}$$

where A_1 and A_2 are members of Σ_1 and Σ_2. Then $\Sigma = \Sigma_1 \times \Sigma_2$ is *defined* to be the *smallest* sigma-algebra containing all these rectangles, i.e., the sigma-algebra generated by all these rectangles. We shall see that the fact that Σ is the *smallest* sigma-algebra is important for Fubini's theorem (see Sects. 1.10 and 1.12).

Next suppose that $(\Omega_1, \Sigma_1, \mu_1)$ and $(\Omega_2, \Sigma_2, \mu_2)$ are two measure spaces. It is a basic and nontrivial fact that there exists a unique measure μ on the product sigma-algebra Σ of Ω with the 'product property' that

$$\mu(A_1 \times A_2) = \mu_1(A_1)\mu_2(A_2)$$

for all rectangles. This measure μ is called the *product measure* and is denoted by $\mu_1 \times \mu_2$. It will be constructed in Theorem 1.10 (product measure). The sigma-algebra Σ has the **section property** that if we take an arbitrary $A \in \Sigma$ and form the set $A_1(x_2) \subset \Omega_1$ defined by $A_1(x_2) = \{x_1 \in \Omega_1 : (x_1, x_2) \in A\}$, then $A_1(x_2)$ is in Σ_1 for *every* choice of x_2. An analogous property holds with 1 and 2 interchanged.

The section property depends crucially on the fact that Σ is defined to be the *smallest sigma-algebra that contains all rectangles*. To prove the section property one reasons as follows. Let $\Sigma' \subset \Sigma$ be the set of all those measurable sets $A \in \Sigma$ that *do have* the section property. Certainly, $\emptyset$ is in Σ' and $\Omega_1 \times \Omega_2$ is also in Σ'. Moreover, all rectangles are in Σ'. From the identity

$$\left(\bigcup_i A_2^i \right)(x_1) = \left(\bigcup_i A_2^i(x_1) \right)$$

which holds for any family of sets it follows that countable unions of sets in Σ' also have the section property. And from $A_2^c(x_1) = (A_2(x_1))^c$ one infers that if $A \in \Sigma'$, then $A^c \in \Sigma'$. Hence $\Sigma' \subset \Sigma$ is a sigma-algebra and since it contains the rectangles it must be equal to the minimal sigma-algebra Σ. This way of reasoning will be used again in the proof of Theorem 1.10.

In the same fashion one easily proves that for any three sigma-algebras $\Sigma_1, \Sigma_2, \Sigma_3$ the smallest sigma-algebra $\Sigma = \Sigma_1 \times \Sigma_2 \times \Sigma_3$ that contains all cubes also has the section property, i.e., for $A \in \Sigma$,

$$A_{23}(x_1) = \{(x_2, x_3) : (x_1, x_2, x_3) \in A\} \in \Sigma_2 \times \Sigma_3$$

for every $x_1 \in \Omega_1$, etc. By cubes we understand sets of the form $A_1 \times A_2 \times A_3$ where $A_i \in \Sigma_i$, $i = 1, 2, 3$.

If we turn to Lebesgue measure, then we find that if $\mathcal{B}^m$ is the Borel sigma-algebra of $\mathbb{R}^m$ then $\mathcal{B}^m \times \mathcal{B}^n = \mathcal{B}^{m+n}$. Note, however, that if we first extend Lebesgue measure to the nonmeasurable sets contained in Borel sets of measure zero, as described above, then the section property does not hold. A counterexample was mentioned earlier, namely a nonmeasurable subset of the real line is, when viewed as a subset of the plane, a subset of a set of measure zero. This failure of the section property is our chief reason for restricting the Lebesgue measure to the Borel sigma-algebra. It also shows that the product of the completion of the Borel sigma-algebra with itself is not complete; if it were complete it would contain the set mentioned above, but then it would fail to have the section property which, as we proved above, the product always has. On the other hand, if we take the completion of the product, then the section property can be shown to hold for *almost* every section.

● Up to now we have avoided proving any difficult theorems in measure theory. The following Theorem 1.3, however, is central to the subject and will be needed later in Sect. 1.10 on the product measure and for the proof of Fubini's theorem in 1.12. Because of its importance, and as an example of a 'pure measure theory' proof, we give it in some detail. The proof, but not the content, of Theorem 1.3 can be skipped on a first reading.

A **monotone class** $\mathcal{M}$ is a collection of sets with two properties:

if $A_i \in \mathcal{M}$ for $i = 1, 2, \ldots$, and if $A_1 \subset A_2 \subset \cdots$, then $\bigcup_i A_i \in \mathcal{M}$;

if $B_i \in \mathcal{M}$ for $i = 1, 2, \ldots$, and if $B_1 \supset B_2 \supset \cdots$, then $\bigcap_i B_i \in \mathcal{M}$.

Obviously any sigma-algebra is a monotone class, and the collection of all subsets of a set Ω is again a monotone class. Thus any collection of subsets is *contained* in a monotone class.

A collection of sets, $\mathcal{A}$, is said to form an **algebra of sets** if for every A and B in $\mathcal{A}$ the differences $A \sim B$, $B \sim A$ and the union $A \cup B$ are in $\mathcal{A}$. A sigma-algebra is then an algebra that is closed under countably many operations of this kind. Note that passage from an algebra, $\mathcal{A}$, to a sigma-algebra amounts to incorporation of countable unions of subsets of $\mathcal{A}$, thereby yielding some collection of sets, $\mathcal{A}_1$, which is no longer closed under taking intersections. Next, we incorporate countable intersections of sets in $\mathcal{A}_1$. This yields a collection of sets $\mathcal{A}_2$ which is not closed under taking unions. Proceeding this way one can arrive at a sigma-algebra by 'transfinite induction', which is enough to cause goose-bumps. The following theorem avoids this and simply states that sigma-algebras are monotone 'limits' of algebras. The key word in the following is 'sigma-algebra'.

1.3 THEOREM (Monotone class theorem)

Let Ω be a set and let $\mathcal{A}$ be an algebra of subsets of Ω such that Ω is in $\mathcal{A}$ and the empty set $\emptyset$ is also in $\mathcal{A}$. Then there exists a smallest monotone class $\mathcal{S}$ that contains $\mathcal{A}$. That class, $\mathcal{S}$, is also the smallest sigma-algebra that contains $\mathcal{A}$.

PROOF. Let $\mathcal{S}$ be the intersection of all monotone classes that contain $\mathcal{A}$, i.e., $Y \in \mathcal{S}$ if and only if Y is in every monotone class containing $\mathcal{A}$. We leave it as an exercise to the reader to show that $\mathcal{S}$ is a monotone class containing $\mathcal{A}$. By definition, it is then the smallest such monotone class.

We first note that it suffices to show that $\mathcal{S}$ is closed under forming complements and finite unions. Assuming this closure for the moment, we have, with $A_1, A_2, \ldots$ in $\mathcal{S}$, that $B_n := \bigcup_{i=1}^n A_i$ is a monotone increasing sequence of sets in $\mathcal{S}$. Since $\mathcal{S}$ is a monotone class $\bigcup_{i=1}^\infty A_i$ is in $\mathcal{S}$. Thus $\mathcal{S}$

is necessarily closed under forming countable unions. The formula

$$\left(\bigcap_{i=1}^{\infty} A_n\right)^c = \left(\bigcup_{i=1}^{\infty} A_i^c\right)$$

implies that $\mathcal{S}$, being closed under forming complements, contains also countable intersections of its members. Thus $\mathcal{S}$ is a sigma-algebra and since any sigma-algebra is a monotone class, $\mathcal{S}$ is the smallest sigma-algebra that contains $\mathcal{A}$.

Next, we show that $\mathcal{S}$ is indeed closed under finite unions. Fix a set $A \in \mathcal{A}$ and consider the collection $\mathcal{C}(A) = \{B \in \mathcal{S} : B \cup A \in \mathcal{S}\}$. Since $\mathcal{A}$ is an algebra, $\mathcal{C}(A)$ contains $\mathcal{A}$. For any increasing sequence of sets B_n in $\mathcal{C}(A)$, $A \cup B_i$ is an increasing sequence of sets in $\mathcal{S}$. Since $\mathcal{S}$ is a monotone class,

$$A \cup \left(\bigcup_{i=1}^{\infty} B_i\right) = \bigcup_{i=1}^{\infty} A \cup B_i$$

is in $\mathcal{S}$ and therefore $\bigcup_{i=1}^{\infty} B_i$ is in $\mathcal{C}(A)$. The reader can show that $\mathcal{C}(A)$ is closed under countable intersections of decreasing sets, and we then conclude that $\mathcal{C}(A)$ is a monotone class containing $\mathcal{A}$. Since $\mathcal{C}(A) \subset \mathcal{S}$ and $\mathcal{S}$ is the smallest monotone class that contains $\mathcal{A}$, $\mathcal{C}(A) = \mathcal{S}$.

Again, fix a set A, but this time an arbitrary one in $\mathcal{S}$, and consider the collection $\mathcal{C}(A) = \{B \in \mathcal{S} : B \cup A \in \mathcal{S}\}$. From the previous argument we know that $\mathcal{A}$ is a subset of $\mathcal{C}(A)$. A verbatim repetition of that argument to this new collection $\mathcal{C}(A)$ will convince the reader that $\mathcal{C}(A)$ is a monotone class and hence $\mathcal{C}(A) = \mathcal{S}$. Thus $\mathcal{S}$ is closed under finite unions, as claimed.

Finally, we address the complementation question. Let $\mathcal{C} = \{B \in \mathcal{S} : B^c \in \mathcal{S}\}$. This set contains $\mathcal{A}$ since $\mathcal{A}$ is an algebra. For any increasing sequence of sets $B_i \in \mathcal{C}$, $i = 1, 2, \ldots$, B_i^c is a decreasing sequence of sets in $\mathcal{S}$. Since $\mathcal{S}$ is a monotone class,

$$\left(\bigcup_{i=1}^{\infty} B_i\right)^c = \bigcap_{i=1}^{\infty} B_i^c$$

is in $\mathcal{S}$. Similarly for any decreasing sequence of sets $B_i \in \mathcal{C}$, $i = 1, 2, \ldots$, B_i^c is an increasing sequence of sets in $\mathcal{S}$ and hence

$$\left(\bigcap_{i=1}^{\infty} B_i\right)^c = \bigcup_{i=1}^{\infty} B_i^c$$

is in $\mathcal{S}$. Again $\mathcal{C} = \mathcal{S}$.

Thus $\mathcal{S}$ is closed under finite intersections and complementation. ∎

As an application of the monotone class theorem we present a uniqueness theorem for measures. It demonstrates a typical way of using the monotone class theorem and it will be handy in Sect. 1.10 on product measures.

1.4 THEOREM (Uniqueness of measures)

Let Ω be a set, $\mathcal{A}$ an algebra of subsets of Ω and Σ the smallest sigma-algebra that contains $\mathcal{A}$. Let μ_1 be a sigma-finite measure in the stronger sense that there exists a sequence of sets $A_i \in \mathcal{A}$ (and not merely $A_i \in \Sigma$), $i = 1, 2, \ldots$, each having finite μ_1 measure, such that $\bigcup_{i=1}^\infty A_i = \Omega$. If μ_2 is a measure that coincides with μ_1 on $\mathcal{A}$, then $\mu_1 = \mu_2$ on all of Σ.

PROOF. First we prove the theorem under the assumption that μ_1 is a finite measure on Ω. consider the set

$$\mathcal{M} = \{A \in \Sigma : \mu_1(A) = \mu_2(A)\}.$$

Clearly this collection of sets contains $\mathcal{A}$ and we shall show that $\mathcal{M}$ is a monotone class. By the previous Theorem 1.3 we then conclude that $\mathcal{M} = \Sigma$. Let $A_1 \subset A_2 \subset \cdots$ be an increasing sequence of sets in $\mathcal{M}$. Define $B_1 = A_1, B_2 = A_1 \sim A_2, \ldots, B_n = A_n \sim A_{n-1}, \ldots$. These sets are mutually disjoint and $\bigcup_{i=1}^n B_i = A_n$, in particular

$$\bigcup_{i=1}^\infty B_i = \bigcup_{i=1}^\infty A_i.$$

By the countable additivity of measures,

$$\mu_1\left(\bigcup_{i=1}^\infty A_i\right) = \sum_{i=1}^\infty \mu_1(B_i) = \lim_{n \to \infty} \sum_{i=1}^n \mu_1(B_i)$$

$$= \lim_{n \to \infty} \mu_1(A_n) = \lim_{n \to \infty} \mu_2(A_n) = \mu_2\left(\bigcup_{i=1}^\infty A_i\right).$$

Hence $\bigcup_{i=1}^\infty A_i$ is in $\mathcal{M}$. Now, with $A \in \mathcal{M}$, its complement A^c is also in $\mathcal{M}$, which follows from the fact that $\mu_i(A^c) = \mu_i(\Omega) - \mu_i(A)$, $i = 1, 2$, and that $\mu_1(\Omega) = \mu_2(\Omega) < \infty$. From this, it is easy to show that $\mathcal{M}$ is a monotone class. We leave the details to the reader.

Next, we return to the sigma-finite case. The theorem for the finite case implies that $\mu_1(B \cap A_0) = \mu_2(B \cap A_0)$ for every $A_0 \subset \mathcal{A}$ and every $B \subset \Sigma$. To see this, simply note that $A_0 \cap \Sigma$ is a sigma-algebra on A_0 which is the smallest one that contains the algebra $\mathcal{A}_0 \cap \mathcal{A}$. (Why?) Recall that, by assumption, there exists a sequence of sets $A_i \subset \mathcal{A}$, $i = 1, 2, \ldots$, each having finite μ_1 measure, such that $\bigcup_{i=1}^{\infty} A_i = \Omega$. Without loss of generality we may assume that these sets are disjoint. (Why?) Now for $B \subset \Sigma$,

$$\mu_1(B) = \mu_1\left(\bigcup_{i=1}^{\infty}(A_i \cap B)\right) = \sum_{i=1}^{\infty} \mu_1(A_i \cap B) = \sum_{i=1}^{\infty} \mu_2(A_i \cap B) = \mu_2(B). \quad\blacksquare$$

1.5 DEFINITION OF MEASURABLE FUNCTIONS AND INTEGRALS

Suppose that $f : \Omega \to \mathbb{R}$ is a real-valued function on Ω. Given a sigma-algebra Σ, we say that f is a **measurable function** (with respect to Σ) if for every number t the **level set**

$$S_f(t) := \{x \in \Omega : f(x) > t\} \qquad (1)$$

is measurable, i.e., $S_f(t) \in \Sigma$. Note that measurability does not require a measure!

More generally, if $f : \Omega \to \mathbb{C}$ is complex-valued, we say that f is measurable if its real and imaginary parts, $\mathrm{Re}\, f$ and $\mathrm{Im}\, f$, are measurable.

REMARK. Instead of the $>$ sign in (1) we could have chosen $\geq$, $\leq$ or $<$. All these definitions are in fact equivalent. To see this, one notes, for example, that

$$\{x \in \Omega : f(x) > t\} = \bigcup_{j=1}^{\infty}\{x \in \Omega : f(x) \geq t + 1/j\}.$$

If Σ is the Borel sigma-algebra $\mathcal{B}$ on $\mathbb{R}^n$, it is evident that every continuous function is Borel measurable, in fact $S_f(t)$ is then open. Other examples of Borel measurable functions are upper and lower semicontinuous functions. Recall that a real-valued function f is **lower semicontinuous** if $S_f(t)$ is open and it is **upper semicontinuous** if $\{x \in \Omega : f(x) < t\}$ is open. f is continuous if it is both upper and lower semicontinuous. To prove measurability, note that when f is lower semicontinuous then the set $\{x : f(x) < t + 1/j\}$ is measurable. Since the set

$$\{x \in \Omega : f(x) \leq t\} = \bigcap_{j=1}^{\infty}\{x : f(x) < t + 1/j\},$$

the set $\{x : f(x) \leq t\}$ is measurable. Therefore $S_f(t) = \Omega \sim \{x : f(x) \leq t\}$ is also measurable.

By pursuing the above reasoning a little further, one can show that for *any* Borel set $A \subset \mathbb{R}$ the set $\{x : f(x) \in A\}$ is Σ-measurable whenever f is Σ-measurable.

An amusing exercise (see Exercises 3, 4, 18) is to prove the facts that whenever f and g are measurable functions then so are the functions $x \mapsto \lambda f(x) + \gamma g(x)$ for λ and $\gamma \in \mathbb{C}$, $x \mapsto f(x)g(x)$, $x \mapsto |f(x)|$ and $x \mapsto \phi(f(x))$, where ϕ is any Borel measurable function from $\mathbb{C}$ to $\mathbb{C}$. In the same vein $x \mapsto \max\{f(x), g(x)\}$ and $x \mapsto \min\{f(x), g(x)\}$ are measurable functions. Moreover, when $f^1, f^2, f^3, \ldots$ is a sequence of measurable functions then the functions $\limsup_{j \to \infty} f^j(x)$ and $\liminf_{j \to \infty} f^j(x)$ are measurable.

Hence, *if a sequence $f^j(x)$ has a limit $f(x)$ for μ-almost every x, then f is a measurable function.* (More precisely, f can be redefined on a set of measure zero so that it becomes measurable.) The reader is urged to prove all these assertions or at least look them up in any standard text.

That a measurable function is defined only almost everywhere can cause some difficulties with some concepts, e.g., with the notion of strict positivity of a function. To remedy this we say that a nonnegative measurable function f is a **strictly positive measurable function** on a measurable set A, if the set $\{x \in A : f(x) = 0\}$ has zero measure.

Similar difficulties arise in the definition of the support of a measurable function. For a given Borel measure μ let f be a Borel measurable function on $\mathbb{R}^n$, or on any topological space for that matter. Recall that the open sets are measurable, i.e., they are members of the sigma-algebra. Consider the collection Ω of *open* subsets ω with the property that $f(x) = 0$ for μ-almost every $x \in \omega$ and let the open set ω^* be the union of all the ω's in Ω. Note that Ω and ω^* might be empty. Now we define the **essential support of** f, ess supp$\{f\}$, to be the complement of ω^*. Thus, ess supp$\{f\}$ is a closed, and hence measurable, set. Consider, e.g., the function f on $\mathbb{R}$, defined by $f(x) = 1$, x rational, and $f(x) = 0$, x not rational, and with μ being Lebesgue measure. Obviously $f(x) = 0$ for a.e. $x \in \mathbb{R}$, and hence ess supp$\{f\} = \emptyset$. Note also that ess supp$\{f\}$ depends on the measure μ and not just on the sigma-algebra. It is a simple exercise to verify that for μ being Lebesgue measure and f continuous, ess supp$\{f\}$ coincides with supp$\{f\}$, defined in Sect. 1.1.

In the remainder of this book we shall, for simplicity, use supp$\{f\}$ to mean ess supp$\{f\}$.

Our next task is to use a measure μ to define integrals of measurable functions. (Recall that the concept of measurability has nothing to do with a measure.)

First, suppose that $f : \Omega \to \mathbb{R}^+$ is a nonnegative real-valued, Σ-measurable function on Ω. (Our notation throughout will be that $\mathbb{R}^+ = \{x \in \mathbb{R} : x \geq 0\}$.) We then define
$$F_f(t) = \mu(S_f(t)),$$
i.e., $F_f(t)$ is the measure of the set on which $f > t$. Evidently $F_f(t)$ is a nonincreasing function of t since $S_f(t_1) \subset S_f(t_2)$ for $t_1 \geq t_2$. Thus $F_f(t) : \mathbb{R}^+ \to \mathbb{R}^+$ is a monotone nonincreasing function and it is an elementary calculus exercise (and a fundamental part of the theory of Riemann integration) to verify that the Riemann integral of such functions is always well defined (although its value might be $+\infty$). This Riemann integral *defines* **the integral of f over** Ω, i.e.,
$$\int_\Omega f(x)\mu(\mathrm{d}x) := \int_0^\infty F_f(t)\,\mathrm{d}t. \tag{2}$$
(Notation: sometimes we abbreviate this integral as $\int f$ or $\int f\,\mathrm{d}\mu$. The symbol $\mu(\mathrm{d}x)$ is intended to display the underlying measure, μ. Some authors use $\mathrm{d}\mu(x)$ while others use just $\mathrm{d}\mu x$. When μ is Lebesgue measure, $\mathrm{d}x$ is used in place of $\mathcal{L}^n\,\mathrm{d}x$.) A heuristic verification of the reason that (2) agrees with the usual definition can be given by introducing Heaviside's step-function $\Theta(s) = 1$ if $s > 0$ and $\Theta(s) = 0$ otherwise. Then, *formally*,
$$\begin{aligned}\int_0^\infty F_f(t)\,\mathrm{d}t &= \int_0^\infty \left\{ \int_\Omega \Theta(f(x) - t)\mu(\mathrm{d}x) \right\} \mathrm{d}t \\ &= \int_\Omega \left\{ \int_0^{f(x)} \mathrm{d}t \right\} \mu(\mathrm{d}x) = \int_\Omega f(x)\mu(\mathrm{d}x).\end{aligned} \tag{3}$$

If f is measurable and nonnegative and if $\int f\,\mathrm{d}\mu < \infty$, we say that f is a **summable** (or **integrable**) **function**.

It is an important fact (which we shall not need, and therefore not prove here) that if the function f is **Riemann integrable**, then its **Riemann integral** coincides with the value given in (2). See, however, Exercise 21 for a special case which will be used in Chapter 6.

More generally, suppose $f : \Omega \to \mathbb{C}$ is a complex-valued function on Ω. Then f consists of two real-valued functions, because we can write $f(x) = g(x) + ih(x)$, with g and h real-valued. In turn, each of these two functions can be thought of as the difference of two nonnegative functions, e.g.,
$$g(x) = g_+(x) - g_-(x) \quad \text{where} \tag{4}$$
$$g_+(x) = \begin{cases} g(x) & \text{if } g(x) > 0, \\ 0 & \text{if } g(x) \leq 0. \end{cases} \tag{5}$$

Alternatively, $g_+(x) = \max(g(x), 0)$ and $g_-(x) = -\min(g(x), 0)$. These are called the **positive and negative parts** of g. If f is measurable, then all four functions are measurable by the earlier remark. If all four functions g_+, g_-, h_+, h_- are summable, we say that f is summable and we define

$$\int f := \int g_+ - \int g_- + i \int h_+ - i \int h_-. \tag{6}$$

Equivalently, f is summable if and only if $x \mapsto |f(x)| \in \mathbb{R}^+$ is a summable function. It is to be emphasized that the integral of f can be defined *only* if f is summable. To attempt to integrate a function that is not summable is to open a Pandora's box of possibly false conclusions and paradoxes. There is, however, a noteworthy exception to this rule: If f is nonnegative we shall often abuse notation slightly by writing $\int f = +\infty$ when f is not summable. With this convention a relation such as $\int g < \int f$ (for $f \geq 0$ and $g \geq 0$) is meant to imply that when g is not summable, then f is also not summable. This convention saves some pedantic verbiage.

Another amusing (and not so trivial) exercise (see Exercise 9) is the verification of the linearity of integration. If f and g are summable, then $\lambda f + \gamma g$ are summable (for any λ and $\gamma \in \mathbb{C}$) and

$$\int_\Omega (\lambda f + \gamma g) \, d\mu = \lambda \int_\Omega f \, d\mu + \gamma \int_\Omega g \, d\mu. \tag{7}$$

The difficulty here lies in computing the level sets of linear combinations of summable functions.

An important class of measurable functions consists of the characteristic functions of measurable sets, as defined in 1.1(1). Clearly,

$$\int_\Omega \chi_A \, d\mu = \mu(A)$$

and hence χ_A is summable if and only if $\mu(A) < \infty$.

Sometimes we shall use the notation $\chi_{\{\cdots\}}$, where $\{\cdots\}$ denotes a set that is specified by condition $\cdots$. For example, if f is a measurable function, $\chi_{\{f>t\}}$ is the characteristic function of the set $S_f(t)$, whence $\int \chi_{\{f>t\}}$ is precisely $F_f(t)$ for $t \geq 0$.

For later use we now show that $\chi_{\{f>t\}}$ is a *jointly* measurable function of x and t. We have to show that the level sets of $\chi_{\{f>t\}}$ are $\Sigma \times \mathcal{B}^1$-measurable, where $\mathcal{B}^1$ is the Borel measure on the half line $\mathbb{R}^+$. The level sets in (x, t)-space are parametrized by $s \geq 0$ and have the form

$$\{(x, t) \in \Omega \times \mathbb{R}^+ : \chi_{\{f>t\}}(x) > s\}.$$

If $s \geq 1$, then the level set is empty and hence measurable. For $0 \leq s < 1$ the level set does not depend on s since $\chi_{\{f>t\}}$ takes only the values zero or one. In fact it is the set 'under the graph of f', i.e., the set $G = \{(x,t) \in \Omega \times \mathbb{R}^+ : 0 \leq t < f(x)\}$. This set is the union of sets of the form $S_f(r) \times [0,r]$ for rational r. (Recall that $[a,b]$ denotes the closed interval $a \leq x \leq b$ while (a,b) denotes the open interval $a < x < b$.) Since the rationals are countable we see that G is the countable union of rectangles and hence is measurable. Another way to prove that $G \subset \mathbb{R}^{n+1}$ is measurable, but which is secretly the same as the previous proof, is to note that

$$G = \{(x,t) : f(x) - t \geq 0\} \cap \{t : t > 0\},$$

and this is a measurable set since the set on which a measurable function ($f(x) - t$, in this case) is nonnegative is measurable by definition. (Why is $f(x) - t$ $\mathcal{L}^{n+1}$-measurable?)

Our definition of the integral suggests that it should be interpreted as the '$\mu \times \mathcal{L}^1$' measure of the set G which is in $\Sigma \times \mathcal{B}^1$. It is reasonable to *define*

$$(\mu \times \mathcal{L}^1)(G) := \int_0^\infty \int_\Omega \chi_{\{f>a\}}(x)\mu(\mathrm{d}x) = \int_\Omega f(x)\mu(\mathrm{d}x). \qquad (8)$$

A necessary condition for this to be a good definition is that it should not matter whether we integrate first over a or over x. In fact, since for every $x \in \Omega$, $\int_0^\infty \chi_{\{f>a\}}(x)\,\mathrm{d}a = f(x)$ (even for nonmeasurable functions), we have (recalling the definition of the integral) that

$$\int_0^\infty \int_\Omega \chi_{\{f>a\}}(x)\mu(\mathrm{d}x)\,\mathrm{d}a = \int_\Omega \int_0^\infty \chi_{\{f>a\}}(x)\,\mathrm{d}a\mu(\mathrm{d}x). \qquad (9)$$

This is a first elementary instance of **Fubini's theorem** about the interchange of integration. We shall see later in Theorem 1.10 that this interchange of integration is valid for *any* set $A \in \Sigma \times \mathcal{B}^1$ and we shall *define* $(\mu \times \mathcal{L}^1)(A)$ to be $\int_\mathbb{R} \mu(\{x : (x,a) \in A\})\,\mathrm{d}a$. We shall also see that $\mu \times \mathcal{L}^1$ defined this way is a measure on $\Sigma \times \mathcal{B}^1$.

• With this brief sketch of the fundamentals behind us, we are now ready to prove one of the basic convergence theorems in the subject. It is due to Beppo-Levi and Lebesgue. (Here and in the following the measure space (Ω, Σ, μ) will be understood.)

Suppose that $f^1, f^2, f^3, \ldots$ is an increasing sequence of nonnegative, summable functions on (Ω, Σ, μ), i.e., for each j, $f^{j+1}(x) \geq f^j(x)$ for Σ-almost every $x \in \Omega$. Because a countable union of sets of measure zero

also has measure zero, it then follows that the sequence of numbers $f^1(x)$, $f^2(x), \ldots$ is nondecreasing for almost every x. This monotonicity allows us to define
$$f(x) := \lim_{j \to \infty} f^j(x)$$
for almost every x, and we can define $f(x) := 0$ on the set of x's for which the above limit does not exist. This limit can, of course, be $+\infty$, but it is well defined a.e. It is also clear that the numbers $I_j := \int_\Omega f^j \, d\mu$ are also nondecreasing and we can define
$$I := \lim_{j \to \infty} I_j.$$

1.6 THEOREM (Monotone convergence)

Let $f^1, f^2, f^3, \ldots$ be an increasing sequence of nonnegative, summable functions on (Ω, Σ, μ), with f and I as defined above. Then f is measurable and, moreover, I is finite if and only if f is summable, in which case $I = \int_\Omega f \, d\mu$. In other words,
$$\lim_{j \to \infty} \int_\Omega f^j(x) \mu(dx) = \int_\Omega \lim_{j \to \infty} f^j(x) \mu(dx), \tag{1}$$
with the understanding that the left side of (1) is $+\infty$ when f is not summable.

▶ *Caution*: The word 'nonnegative' in the hypothesis is crucial!

PROOF. To compute $\int f^j$ we must first compute
$$F_{f^j}(t) = \mu(\{x : f^j(x) > t\}).$$
Note that, by definition, the set $\{x : f(x) > t\}$ equals the union of the increasing, countable family of sets $\{x : f^j(x) > t\}$. Hence, by 1.2(4), $\lim_{j \to \infty} F_{f^j}(t) = F_f(t)$ for every t. Moreover, this convergence is plainly monotone.

To prove our theorem, it then suffices to prove the corresponding theorem for the Riemann integral of monotone functions. That is,
$$\lim_{j \to \infty} \int_0^\infty F_{f^j}(t) \, dt = \int_0^\infty F_f(t) \, dt \tag{2}$$
given that each function $F_{f^j}(t)$ is monotone (in t), and the family is monotone in the index j, with the pointwise limit $F_f(t)$. This is an easy exercise; all that is needed is to note that the upper and lower Riemann sums converge. ∎

● The previous theorem can be paraphrased as saying that the functional $f \mapsto \int f$ on nonnegative functions behaves like a continuous functional with respect to sequences that converge pointwise and *monotonically*. It is easy to see that $f \mapsto \int f$ is not continuous in general, i.e., if f^j is a sequence of positive functions and if $f^j \to f$ pointwise a.e. it is not true in general that $\lim_{j \to \infty} \int f^j = \int f$, or even that the limit exists (see the Remark after the next lemma). What is true, however, is that $f \mapsto \int f$ is pointwise lower semicontinuous, i.e., $\liminf_{j \to \infty} \int f^j \geq \int f$ if $f^j \to f$ pointwise. The precise enunciation of that fact is the lemma of Fatou.

1.7 LEMMA (Fatou's lemma)

Let $f^1, f^2, \ldots$ be a sequence of nonnegative, summable functions on (Ω, Σ, μ). Then $f(x) := \liminf_{j \to \infty} f^j(x)$ is measurable and

$$\liminf_{j \to \infty} \int_\Omega f^j(x) \mu(\mathrm{d}x) \geq \int_\Omega f(x) \mu(\mathrm{d}x).$$

in the sense that the finiteness of the left side implies that f is summable.

▶ *Caution*: The word 'nonnegative' is crucial.

PROOF. Define $F^k(x) = \inf_{j \geq k} f^j(x)$. Since

$$\{x : F^k(x) \geq t\} = \bigcap_{j \geq k} \{x : f^j(x) \geq t\},$$

we see that $F^k(x)$ is measurable for all $k = 1, 2, \ldots$ by the Remark in 1.5. Moreover $F^k(x)$ is summable since $F^k(x) \leq f^k(x)$. The sequence F^k is obviously increasing and its limit is given by $\sup_{k \geq 1} \inf_{j \geq k} f^j(x)$ which is, by definition, $\liminf_{j \to \infty} f^j(x)$. We have that

$$\liminf_{j \to \infty} \int_\Omega f^j(x) \mu(\mathrm{d}x) := \sup_{k \geq 1} \inf_{j \geq k} \int_\Omega f^j(x) \mu(\mathrm{d}x)$$

$$\geq \lim_{k \to \infty} \int_\Omega F^k(x) \mu(\mathrm{d}x) = \int_\Omega f(x) \mu(\mathrm{d}x).$$

The last equality holds by monotone convergence and shows that f is summable if the left side is finite. The first equality is a definition. The middle inequality comes from the general fact that $\inf_j \int h^j \geq \inf_j \int (\inf_j h^j) = \int (\inf_j h^j)$, since $(\inf_j h^j)$ does not depend on j. ∎

REMARK. In case $f^j(x)$ converges to $f(x)$ for almost every $x \in \Omega$ the lemma says that

$$\liminf_j \int_\Omega f^j(x)\mu(\mathrm{d}x) \geq \int_\Omega f(x)\mu(\mathrm{d}x).$$

Even in this case the inequality can be strict. To give an example, consider on $\mathbb{R}$ the sequence of functions $f^j(x) = 1/j$ for $|x| \leq j$ and $f^j(x) = 0$ otherwise. Obviously $\int_\mathbb{R} f^j(x)\,\mathrm{d}x = 2$ for all j but $f^j(x) \to 0$ pointwise for all x.

● So far we have only considered the interchange of limits and integrals for nonnegative functions. The following theorem, again due to Lebesgue, is the one that is usually used for applications and takes care of this limitation. It is one of the most important theorems in analysis. It is equivalent to the monotone convergence theorem in the sense that each can be simply derived from the other.

1.8 THEOREM (Dominated convergence)

Let $f^1, f^2, \ldots$ be a sequence of complex-valued summable functions on (Ω, Σ, μ) and assume that these functions converge to a function f pointwise a.e. If there exists a summable, nonnegative function $G(x)$ on (Ω, Σ, μ) such that $|f^j(x)| \leq G(x)$ for all $j = 1, 2, \ldots$, then $|f(x)| \leq G(x)$ and

$$\lim_{j \to \infty} \int_\Omega f^j(x)\mu(\mathrm{d}x) = \int_\Omega f(x)\mu(\mathrm{d}x).$$

▶ *Caution*: The existence of the dominating G is crucial!

PROOF. It is obvious that the real and imaginary parts of f^j, R^j and I^j, satisfy the same assumptions as f^j itself. The same is true for the positive and negative parts of R^j and I^j. Thus it suffices to prove the theorem for nonnegative functions f^j and f. By Fatou's lemma

$$\liminf_{j \to \infty} \int_\Omega f^j \geq \int_\Omega f.$$

Again by Fatou's lemma

$$\liminf_{j \to \infty} \int_\Omega (G(x) - f^j(x))\mu(\mathrm{d}x) \geq \int_\Omega (G(x) - f(x))\mu(\mathrm{d}x),$$

since $G(x) - f^j(x) \geq 0$ for all j and all $x \in \Omega$. Summarizing these two inequalities we obtain

$$\liminf_{j \to \infty} \int_\Omega f^j(x)\mu(\mathrm{d}x) \geq \int_\Omega f(x)\mu(\mathrm{d}x) \geq \limsup_{j \to \infty} \int_\Omega f^j(x)\mu(\mathrm{d}x),$$

which proves the theorem. ∎

REMARK. The previous theorem allows a slight, but useful, generalization in which the dominating function $G(x)$ is replaced by a *sequence* $G^j(x)$ with the property that there exists a summable G such that

$$\int_\Omega |G(x) - G^j(x)|\mu(\mathrm{d}x) \to 0 \quad \text{as } j \to \infty$$

and such that $0 \leq |f^j(x)| \leq G^j(x)$. Again, if $f^j(x)$ converges pointwise a.e. to f the limit and the integral can be interchanged, i.e.,

$$\lim_{j \to \infty} \int_\Omega f^j(x)\mu(\mathrm{d}x) = \int_\Omega f(x)\mu(\mathrm{d}x).$$

To see this assume first that $f^j(x) \geq 0$ and note that

$$\int (G - f^j)_+ \to \int (G - f)_+ \quad \text{as } j \to \infty$$

since $(G - f^j)_+ \leq G$, using dominated convergence. Next observe that

$$\int (G - f^j)_- = \int (G - G^j + G^j - f^j)_- \leq \int (G - G^j)_-$$

since $G^j - f^j \geq 0$. See 1.5(5). The last integral however tends to zero as $j \to \infty$, by assumption. Thus we obtain

$$\lim_{j \to \infty} \int (G - f^j) = \int (G - f)_+ = \int (G - f)$$

since clearly $f(x) \leq G(x)$. The generalization in which f takes complex values is straightforward.

● Theorem 1.8 was proved using Fatou's lemma. It is interesting to note that Theorem 1.8 can be used, in turn, to prove the following generalization of Fatou's lemma. Suppose that f^j is a sequence of nonnegative functions that converges pointwise to a function f. As we have seen in the Remark after Lemma 1.7, limit and integral cannot be interchanged since, intuitively, the sequence f^j might 'leak out to infinity'. The next theorem taken from [Brézis-Lieb] makes this intuition precise and provides us with a correction term that changes Fatou's lemma from an inequality to an equality. While it is not going to be used in this book, it is of intrinsic interest as a theorem in measure theory and has been used effectively to solve some problems in the calculus of variations. We shall state a simple version of the theorem; the reader can consult the original paper for the general version in which, among other things, $f \mapsto |f|^p$ is replaced by a larger class of functions, $f \mapsto j(f)$.

1.9 THEOREM (Missing term in Fatou's lemma)

Let f^j be a sequence of complex-valued functions on a measure space that converges pointwise a.e. to a function f (which is measurable by the remarks in 1.5). Assume, also, that the f^j's are uniformly p^{th} power summable for some fixed $0 < p < \infty$, i.e.,

$$\int_\Omega |f^j(x)|^p \mu(\mathrm{d}x) < C \quad \text{for } j = 1, 2, \ldots$$

and for some constant C. Then

$$\lim_{j \to \infty} \int_\Omega \bigl||f^j(x)|^p - |f^j(x) - f(x)|^p - |f(x)|^p\bigr| \mu(\mathrm{d}x) = 0. \tag{1}$$

REMARKS. (1) By Fatou's lemma, $\int |f|^p \leq C$.

(2) By applying the triangle inequality to (1) we can conclude that

$$\int |f^j|^p = \int |f|^p + \int |f - f^j|^p + o(1), \tag{2}$$

where $o(1)$ indicates a quantity that vanishes as $j \to \infty$. Thus the correction term is $\int |f - f^j|^p$, which measures the 'leakage' of the sequence f^j. One obvious consequence of (2), for all $0 < p < \infty$, is that if $\int |f - f^j|^p \to 0$ and if $f^j \to f$ a.e., then

$$\int |f^j|^p \to \int |f|^p.$$

(In fact, this can be proved directly under the sole assumption that $\int |f - f^j|^p \to 0$. When $1 \leq p < \infty$ this a trivial consequence of the triangle inequality in 2.4(2). When $0 < p < 1$ it follows from the elementary inequality $|a + b|^p \leq |a|^p + |b|^p$ for all complex a and b.) Another consequence of (2), for all $0 < p < \infty$, is that if $\int |f^j|^p \to \int |f|^p$ and $f^j \to f$ a.e., then

$$\int |f - f^j|^p \to 0.$$

PROOF. Assume, for the moment, that the following family of inequalities, (3), is true: For any $\varepsilon > 0$ there is a constant C_ε such that for all numbers $a, b \in \mathbb{C}$
$$\big||a+b|^p - |b|^p\big| \leq \varepsilon |b|^p + C_\varepsilon |a|^p. \tag{3}$$

Next, write $f^j = f + g^j$ so that $g^j \to 0$ pointwise a.e. by assumption. We claim that the quantity
$$G^j_\varepsilon = \big(\big||f+g^j|^p - |g^j|^p - |f|^p\big| - \varepsilon |g^j|^p\big)_+ \tag{4}$$

satisfies $\lim_{j \to \infty} \int G^j_\varepsilon = 0$. Here $(h)_+$ denotes as usual the positive part of a function h. To see this, note first that
$$\big||f+g^j|^p - |g^j|^p - |f|^p\big|$$
$$\leq \big||f+g^j|^p - |g^j|^p\big| + |f|^p \leq \varepsilon |g^j|^p + (1+C_\varepsilon)|f|^p$$

and hence $G^j_\varepsilon \leq (1+C_\varepsilon)|f|^p$. Moreover $G^j_\varepsilon \to 0$ pointwise a.e. and hence the claim follows by Theorem 1.8 (dominated convergence). Now
$$\int \big||f+g^j|^p - |g^j|^p - |f|^p\big| \leq \varepsilon \int |g^j|^p + \int G^j_\varepsilon.$$

We have to show $\int |g^j|^p$ is uniformly bounded. Indeed,
$$\int |g^j|^p = \int |f - f^j|^p \leq 2^p \int (|f|^p + |f^j|^p) \leq 2^{p+1} C.$$

Therefore,
$$\limsup_{j \to \infty} \int \big||f+g^j|^p - |g^j|^p - |f|^p\big| \leq \varepsilon D.$$

Since ε was arbitrary the theorem is proved.

It remains to prove (3). The function $t \mapsto |t|^p$ is convex if $p > 1$. Hence $|a+b|^p \leq (|a|+|b|)^p \leq (1-\lambda)^{1-p}|a|^p + \lambda^{1-p}|b|^p$ for any $0 < \lambda < 1$. The choice $\lambda = (1+\varepsilon)^{-1/(p-1)}$ yields (3) in the case where $p > 1$. If $0 < p \leq 1$ we have the simple inequality $|a+b|^p - |b|^p \leq |a|^p$ whose proof is left to the reader. ∎

• With these convergence tools at our disposal we turn to the question of proving Fubini's theorem, 1.12. Our strategy to prove Fubini's theorem in full generality will be the following: First, we prove the 'easy' form in Theorem 1.10; this will imply 1.5(9). Then we use a small generalization of Theorem 1.10 to establish the general case in Theorem 1.12.

1.10 THEOREM (Product measure)

Let $(\Omega_1, \Sigma_1, \mu_1), (\Omega_2, \Sigma_2, \mu_2)$ be two sigma-finite measure spaces. Let A be a measurable set in $\Sigma_1 \times \Sigma_2$ and, for every $x \in \Omega_2$, set $f(x) := \mu_1(A_1(x))$ and, for every $y \in \Omega_1$, $g(y) := \mu_2(A_2(y))$. (Note that by the considerations at the end of Sect. 1.2 the sections are measurable and hence these quantities are defined). Then f is μ_2-measurable, g is μ_1-measurable and

$$(\mu_1 \times \mu_2)(A) := \int_{\Omega_2} f(x) \mu_2(dx) = \int_{\Omega_1} g(y) \mu_1(dy). \qquad (1)$$

Moreover, $\mu_1 \times \mu_2$, the product of the measures μ_1 and μ_2, defined in (1), is a sigma-finite measure on $\Sigma_1 \times \Sigma_2$.

PROOF. The measurability of f and g parallels the proof of the section property in Sect. 1.2 and is left to Exercise 22.

Consider any collection of disjoint sets A^i, $i = 1, 2, \ldots$, in $\Sigma_1 \times \Sigma_2$. Clearly their sections $A_1^i(x)$, $i = 1, 2, \ldots$, which are measurable (see Sect. 1.2), are also disjoint and hence

$$\mu_1\left(\left(\bigcup_{i=1}^{\infty} A_1^i\right)(x)\right) = \sum_{i=1}^{\infty} \mu_1(A_1^i(x)).$$

The monotone convergence theorem then yields the countable additivity of $\mu_1 \times \mu_2$. Similarly, the second integral in (1) also defines a countably additive measure.

We now verify the assumptions of Theorem 1.4 (uniqueness of measures). Define $\mathcal{A}$ to be the set of finite unions of rectangles, with $\Omega_1 \times \Omega_2$ and the empty set included. It is easy to see that this set is an algebra since the difference of two sets in $\mathcal{A}$ can be written again as a union of rectangles. Simply use the identities

$$(A_1 \times B_1) \cap (A_2 \times B_2) = (A_1 \cap A_2) \times (B_1 \cap B_2)$$

and

$$(A_1 \times B_1) \sim (A_2 \times B_2) = [(A_1 \sim A_2) \times B_1] \cup [A_2 \times (B_1 \sim B_2)].$$

By assumption there exists a collection of sets $A_i \subset \Omega_1$ with $\mu_1(A_i) < \infty$ for $i = 1, 2, \ldots$ and with

$$\bigcup_{i=1}^{\infty} A_i = \Omega_1.$$

Similarly, there exists a collection $B_j \subset \Omega_2$ with $\mu_2(B_j) < \infty$ for $j = 1, 2, \ldots$ and with
$$\bigcup_{j=1}^{\infty} B_j = \Omega_2.$$
Clearly the collection of rectangles $A_i \times B_j$ is countable, covers $\Omega_1 \times \Omega_2$ and
$$(\mu_1 \times \mu_2)(A_i \times B_j) = \mu_1(A_i)\mu_2(B_j) < \infty.$$
Thus, the two measures defined by the two integrals in (1) are sigma-finite. Now, note that the two integrals in (1) coincide on $\mathcal{A}$. Since, by definition, $\Sigma_1 \times \Sigma_2$ is the smallest sigma-algebra that contains $\mathcal{A}$, Theorem 1.4 yields (1) on all of $\Sigma_1 \times \Sigma_2$. ∎

● The following generalization of the previous theorem is useful and is an important step in proving Fubini's theorem.

1.11 COROLLARY (Commutativity and associativity of product measures)

Let $(\Omega_i, \Sigma_i, \mu_i)$ for $i = 1, 2, 3$ be sigma-finite measure spaces. For $A \in \Sigma_1 \times \Sigma_2$ define the reflected set
$$RA := \{(x, y) : (y, x) \in A\}.$$
This defines a one-to-one correspondence between $\Sigma_1 \times \Sigma_2$ and $\Sigma_2 \times \Sigma_1$. Then the formation of the product measure $\mu_1 \times \mu_2$ is commutative in the sense that
$$(\mu_2 \times \mu_1)(RA) = (\mu_1 \times \mu_2)(A)$$
for every $A \in \Sigma_1 \times \Sigma_2$. Moreover, the formation of product measures is associative, i.e.
$$(\mu_1 \times \mu_2) \times \mu_3 = \mu_1 \times (\mu_2 \times \mu_3). \tag{1}$$

PROOF. The proof of the commutativity is an obvious consequence of the previous theorem. To see the associativity, simply note that the sigma-algebras associated with $(\mu_1 \times \mu_2) \times \mu_3$ and $\mu_1 \times (\mu_2 \times \mu_3)$ are the smallest monotone classes that contain unions of cubes. Hence (1) follows, since the two measures coincide on cubes. ∎

1.12 THEOREM (Fubini's theorem)

Consider two sigma-finite *measure spaces* $(\Omega_i, \Sigma_i, \mu_i)$, $i = 1, 2$, *and let f be a $\Sigma_1 \times \Sigma_2$ measurable function on $\Omega_1 \times \Omega_2$. If $f \geq 0$, then the following three integrals are equal (in the sense that all three can be infinite):*

$$\int_{\Omega_1 \times \Omega_2} f(x,y)(\mu_1 \times \mu_2)(\mathrm{d}x\,\mathrm{d}y), \tag{1}$$

$$\int_{\Omega_1} \left(\int_{\Omega_2} f(x,y) \mu_2(\mathrm{d}y) \right) \mu_1(\mathrm{d}x), \tag{2}$$

$$\int_{\Omega_2} \left(\int_{\Omega_1} f(x,y) \mu_1(\mathrm{d}x) \right) \mu_2(\mathrm{d}y). \tag{3}$$

If f is complex-valued, then the above holds if one assumes in addition that

$$\int_{\Omega_1 \times \Omega_2} |f(x,y)|(\mu_1 \times \mu_2)(\mathrm{d}x\,\mathrm{d}y) < \infty. \tag{4}$$

REMARK. Sigma-finiteness is essential! In Exercise 19 we ask the reader to construct a counterexample.

PROOF. The second part of the statement follows from the first applied to the positive and negative parts of the $\operatorname{Re} f$ and $\operatorname{Im} f$ separately. As for (1), (2), (3), recall that by Theorem 1.10 (product measure) and the considerations at the end of Sect. 1.5 the value of the integral in (1) is given by

$$(\mu_1 \times \mu_2 \times \mathcal{L}^1)(G) \tag{5}$$

where $G = \{(x, y, t) \in \Omega_1 \times \Omega_2 \times \mathbb{R} : 0 \leq t < f(x, y)\}$, i.e., G is the set under the graph of f. Note that by the previous corollary the sequence of the factors in (5) is of no concern. Hence one can interpret (5) in three ways as

$$(\mathcal{L}^1 \times (\mu_1 \times \mu_2))(G), \qquad (\mu_1 \times (\mathcal{L}^1 \times \mu_2))(R_1 G)$$

and

$$(\mu_2 \times (\mathcal{L}^1 \times \mu_1))(R_2 G)$$

where R_1 and R_2 are the appropriate reflections. By the previous corollary these numbers are all equal and thus the theorem follows from the definitions

$$\int_{\Omega_1 \times \Omega_2} f(x,y)(\mu_1 \times \mu_2)(\mathrm{d}x\,\mathrm{d}y) = \int_0^\infty (\mu_1 \times \mu_2)(\chi_{f>t})\,\mathrm{d}t,$$

$$\int_{\Omega_1} \mu_1(\mathrm{d}x) \int_{\Omega_2} f(x,y)\mu(\mathrm{d}y) = \int_{\Omega_1} \mu_1(\mathrm{d}x) \int_0^\infty \mu_2(\chi_{f(x,\cdot)>t})\,\mathrm{d}t,$$

and similarly with μ_1 and μ_2 interchanged. ∎

• The next theorem is an elementary illustration of the use of Fubini's theorem. It is also extremely useful in practice because it permits us, in many cases, to reduce a problem about an integral of a general function to a problem about the integration of characteristic functions, i.e., functions that take only the values 0 or 1.

1.13 THEOREM (Layer cake representation)

Let ν be a measure on the Borel sets of the positive real line $[0, \infty)$ such that

$$\phi(t) := \nu([0, t)) \tag{1}$$

is finite for every $t > 0$. (Note that $\phi(0) = 0$ and that ϕ, being monotone, is Borel measurable.) Now let (Ω, Σ, μ) be a measure space and f any nonnegative measurable function on Ω. Then

$$\int_\Omega \phi(f(x))\mu(\mathrm{d}x) = \int_0^\infty \mu(\{x : f(x) > t\})\nu(\mathrm{d}t). \tag{2}$$

In particular, by choosing $\nu(\mathrm{d}t) = pt^{p-1}\,\mathrm{d}t$ for $p > 0$, we have

$$\int_\Omega f(x)^p \mu(\mathrm{d}x) = p \int_0^\infty t^{p-1} \mu(\{x : f(x) > t\})\,\mathrm{d}t. \tag{3}$$

By choosing μ to be the Dirac measure at some point $x \in \mathbb{R}^n$ and $p = 1$ we have

$$f(x) = \int_0^\infty \chi_{\{f>t\}}(x)\,\mathrm{d}t. \tag{4}$$

REMARKS. (1) It is formula (4) that we call the layer cake representation of f. (Approximate the $\mathrm{d}t$ integral by a Riemann sum and the allusion will be obvious.)

(2) The theorem can easily be generalized to the case in which ν is

replaced by the difference of two (positive) measures, i.e., $\nu = \nu_1 - \nu_2$. Such a difference is called a **signed measure**. The functions ϕ that can be written as in (1) with this ν are called functions of **bounded variation**. The additional assumption needed for the theorem is that for the given f, and each of the measures ν_1 and ν_2, one of the integrands in (2) is summable. As an example,

$$\int_\Omega \sin[f(x)]\mu(\mathrm{d}x) = \int_0^\infty (\cos t)\mu(\{x : f(x) > t\})\,\mathrm{d}t.$$

(3) In the case where $\phi(t) = t$, equation (2) is just the *definition* of the integral of f.

(4) Our proof uses Fubini's theorem, but the theorem can also be proved by appealing to the original definition of the integral and computing the μ-measure of the set $\{x : \phi(f(x)) > t\}$. This can be tedious (we leave this to the reader) in case ϕ is not strictly monotone.

PROOF. Recall that

$$\int_0^\infty \mu(\{x : f(x) > t\})\nu(\mathrm{d}t) = \int_0^\infty \int_\Omega \chi_{\{f>t\}}(x)\mu(\mathrm{d}x)\nu(\mathrm{d}t)$$

and that $\chi_{\{f>t\}}(x)$ is jointly measurable as discussed in Sect. 1.5. By applying Theorem 1.12 (Fubini's theorem) the right side equals

$$\int_\Omega \left(\int_0^\infty \chi_{\{f>t\}}(x)\nu(\mathrm{d}t)\right)\mu(\mathrm{d}x).$$

The result follows by observing that

$$\int_0^\infty \chi_{\{f>t\}}(x)\nu(\mathrm{d}t) = \int_0^{f(x)} \nu(\mathrm{d}t) = \phi(f(x)). \qquad \blacksquare$$

● Another application of the notion of level sets is the 'bathtub principle'. It solves a simple minimization problem — one that arises from time to time, but which sometimes appears confusing until the problem is viewed in the correct light. The proof, which we leave to the reader, is an easy exercise in manipulating level sets.

1.14 THEOREM (Bathtub principle)

Let (Ω, Σ, μ) be a measure space and let f be a real-valued, measurable function on Ω such that $\mu(\{x : f(x) < t\})$ is finite for all $t \in \mathbb{R}$. Let the number $G > 0$ be given and define a class of measurable functions on Ω by

$$\mathcal{C} = \left\{ g : 0 \leq g(x) \leq 1 \text{ for all } x \text{ and } \int_\Omega g(x)\mu(\mathrm{d}x) = G \right\}.$$

Then the minimization problem

$$I = \inf_{g \in \mathcal{C}} \int_\Omega f(x)g(x)\mu(\mathrm{d}x) \qquad (1)$$

is solved by

$$g(x) = \chi_{\{f<s\}}(x) + c\chi_{\{f=s\}} \qquad (2)$$

and

$$I = \int_{f<s} f(x)\mu(\mathrm{d}x) + cs\mu(\{x : f(x) = s\}), \qquad (3)$$

where

$$s = \sup\{t : \mu(\{x : f(x) < t\}) \leq G\}, \qquad (4)$$

$$c\mu(\{x : f(x) = s\}) = G - \mu(\{x : f(x) < s\}). \qquad (5)$$

The minimizer given in (2) is unique if $G = \mu(\{x : f(x) < s\})$ or if $G = \mu(\{x : f(x) \leq s\})$.

In order to understand why this is like filling a bathtub (and also for the purpose of constructing a proof of Theorem 1.14) think of the graph of f as a bathtub, take μ to be Lebesgue measure, and think of filling this bathtub with a fluid whose density g is not allowed to be greater than 1, but whose total mass, G, is given.

● The following theorem can be skipped at first reading for it will not be needed until Chapter 6 in the proof of Theorem 6.22 (positive distributions are measures). It provides a tool for constructing measures. Usually one is given a 'measure' on some collection of sets that is only finitely additive. The first step is to extend this 'measure' to an outer measure (defined by (i), (ii) and (iii) in Theorem 1.15 below) on *all* subsets. (Note: an outer measure is not necessarily finitely additive.) The second step is to restrict this outer measure to a class of sets that form a sigma-algebra in such a way that it is countably additive there. This construction is very general and the idea is due to Carathéodory.

1.15 THEOREM (Constructing a measure from an outer measure)

*Let Ω be a set and let μ be an **outer measure** on the collection of subsets of Ω, i.e., a nonnegative set function satisfying*
  (i) $\mu(\varnothing) = 0$,
  (ii) $\mu(A) \leq \mu(B)$ if $A \subset B$,
  (iii)
$$\mu\left(\bigcup_{i=1}^{\infty} A_i\right) \leq \sum_{i=1}^{\infty} \mu(A_i)$$

for any countable collection of subsets of Ω.

*Define Σ to be the collection of sets satisfying **Carathéodory's criterion**, namely $A \in \Sigma$ if*

$$\mu(E) = \mu(E \cap A) + \mu(E \cap A^c) \qquad (1)$$

for every set $E \subset \Omega$.

Then Σ is a sigma-algebra and the restriction of μ to Σ is a countably additive measure. The sets in Σ are called the measurable sets.

PROOF. Clearly Σ is not empty since $\varnothing \in \Sigma$ and $\Omega \in \Sigma$. Obviously with $A \in \Sigma$, $A^c \in \Sigma$. It is an instructive exercise for the reader to show that any finite union and any finite intersection of measurable sets is measurable. Thus Σ is an algebra.

Let E be any set in Ω and let $B_1, B_2, \ldots$ be a countable collection of disjoint measurable sets. Certainly

$$\mu\left(E \cap \left(\bigcup_{i=1}^{\infty} B_i\right)\right) = \mu\left(\bigcup_{i=1}^{\infty} (E \cap B_i)\right) \leq \sum_{i=1}^{\infty} \mu(E \cap B_i)$$

by (iii). Thus, by (ii),

$$\mu\left(E \cap \bigcup_{i=1}^{m} B_i\right)$$

is an increasing sequence and

$$\lim_{m \to \infty} \mu\left(E \cap \left(\bigcup_{i=1}^{m} B_i\right)\right) \leq \mu\left(E \cap \left(\bigcup_{i=1}^{\infty} B_i\right)\right) \leq \sum_{i=1}^{\infty} \mu(E \cap B_i).$$

We show that there must be equality in the above chain of inequalities. To see this, note that by (iii)

$$\mu(E) \le \mu\left(E \cap \left(\bigcup_{i=1}^{m} B_i\right)\right) + \mu\left(E \cap \left(\bigcup_{i=1}^{m} B_i\right)^c\right)$$

$$\le \sum_{i=1}^{m} \mu(E \cap B_i) + \mu\left(E \cap \left(\bigcap_{j=1}^{m} B_j^c\right)\right). \tag{2}$$

Since the B_i's are disjoint, we have for every $i = 1, 2, \ldots,$

$$E \cap B_i = E \cap \left(\bigcap_{j<i} B_j^c\right) \cap B_i$$

and hence the right side of (2) equals

$$\sum_{i=1}^{m-1} \mu\left(E \cap \left(\bigcap_{j<i} B_j^c\right) \cap B_i\right) + \mu\left(E \cap \left(\bigcap_{j<m} B_j^c\right) \cap B_m\right)$$
$$+ \mu\left(E \cap \left(\bigcap_{j=1}^{m} B_j^c\right)\right). \tag{3}$$

By the measurability of B_m the last term in (3) equals

$$\mu\left(E \cap \left(\bigcap_{j=1}^{m-1} B_j^c\right)\right),$$

and hence the right side of (2) is not changed when m is replaced by $m-1$. Repeating this step we conclude that the right side of (2) equals $\mu(E)$. Hence,

$$\mu(E) = \mu\left(E \cap \left(\bigcup_{i=1}^{m} B_i\right)\right) + \mu\left(E \cap \left(\bigcup_{i=1}^{m} B_i\right)^c\right), \tag{4}$$

i.e., $\bigcup_{i=1}^{m} B_i$ is measurable and

$$\mu\left(E \cap \left(\bigcup_{i=1}^{m} B_i\right)\right) = \sum_{i=1}^{m} \mu(E \cap B_i). \tag{5}$$

From this we conclude that

$$\lim_{m \to \infty} \mu\left(E \cap \left(\bigcup_{i=1}^{m} B_i\right)\right) = \mu\left(E \cap \left(\bigcup_{i=1}^{\infty} B_i\right)\right)$$
$$= \sum_{i=1}^{\infty} \mu(E \cap B_i). \tag{6}$$

Since

$$\mu\left(E \cap \left(\bigcup_{i=1}^{m} B_i\right)^c\right) \geq \mu\left(E \cap \left(\bigcup_{i=1}^{\infty} B_i\right)^c\right),$$

(6) and (4) yield

$$\mu(E) \geq \mu\left(E \cap \left(\bigcup_{i=1}^{\infty} B_i\right)\right) + \mu\left(E \cap \left(\bigcup_{i=1}^{\infty} B_i\right)^c\right) \tag{7}$$

and hence, by (iii), $\bigcup_{i=1}^{\infty} B_i$ is measurable. By setting $E = \Omega$ in (6) we obtain the countable additivity, i.e.,

$$\mu\left(\bigcup_{i=1}^{\infty} B_i\right) = \sum_{i=1}^{\infty} \mu(B_i). \tag{8}$$

Having established that countable unions of disjoint measurable sets are measurable, it is straightforward to show that Σ is a monotone class. By Theorem 1.3 (monotone class theorem) the smallest monotone class containing the algebra Σ is also the smallest sigma-algebra containing Σ. Hence Σ is a sigma-algebra and μ is a countably additive measure on Σ. ∎

Exercises for Chapter 1

1. Complete the proof of Theorem 1.3 (monotone class theorem).

2. With regard to the remark about continuous functions in Sect. 1.5, show that f is continuous (in the sense of the usual ε, δ definition) if and only if f is both upper and lower semicontinuous. Show that f is upper semicontinuous at x if and only if, for every sequence $x_1, x_2, \ldots$ converging to x, we have $f(x) \geq \limsup_{n\to\infty} f(x_n)$.

3. Prove the assertion made in Sect. 1.5 that for any Borel set $A \subset \mathbb{R}$ and any sigma-algebra Σ the set $\{x : f(x) \in A\}$ is Σ-measurable whenever the function f is Σ-measurable.

4. (Continuation of Problem 3): Let $\phi : \mathbb{C} \to \mathbb{C}$ be a Borel measurable function and let the complex-valued function f be Σ-measurable. Prove that $\phi(f(x))$ is Σ-measurable.

5. Prove equation (2) in Theorem 1.6 (monotone convergence).

6. Give the alternative proof of the layer cake representation, alluded to in Remark (3) of 1.13, that does not make use of Fubini's theorem.

7. Prove Theorem 1.14 (bathtub principle).

8. Prove the statement about finite unions and intersections in the first paragraph of the proof of Theorem 1.15 (constructing a measure from an outer measure).

    ▶ *Hint.* For any two measurable sets A, B and E arbitrary, show that
    $$\mu(E) = \mu(E \cap A \cap B) + \mu(E \cap A^c \cap B) + \mu(E \cap A \cap B^c) \\ + \mu(E \cap A^c \cap B^c).$$

    Use this to prove that $A \cap B$ is measurable.

9. Verify the linearity of the integral as given in 1.5(7) by completing the steps outlined below. In what follows, f and g are nonnegative summable functions.
    a) Show that $f + g$ is also summable. In fact, by a simple argument $\int (f+g) \leq 2(\int f + \int g)$.
    b) For any integer N find two functions f_N and g_N that take only finitely many values, such that $|\int f - \int f_N| \leq C/N$, $|\int g - \int g_N| \leq C/N$ and

$|\int (f+g) - \int (f_N + g_N)| \leq C/N$ for some constant C independent of N.

c) Show that for f_N and g_N as above $\int (f_N + g_N) = \int f_N + \int g_N$, thus proving the additivity of the integral for nonnegative functions.

d) In a similar fashion, show that for $f, g \geq 0$, $\int (f - g) = \int f - \int g$.

e) Now use c) and d) to prove the linearity of the integral.

10. Prove that when we add and subtract the subsets of sets of zero measure to the sets of a sigma-algebra then the result is again a sigma-algebra and the extended measure is again a measure.

11. Prove that the measure constructed in Theorem 1.15 is complete, i.e., every subset of a measurable set that has measure zero is measurable.

12. Find a simple condition on $f_n(x)$ so that

$$\sum_{n=0}^{\infty} \int_\Omega f_n(x) \mu(\mathrm{d}x) = \int_\Omega \left\{ \sum_{n=0}^{\infty} f_n(x) \right\} \mu(\mathrm{d}x).$$

13. Let f be the function on $\mathbb{R}^n$ defined by $f(x) = |x|^{-p} \chi_{\{|x|<1\}}(x)$. Compute $\int f \, \mathrm{d}\mathcal{L}^n$ in two ways: (i) Use polar coordinates and compute the integral by the standard calculus method. (ii) Compute $\mathcal{L}^n(\{x : f(x) > a\})$ and then use Lebesgue's definition.

14. Prove that $j(x)$, defined in 1.1(2), is infinitely differentiable.

15. **Urysohn's lemma.** Let $\Omega \subset \mathbb{R}^n$ be open and let $K \subset \Omega$ be compact. Prove that there is a $\psi \in C_c^\infty(\Omega)$ with $\psi(x) = 1$ for all $x \in K$.

    ▶ *Hints.* (a) Replace K by a slightly larger compact set K_ε, i.e., $K \subset K_\varepsilon \subset \Omega$; (b) Using the distance function $d(x, K_\varepsilon) = \inf\{|x - y| : y \in K_\varepsilon\}$, construct a function $\psi \in C_c^0(\Omega)$ with $\psi_\varepsilon = 1$ on K_ε and $\psi_\varepsilon(x) = 0$ for $x \notin K_{2\varepsilon} \subset \Omega$; (c) Take $j_\varepsilon(x) = \varepsilon^{-n} j(x/\varepsilon)$, with j given in Exercise 14 and $\int j = 1$ (here $\int$ denotes the *Riemann* integral from elementary calculus). Define $\psi(x) = \int j_\varepsilon(x - y) \psi_\varepsilon(y) \, \mathrm{d}y$ (again, the Riemann integral); (d) Verify that ψ has the correct properties. To show that $\psi \in C_c^\infty(\Omega)$ it will be necessary to differentiate 'under the integral sign', a process that can be justified with standard theorems from calculus.

16. Let $\Omega \subset \mathbb{R}^n$ be open and $\phi \in C_c^\infty(\Omega)$. Show that there exist nonnegative functions ϕ_1 and ϕ_2, both in $C_c^\infty(\Omega)$, such that $\phi = \phi_1 - \phi_2$.

17. Show that the infimum of a family of continuous functions is upper semi-continuous.

18. Simple facts about measure:
    a) Show that the condition $\{x : f(x) > a\}$ is measurable for all $a \in \mathbb{R}$ holds if and only if it holds for all rational a.
    b) For rational a, show that
    $$\{x : f(x) + g(x) > a\} = \bigcup_{b \text{ rational}} \left(\{x : f(x) > b\} \cap \{x : g(x) > a - b\}\right).$$
    c) In a similar way, prove that fg is measurable if f and g are measurable.

19. Give a 'counterexample' to Fubini's theorem in the absence of sigma-finiteness.
    ▶ *Hint.* Take Lebesgue measure on $[0, 1]$ as one space and counting measure on $[0, 1]$ as the other. (**The counting measure** of a set is just the number of elements in the set.)

20. If f and g are two continuous functions on a common open set in $\mathbb{R}^n$ that agree everywhere on the complement of a set of zero Lebesgue measure, then, in fact, f and g agree everywhere.

21. Prove that if $f : \mathbb{R}^n \to \mathbb{C}$ is uniformly continuous and summable, then the Riemann integral of f equals its Lebesgue integral.

22. Theorem 1.10 (product measure) asserts that f and g are measurable functions. Prove this by imitating the proof of the section property in Sect. 1.2.

Chapter 2

L^p-Spaces

This and the next two chapters contain basic facts about functions, the objects of principal interest in the rest of the book. The main topic is the definition and properties of pth-power summable functions.

This topic does not utilize any metric properties of the domain, e.g., the Euclidean structure of $\mathbb{R}^n$, and therefore can be stated in greater generality than we shall actually need later. This generality is sometimes useful in other contexts, however. On a first reading it may be simplest to replace the measure $\mu(\mathrm{d}x)$ on the space Ω by Lebesgue measure $\mathrm{d}x$ on $\mathbb{R}^n$ and to regard Ω as a Lebesgue measurable subset of $\mathbb{R}^n$.

2.1 DEFINITION OF L^p-SPACES

Let Ω be a measure space with a (positive) measure μ and let $1 \leq p < \infty$. We define $L^p(\Omega, \mathrm{d}\mu)$ to be the following class of measurable functions:

$$L^p(\Omega, \mathrm{d}\mu) = \{f \ : \ f : \Omega \to \mathbb{C}, \ f \text{ is } \mu\text{-measurable} \ \text{ and } \ |f|^p \text{ is } \mu\text{-summable}\}. \tag{1}$$

Usually we omit μ in the notation and write instead $L^p(\Omega)$ if there is no ambiguity. Most of the time we have in mind that Ω is a Lebesgue measurable subset of $\mathbb{R}^n$ and μ is Lebesgue measure.

The reason we exclude $p < 1$ is that 3(c) below fails when $p < 1$.

On account of the inequality $|\alpha + \beta|^p \leq 2^{p-1}(|\alpha|^p + |\beta|^p)$ we see that for arbitrary complex numbers a and b, $af + bg$ is in $L^p(\Omega)$ if f and g are. Thus $L^p(\Omega)$ is a vector space.

For each $f \in L^p(\Omega)$ we define the **norm** to be

$$\|f\|_p = \left(\int_\Omega |f(x)|^p \mu(\mathrm{d}x)\right)^{1/p}. \tag{2}$$

Sometimes we shall write this as $\|f\|_{L^p(\Omega)}$ if there is possibility of confusion. This norm has the following *three crucial properties* that make it truly a norm:

(a) $\|\lambda f\|_p = |\lambda| \|f\|_p$ for $\lambda \in \mathbb{C}$.

(b) $\|f\|_p = 0$ if and only if $f(x) = 0$ for μ-almost every point x. (3)

(c) $\|f + g\|_p \leq \|f\|_p + \|g\|_p$.

(Technically, (2) only defines a semi-norm because of the 'almost every' caveat in 3(b), i.e., $\|f\|_p$ can be zero without $f \equiv 0$. Later on, when we define equivalence classes, (2) will be an honest norm on these classes.) Property (a) is obvious and (b) follows from the definition of the integral. Less trivial is property (c) which is called the **triangle inequality**. It will follow immediately from Theorem 2.4 (Minkowski's inequality). The triangle inequality is the same thing as **convexity of the norm**, i.e., if $0 \leq \lambda \leq 1$, then

$$\|\lambda f + (1-\lambda)g\|_p \leq \lambda \|f\|_p + (1-\lambda)\|g\|_p.$$

We can also define $L^\infty(\Omega, \mathrm{d}\mu)$ by

$$L^\infty(\Omega, \mathrm{d}\mu) = \{f\ :\ f : \Omega \to \mathbb{C},\ f \text{ is } \mu\text{-measurable and there exists} \\ \text{a finite constant } K \text{ such that } |f(x)| \leq K \text{ for } \mu\text{-a.e. } x \in \Omega\}. \tag{4}$$

For $f \in L^\infty(\Omega)$ we define the norm

$$\|f\|_\infty = \inf\{K\ :\ |f(x)| \leq K \text{ for } \mu\text{-almost every } x \in \Omega\}. \tag{5}$$

Note that the norm depends on μ. This quantity is also called the **essential supremum** of f and is denoted by $\operatorname{ess\,sup}_x |f(x)|$. (Do not confuse this with ess supp—which has one more p.) Unlike the usual supremum, ess sup ignores sets of μ-measure zero. E.g., if $\Omega = \mathbb{R}$ and $f(x) = 1$ if x is rational and $f(x) = 0$ otherwise, then (with respect to Lebesgue measure) $\operatorname{ess\,sup}_x |f(x)| = 0$, while $\sup_x |f(x)| = 1$.

One can easily verify that the L^∞ norm has the same properties (a), (b) and (c) as above. Note that property (b) would fail if ess sup is replaced by sup. Also note that $|f(x)| \leq \|f\|_\infty$ for almost every x.

We leave it as an exercise to the reader to prove that when $f \in L^\infty(\Omega) \cap L^q(\Omega)$ for some q then $f \in L^p(\Omega)$ for all $p > q$ and

$$\|f\|_\infty = \lim_{p \to \infty} \|f\|_p. \tag{6}$$

This equation is the reason for denoting the space defined in (4) by $L^\infty(\Omega)$.

An important concept, whose meaning will become clear later, is the **dual index** to p (for $1 \leq p \leq \infty$, of course). This is often denoted by p', but we shall often use q, and it is given by

$$\frac{1}{p} + \frac{1}{p'} = 1. \tag{7}$$

Thus, 1 and ∞ are dual, while the dual of 2 is 2.

Unfortunately, the norms we have defined do not serve to distinguish all different measurable functions, i.e., if $\|f - g\|_p = 0$ we can only conclude that $f(x) = g(x)$ μ-almost everywhere. To deal with this nuisance we can redefine $L^p(\Omega, \mathrm{d}\mu)$ so that its elements are not functions but *equivalence classes* of functions. That is to say, if we pick an $f \in L^p(\Omega)$ we can define $\widetilde{f}$ to be the set of all those functions that differ from f only on a set of μ-measure zero. If h is such a function we write $f \sim h$; moreover if $f \sim h$ and $h \sim g$, then $f \sim g$. Consequently, two such sets $\widetilde{f}$ and $\widetilde{k}$ are either identical or disjoint. We can now define

$$\|\widetilde{f}\|_p := \|f\|_p$$

for some $f \in \widetilde{f}$. The point is that this definition does not depend on the choice of $f \in \widetilde{f}$.

Thus we have two vector spaces. The first consists of functions while the second consists of equivalence classes of functions. (It is left to the reader to understand how to make the set of equivalence classes into a vector space.) For the first, $\|f - g\|_p = 0$ does not imply $f = g$, but for the second space it does. Some authors distinguish these spaces by different symbols, but all agree that it is the second space that should be called $L^p(\Omega)$. Nevertheless most authors will eventually slip into the tempting trap of saying 'let f be a function in $L^p(\Omega)$' which is technically nonsense in the context of the second definition. Let the reader be warned that we will generally commit this sin. Thus when we are talking about L^p-functions and we write $f = g$ we really have in mind that f and g are two functions that agree μ-almost everywhere. If the context is changed to, say, continuous functions, then $f = g$ means $f(x) = g(x)$ for all x. In particular, we note that it makes no sense to ask for the value $f(0)$, say, if f is an L^p-function.

- A **convex set** $K \subset \mathbb{R}^n$ is one for which $\lambda x + (1-\lambda)y \in K$ for all $x, y \in K$ and all $0 \leq \lambda \leq 1$. A **convex function**, f, on a convex set $K \subset \mathbb{R}^n$ is a real-valued function satisfying

$$f(\lambda x + (1-\lambda)y) \leq \lambda f(x) + (1-\lambda) f(y) \tag{8}$$

for all $x, y \in K$ and all $0 < \lambda < 1$. If equality *never* holds in (8) when $y \neq x$ and $0 < \lambda < 1$, then f is **strictly convex**. More generally, we say that f is strictly convex at a point $x \in K$ if $f(x) < \lambda f(y) + (1-\lambda) f(z)$ whenever $x = \lambda y + (1-\lambda) z$ for $0 < \lambda < 1$ and $y \neq z$. If the inequality (8) is reversed, f is said to be **concave** (alternatively, f is concave $\iff$ $-f$ is convex). It is easy to prove that if K is an *open* set, then a convex function is continuous.

A **support plane** to a graph of a function $f : K \to \mathbb{R}$ at a point $x \in K$ is a plane (in $\mathbb{R}^{n+1}$) that touches the graph at $(x, f(x))$ and that nowhere lies above the graph. In general, a support plane might not exist at x, but if f is convex on K, its graph has at least one support plane at each point of the interior of K. Thus there exists a vector $V \in \mathbb{R}^n$ (which depends on x) such that

$$f(y) \geq f(x) + V \cdot (y - x) \tag{9}$$

for all $y \in K$. If the support plane at x is unique it is called a **tangent plane**. If f is convex, the existence of a tangent plane at x is equivalent to differentiability at x.

If $n = 1$ and if f is convex, f need not be differentiable at x. However, when x is in the interior of the interval K, f always has a **right derivative**, $f'_+(x)$, and a **left derivative**, $f'_-(x)$, at x, e.g.,

$$f'_+(x) := \lim_{\varepsilon \searrow 0}[f(x+\varepsilon) - f(x)]/\varepsilon.$$

See [Hardy–Littlewood–Pólya] and Exercise 18.

2.2 THEOREM (Jensen's inequality)

Let $J : \mathbb{R} \to \mathbb{R}$ be a convex function. Let f be a μ-measurable, real-valued function on Ω. Since J is convex, it is continuous and therefore $(J \circ f)(x) := J(f(x))$ is also a μ-measurable function on Ω. We assume that $\mu(\Omega) = \int_\Omega \mu(\mathrm{d}x)$ is finite.

Suppose now that $f \in L^1(\Omega)$ and let $\langle f \rangle$ be the average of f, i.e.,

$$\langle f \rangle = \frac{1}{\mu(\Omega)} \int_\Omega f \, \mathrm{d}\mu.$$

Then

(i) $[J \circ f]_-$, *the negative part of* $[J \circ f]$, *is in* $L^1(\Omega)$, *whence* $\int_\Omega (J \circ f)(x)\mu(dx)$ *is well defined although it might be* $+\infty$.

(ii) $$\langle J \circ f \rangle \geq J(\langle f \rangle). \tag{1}$$

If J is strictly convex at $\langle f \rangle$ there is equality in (1) if and only if f is a constant function.

PROOF. Since J is convex its graph has at least one support line at each point. Thus, there is a constant $V \in \mathbb{R}$ such that

$$J(t) \geq J(\langle f \rangle) + V(t - \langle f \rangle) \tag{2}$$

for all $t \in \mathbb{R}$. From this we conclude that

$$[J(f)]_-(x) \leq |J(\langle f \rangle)| + |V||\langle f \rangle| + |V||f(x)|,$$

and hence, recalling that $\mu(\Omega) < \infty$, (i) is proved.

If we now substitute $f(x)$ for t in (2) and integrate over Ω we arrive at (1).

Assume now that J is strictly convex at $\langle f \rangle$. Then (2) is a strict inequality either for all $t > \langle f \rangle$ or for all $t < \langle f \rangle$. If f is not a constant, then $f(x) - \langle f \rangle$ takes on both positive and negative values on sets of positive measure. This implies the last assertion of the theorem. ∎

● The importance of the next inequality can hardly be overrated. There are many proofs of it and the one we give is not necessarily the simplest; we give it in order to show how the inequality is related to Jensen's inequality. Another proof is outlined in the exercises.

2.3 THEOREM (Hölder's inequality)

Let p and q be dual indices, i.e., $1/p + 1/q = 1$ with $1 \leq p \leq \infty$. Let $f \in L^p(\Omega)$ and $g \in L^q(\Omega)$. Then the pointwise product, given by $(fg)(x) = f(x)g(x)$, is in $L^1(\Omega)$ and

$$\left| \int_\Omega fg \, d\mu \right| \leq \int_\Omega |f||g| \, d\mu \leq \|f\|_p \|g\|_q. \tag{1}$$

The first inequality in (1) is an equality if and only if

(i) $f(x)g(x) = e^{i\theta}|f(x)||g(x)|$ *for some real constant θ and for μ-almost every x.*

If $f \not\equiv 0$ the second inequality in (1) is an equality if and only if there is a constant $\lambda \in \mathbb{R}$ such that:

(iia) *If $1 < p < \infty$, $|g(x)| = \lambda |f(x)|^{p-1}$ for μ-almost every x.*

(iib) *If $p = 1$, $|g(x)| \le \lambda$ for all x and $|g(x)| = \lambda$ when $f(x) \ne 0$.*

(iic) *If $p = \infty$, $|f(x)| \le \lambda$ for all x and $|f(x)| = \lambda$ when $g(x) \ne 0$.*

REMARKS. (1) The special case $p = q = 2$ is the **Schwarz inequality**

$$\left| \int_\Omega fg \right|^2 \le \int_\Omega |f|^2 \int_\Omega |g|^2 . \qquad (2)$$

(2) If $f_1, \ldots, f_m$ are functions on Ω with $f_i \in L^{p_i}(\Omega)$ and $\sum_{j=1}^m 1/p_i = 1$ then

$$\left| \int_\Omega \prod_{j=1}^m f_i \, d\mu \right| \le \prod_{j=1}^m \|f_i\|_{p_i} . \qquad (3)$$

This generalization is a simple consequence of (1) with $f := f_1$ and $g := \prod_{j=2}^m f_j$. Then use induction on $\int_\Omega |g|^p$.

PROOF. The left inequality in (1) is a triviality, so we may as well suppose $f \ge 0$ and $g \ge 0$ (note that condition (i) is what is needed for equality here). The cases $p = \infty$ and $q = \infty$ are trivial so we suppose that $1 < p, q < \infty$. Set $A = \{x : g(x) > 0\} \subset \Omega$ and let $B = \Omega \sim A = \{x : g(x) = 0\}$. Since

$$\int_\Omega f^p \, d\mu = \int_A f^p \, d\mu + \int_B f^p \, d\mu,$$

since $\int_\Omega g^p \, d\mu = \int_A g^p \, d\mu$, and since $\int_\Omega fg \, d\mu = \int_A fg \, d\mu$, we see that it suffices—in order to prove (1)—to assume that $\Omega = A$. Introduce a new measure on $\Omega = A$ by $\nu(dx) = g(x)^q \mu(dx)$. Also, set $F(x) = f(x)g(x)^{-q/p}$ (which makes sense since $g(x) > 0$ a.e.). Then, with respect to the measure ν, we have that $\langle F \rangle = \int_\Omega fg \, d\mu / \int_\Omega g^q \, d\mu$. On the other hand, with $J(t) = |t|^p$, $\int_\Omega J \circ F \, d\nu = \int_\Omega f^p \, d\mu$. Our conclusion (1) is then an immediate consequence of Jensen's inequality—as is the condition for equality. ■

2.4 THEOREM (Minkowski's inequality)

Suppose that Ω and Γ are any two spaces with sigma-finite measures μ and ν respectively. Let f be a nonnegative function on $\Omega \times \Gamma$ which is $\mu \times \nu$-measurable. Let $1 \leq p < \infty$. Then

$$\int_\Gamma \left(\int_\Omega f(x,y)^p \mu(\mathrm{d}x)\right)^{1/p} \nu(\mathrm{d}y) \geq \left(\int_\Omega \left(\int_\Gamma f(x,y)\nu(\mathrm{d}y)\right)^p \mu(\mathrm{d}x)\right)^{1/p} \quad (1)$$

in the sense that the finiteness of the left side implies the finiteness of the right side.

Equality and finiteness in (1) for $1 < p < \infty$ imply the existence of a μ-measurable function $\alpha : \Omega \to \mathbb{R}^+$ and a ν-measurable function $\beta : \Gamma \to \mathbb{R}^+$ such that

$$f(x,y) = \alpha(x)\beta(y) \quad \text{for } \mu \times \nu\text{-almost every } (x,y).$$

A special case of this is the **triangle inequality**. *For $f, g \in L^p(\Omega, \mathrm{d}\mu)$ (possibly complex functions)*

$$\|f + g\|_p \leq \|f\|_p + \|g\|_p \quad \text{for } 1 \leq p \leq \infty. \quad (2)$$

If $f \not\equiv 0$ and if $1 \leq p < \infty$, there is equality in (2) if and only if $g = \lambda f$ for some $\lambda \geq 0$.

PROOF. First we note that the two functions

$$\int_\Omega f(x,y)^p \mu(\mathrm{d}x) \quad \text{and} \quad H(x) := \int_\Gamma f(x,y)\nu(\mathrm{d}y)$$

are measurable functions. This follows from Theorem 1.12 (Fubini's theorem) and the assumption that f is $\mu \times \nu$-measurable. We can assume that $f > 0$ on a set of positive $\mu \times \nu$ measure, for otherwise there is nothing to prove. We can also assume that the right side of (1) is finite; if not we can truncate f so that it is finite and then use a monotone convergence argument to remove the truncation. Sigma-finiteness is again used in this step.

The right side of (1) can be written as follows:

$$\int_\Omega H(x)^p \mu(\mathrm{d}x) = \int_\Omega \left(\int_\Gamma f(x,y)\nu(\mathrm{d}y)\right) H(x)^{p-1} \mu(\mathrm{d}x)$$
$$= \int_\Gamma \left(\int_\Omega f(x,y) H(x)^{p-1} \mu(\mathrm{d}x)\right) \nu(\mathrm{d}y).$$

The last equation follows by Fubini's theorem. Using Theorem 2.3 (Hölder's inequality) on the right side we obtain

$$\int_\Omega H(x)^p \mu(\mathrm{d}x) \le \int_\Gamma \left(\int_\Omega f(x,y)^p \mu(\mathrm{d}x)\right)^{1/p} \left(\int_\Omega H(x)^p \mu(\mathrm{d}x)\right)^{\frac{p-1}{p}} \nu(\mathrm{d}y). \tag{3}$$

Dividing both sides of (3) by $\left(\int_\Omega H(x)^p \mu(\mathrm{d}x)\right)^{(p-1)/p}$, which is neither zero nor infinity (by our assumptions about f), yields (1).

The equality sign in the use of Hölder's inequality implies that for ν-almost every y there exists a number $\lambda(y)$ (i.e., independent of x) such that

$$\lambda(y) H(x) = f(x,y) \text{ for } \mu\text{-almost every } x. \tag{4}$$

As mentioned above, H is μ-measurable. To see that λ is ν-measurable we note that

$$\lambda(y) \int_\Omega H(x)^p \mu(\mathrm{d}x) = \int_\Omega f(x,y)^p \mu(\mathrm{d}x),$$

and this yields the desired result since the right side is ν-measurable (by Fubini's theorem).

It remains to prove (2). First, by observing that

$$|f(x) + g(x)| \le |f(x)| + |g(x)|, \tag{5}$$

the problem is reduced to proving (2) for nonnegative functions. Evidently, (5) implies (2) when $p = 1$ or ∞, so we can assume $1 < p < \infty$. We set $F(x,1) = |f(x)|$, $F(x,2) = |g(x)|$ and let ν be the counting measure of the set $\Gamma = \{1,2\}$, namely $\nu(\{1\}) = \nu(\{2\}) = 1$. Then the inequality (2) is seen to be a special case of (1). (Note the use of Fubini's theorem here.)

Equality in (2) entails the existence of constants λ_1 and λ_2 (independent of x) such that

$$|f(x)| = \lambda_1(|f(x)| + |g(x)|) \quad \text{and} \quad |g(x)| = \lambda_2(|f(x)| + |g(x)|). \tag{6}$$

Thus, $|g(x)| = \lambda |f(x)|$ almost everywhere for some constant λ. However, equality in (5) means that $g(x) = \lambda f(x)$ with λ real and nonnegative. ∎

• If $1 < p < \infty$, then $L^p(\Omega)$ possesses another geometric structure that has many consequences, among them the characterization of the dual of $L^p(\Omega)$ (2.14) and, in connection with weak convergence, Mazur's theorem (2.13). This structure is called **uniform convexity** and will be described next. The version we give is optimal and is due to [Hanner]; the proof is in [Ball-Carlen-Lieb]. It improves the triangle (or convexity) inequality

$$\|f + g\|_p \le \|f\|_p + \|g\|_p.$$

2.5 THEOREM (Hanner's inequality)

Let f and g be functions in $L^p(\Omega)$. If $1 \leq p \leq 2$, we have

$$\|f+g\|_p^p + \|f-g\|_p^p \geq (\|f\|_p + \|g\|_p)^p + \big|\|f\|_p - \|g\|_p\big|^p, \tag{1}$$

$$(\|f+g\|_p + \|f-g\|_p)^p + \big|\|f+g\|_p - \|f-g\|_p\big|^p \leq 2^p(\|f\|_p^p + \|g\|_p^p). \tag{2}$$

If $2 \leq p < \infty$, the inequalities are reversed.

REMARK. When $\|f\|_p = \|g\|_p$, (2) improves the triangle inequality $\|f+g\|_p \leq \|f\|_p + \|g\|_p$ because, by convexity of $t \mapsto |t|^p$, the left side of (2) is not smaller than $2\|f+g\|_p^p$. To be more precise, it is easy to prove (Exercise 4) that the left side of (2) is bounded below for $1 \leq p \leq 2$ and for $\|f-g\|_p \leq \|f+g\|_p$ by

$$2\|f+g\|_p^p + p(p-1)\|f+g\|_p^{p-2}\|f-g\|_p^2.$$

The geometric meaning of Theorem 2.5 is explored in Exercise 5.

PROOF. (1) and (2) are identities when $p = 2$ ((1) is then called the **parallelogram identity**) and reduce to the triangle inequality if $p = 1$. (2) is derived from (1) by the replacements $f \to f+g$ and $g \to f-g$. Thus, we concentrate on proving (1) for $p \neq 2$. We can obviously assume that $R := \|g\|_p/\|f\|_p \leq 1$ and that $\|f\|_p = 1$. For $0 \leq r \leq 1$ define

$$\alpha(r) = (1+r)^{p-1} + (1-r)^{p-1}$$

and

$$\beta(r) = [(1+r)^{p-1} - (1-r)^{p-1}]r^{1-p},$$

with $\beta(0) = 0$ for $p < 2$ and $\beta(0) = \infty$ for $p > 2$. We first claim that the function $F_R(r) = \alpha(r) + \beta(r)R^p$ has its maximum at $r = R$ (if $p < 2$) and its minimum at $r = R$ (if $p > 2$). In both cases $F_R(R) = (1+R)^p + (1-R)^p$. To prove this assertion we can use the calculus to compute

$$dF_R(r)/dr = \alpha'(r) + \beta'(r)R^p$$
$$= (p-1)[(1+r)^{p-2} - (1-r)^{p-2}](1 - (R/r)^p),$$

which shows that the derivative of $F_R(r)$ vanishes only at $r = R$ and that the sign of the derivative for $r \neq R$ is such that the point $r = R$ is a maximum or minimum as stated above. Furthermore, for all $0 \leq r \leq 1$ we have that $\beta(r) \leq \alpha(r)$ (if $p < 2$) and $\beta(r) \geq \alpha(r)$ (if $p > 2$) and thus, when $R > 1$,

$$\alpha(r) + \beta(r)R^p \leq \alpha(r)R^p + \beta(r) \text{ (if } p < 2\text{)}$$

and
$$\alpha(r) + \beta(r)R^p \geq \alpha(r)R^p + \beta(r) \text{ (if } p > 2).$$

Thus, in all cases we have for all $0 \leq r \leq 1$ and all nonnegative numbers A and B

$$\alpha(r)|A|^p + \beta(r)|B|^p \leq |A+B|^p + |A-B|^p, \qquad p < 2, \qquad (3)$$

and the reverse if $p > 2$. It is important to note that equality holds if $r = B/A \leq 1$.

In fact, (3) and its reverse for $p > 2$ hold for complex A and B (that is why we wrote (3) with $|A|, |B|$, etc.). To see this note that it suffices to prove it when $A = a$ and $B = be^{i\theta}$ with $a, b > 0$. It then suffices to show that $(a^2 + b^2 + 2ab \cos\theta)^{p/2} + (a^2 + b^2 - 2ab \cos\theta)^{p/2}$ has its minimum when $\theta = 0$ (if $p < 2$) or its maximum when $\theta = 0$ (if $p > 2$). But this follows from the fact that the function $x \mapsto x^r$ is concave (if $0 < r < 1$) or convex (if $r > 1$).

To prove (1) it suffices, then, to prove that when $1 \leq p < 2$

$$\int \{|f+g|^p + |f-g|^p\} \, d\mu \geq \alpha(r) \int |f|^p \, d\mu + \beta(r) \int |g|^p \, d\mu \qquad (4)$$

for every $0 \leq r \leq 1$, and the reverse inequality when $p > 2$. But to prove (4) it suffices to prove it pointwise, i.e., for complex numbers f and g. That is, we have to prove

$$|f+g|^p + |f-g|^p \geq \alpha(r)|f|^p + \beta(r)|g|^p \quad \text{for } p < 2$$

(and the reverse for $p > 2$). But this follows from (3). ∎

- Differentiability of $\|f+tg\|_p^p = \int |f+tg|^p$ with respect to $t \in \mathbb{R}$ will prove to be useful. Note that this function of t is convex and hence always has a left and right derivative. In case $p = 1$ it may not be truly differentiable, however, but it is so for $p > 1$, as we show next.

2.6 THEOREM (Differentiability of norms)

Suppose f and g are functions in $L^p(\Omega)$ with $1 < p < \infty$. The function defined on $\mathbb{R}$ by

$$N(t) = \int_\Omega |f(x) + tg(x)|^p \mu(dx)$$

is differentiable and its derivative at $t = 0$ is given by

$$\left.\frac{d}{dt}N\right|_{t=0} = \frac{p}{2} \int_\Omega |f(x)|^{p-2} \{\overline{f}(x)g(x) + f(x)\overline{g}(x)\} \mu(dx). \qquad (1)$$

REMARKS. (1) Note that $|f|^{p-2}f$ is well defined for $1 < p$, even when $f = 0$, in which case it equals 0. This convention will occur frequently in the sequel. Note also that $|f|^{p-2}f$ and $|f|^{p-2}\overline{f}$ are functions in $L^{p'}(\Omega)$.

(2) This notion of derivative of the norm is called the **Gateaux-** or **directional derivative**.

PROOF. It is an elementary fact from calculus that for complex numbers f and g we have

$$\lim_{t \to 0} [|f + tg|^p - |f|^p]/t = |f|^{p-2}(\overline{f}g + f\overline{g}),$$

i.e., $|f + tg|^p$ is differentiable. Our problem, then, is to interchange differentiation and integration. To do so we use the inequality (for $t \le 1$)

$$|f|^p - |f - g|^p \le \frac{1}{t}\{|f + tg|^p - |f|^p\} \le |f + g|^p - |f|^p,$$

which follows from the convexity of $x \to x^p$ (e.g., $|f + tg|^p \le (1-t)|f|^p + t|f+g|^p$). Since $|f|^p, |f+g|^p$ and $|f-g|^p$ are fixed, summable functions, we can do the necessary interchange thanks to the dominated convergence theorem. ∎

2.7 THEOREM (Completeness of L^p-spaces)

*Let $1 \le p \le \infty$ and let f^i, for $i = 1, 2, 3, \ldots$, be a **Cauchy sequence** in $L^p(\Omega)$, i.e., $\|f^i - f^j\|_p \to 0$ as $i, j \to \infty$. (This means that for each $\varepsilon > 0$ there is an N such that $\|f^i - f^j\|_p < \varepsilon$ when $i > N$ and $j > N$.) Then there exists a unique function $f \in L^p(\Omega)$ such that $\|f^i - f\|_p \to 0$ as $i \to \infty$. We denote this latter fact by*

$$f^i \to f \quad \text{as} \quad i \to \infty,$$

*and we say that f^i **converges strongly to** f in $L^p(\Omega)$.*

Moreover, there exists a subsequence $f^{i_1}, f^{i_2}, \ldots$ (with $i_1 < i_2 < \cdots$, of course) and a nonnegative function F in $L^p(\Omega)$ such that

(i) *Domination :* $|f^{i_k}(x)| \le F(x)$ *for all k and μ-almost every x.* (1)

(ii) *Pointwise convergence:* $\lim_{k \to \infty} f^{i_k}(x) = f(x)$ *for μ-almost every x.* (2)

REMARK. 'Convergence' and 'strong convergence' are used interchangeably. The phrase norm convergence is also used.

PROOF. The first, and most important remark, concerns a strategy that is frequently very useful. Namely, it suffices to show the strong convergence for *some* subsequence. To prove this sufficiency, let f^{i_k} be a subsequence that converges strongly to f in $L^p(\Omega)$ as $k \to \infty$. Since, by the triangle inequality,

$$\|f^i - f\|_p \leq \|f^i - f^{i_k}\|_p + \|f^{i_k} - f\|_p,$$

we see that for any $\varepsilon > 0$ we can make the last term on the right side less than $\varepsilon/2$ by choosing k large. The first term on the right can be made smaller than $\varepsilon/2$ by choosing i and k large enough, since f^i is a Cauchy sequence. Thus, $\|f^i - f\|_p < \varepsilon$ for i large enough and we can conclude convergence for the *whole sequence*, i.e., $f^i \to f$. This also proves, incidentally, that the limit—if it exists—is unique.

To obtain such a convergent subsequence pick a number i_1 such that $\|f^{i_1} - f^n\|_p \leq 1/2$ for all $n \geq i_1$. That this is possible is precisely the definition of a Cauchy sequence. Now choose i_2 such that $\|f^{i_2} - f^n\|_p < 1/4$ for all $n \geq i_2$ and so on. Thus we have obtained a subsequence of the integers, i_k, with the property that $\|f^{i_k} - f^{i_{k+1}}\|_p \leq 2^{-k}$ for $k = 1, 2, \ldots$. Consider the monotone sequence of positive functions

$$F_l(x) := |f^{i_1}(x)| + \sum_{k=1}^{l} |f^{i_k}(x) - f^{i_{k+1}}(x)|. \qquad (3)$$

By the triangle inequality

$$\|F_l\|_p \leq \|f^{i_1}\|_p + \sum_{k=1}^{l} 2^{-k} \leq \|f^{i_1}\|_p + 1.$$

Thus, by the monotone convergence theorem, F_l converges pointwise μ-a.e. to a positive function F which is in $L^p(\Omega)$ and hence is finite almost everywhere. The sequence

$$f^{i_{k+1}}(x) = f^{i_1}(x) + (f^{i_2}(x) - f^{i_1}(x)) + \cdots + (f^{i_{k+1}}(x) - f^{i_k}(x)) \qquad (4)$$

thus converges absolutely for almost every x, and hence it also converges for the same x's to some number $f(x)$. Since $|f^{i_k}(x)| \leq F(x)$ and $F \in L^p(\Omega)$, we know by dominated convergence that f is in $L^p(\Omega)$. Again by dominated convergence $\|f^{i_k} - f\|_p \to 0$ as $k \to \infty$ since $|f^{i_k}(x) - f(x)| \leq F(x) + |f(x)| \in L^p(\Omega)$. Thus, the subsequence f^{i_k} converges strongly in $L^p(\Omega)$ to f. ∎

• An example of the use of uniform convexity, Theorem 2.5, is provided by the following projection lemma, which will be useful later.

2.8 LEMMA (Projection on convex sets)

*Let $1 < p < \infty$ and let K be a convex set in $L^p(\Omega)$ (i.e., $f, g \in K \Rightarrow tf + (1-t)g \in K$ for all $0 \leq t \leq 1$) which is also a **norm closed set** (i.e., if $\{g^i\}$ is a Cauchy sequence in K, then its limit, g, is also in K). Let $f \in L^p(\Omega)$ be any function that is not in K and define the distance as*

$$D = \text{dist}(f, K) = \inf_{g \in K} \|f - g\|_p. \tag{1}$$

Then there is a function $h \in K$ such that $D = \|f - h\|_p$. Every function $g \in K$ satisfies

$$\operatorname{Re} \int_\Omega [(g-h)(\overline{f}-\overline{h})]|f-h|^{p-2}\,d\mu \leq 0. \tag{2}$$

PROOF. We shall prove this for $p \leq 2$ using the uniform convexity result 2.5(2) and shall assume $f = 0$. We leave the rest to the reader. Let h^j, $j = 1, 2, \ldots$ be a minimizing sequence in K, i.e., $\|h^j\|_p \to D$. We shall show that this is a Cauchy sequence. First note that $\|h^j + h^k\|_p \to 2D$ as $j, k \to \infty$ (because $\|h^j + h^k\|_p \leq \|h^j\|_p + \|h^k\|_p$, which converges to $2D$, but $\|h^j + h^k\|_p \geq 2D$ since $\frac{1}{2}(h^j + h^k) \in K$). From 2.5(2) we have that

$$(\|h^j + h^k\|_p + \|h^j - h^k\|_p)^p + \big|\|h^j + h^k\|_p - \|h^j - h^k\|_p\big|^p \leq 2^p\{\|h^j\|_p^p + \|h^k\|_p^p\}.$$

The right side converges as $j, k \to \infty$ to $2^{p+1}D^p$. Suppose that $\|h^j - h^k\|_p$ does not tend to zero, but instead (for infinitely many j's and k's) stays bounded below by some number $b > 0$. Then we would have

$$|2D + b|^p + |2D - b|^p \leq 2^{p+1}D^p,$$

which implies that $b = 0$ (by the *strict* convexity of $x \to |2D+x|^p$, which implies that $|2D+x|^p + |2D-x|^p > 2|2D|^p$ unless $x = 0$). Thus, our sequence is Cauchy and, since K is closed, it has a limit $h \in K$.

To verify (2) we fix $g \in K$ and set $g_t = (1-t)h + tg \in K$ for $0 \leq t \leq 1$. Then (with $f = 0$ as before) $N(t) := \|f - g_t\|_p^p \geq D^p$ while $N(0) = D^p$. Since $N(t)$ is differentiable (Theorem 2.6) we have that $N'(0) \geq 0$, and this is exactly (2) (using 2.6(1)). ∎

2.9 DEFINITION (Continuous linear functionals and weak convergence)

The notion of strong convergence just mentioned in Theorem 2.7 (completeness of L^p-spaces) is not the only useful notion of convergence in $L^p(\Omega)$. The second notion, **weak convergence**, requires continuous linear functionals—which we now define. (Incidentally, what is said here applies to any normed vector space—not just $L^p(\Omega)$.) Weak convergence is often more useful than strong convergence for the following reason. We know that a closed, bounded set, A, in $\mathbb{R}^n$ is compact, i.e., every sequence $x^1, x^2, \ldots$ in A has a subsequence with a limit in A. *The analogous compactness assertion in $L^p(\mathbb{R}^n)$, or even $L^p(\Omega)$ for Ω a compact set in $\mathbb{R}^n$, is false.* Below, we show how to construct a sequence of functions, bounded in $L^p(\mathbb{R}^n)$ for every p, but for which there is no convergent subsequence in *any* $L^p(\mathbb{R}^n)$.

If weak convergence is substituted for strong convergence, the situation improves. *The main theorem here, toward which we are headed, is the Banach–Alaoglu Theorem 2.18 which shows that the bounded sets are compact, with this notion of weak convergence, when $1 < p < \infty$.*

A map, L, from $L^p(\Omega)$ to the complex numbers is a **linear functional** if

$$L(af_1 + bf_2) = aL(f_1) + bL(f_2) \tag{1}$$

for all $f_1, f_2 \in L^p(\Omega)$ and $a, b \in \mathbb{C}$. It is a **continuous linear functional** if, for every strongly convergent sequence, f^i,

$$L(f^i) \to L(f) \quad \text{when} \quad f^i \to f. \tag{2}$$

It is a **bounded linear functional** if

$$|L(f)| \leq K\|f\|_p \tag{3}$$

for some finite number K. We leave it as a very easy exercise for the reader to prove that

$$\text{bounded} \iff \text{continuous} \tag{4}$$

for linear maps.

The set of *continuous* linear functionals (continuity is crucial) on $L^p(\Omega)$ is called the **dual** of $L^p(\Omega)$ and is denoted by $L^p(\Omega)^*$. It is also a vector space over the complex numbers (since sums and scalar multiples of elements of $L^p(\Omega)^*$ are in $L^p(\Omega)^*$). This new space has a norm defined by

$$\|L\| = \sup\{|L(f)| : \|f\|_p \leq 1\}. \tag{5}$$

The reader is asked to check that this definition (5) has the three crucial properties of a norm given in 2.1 (a,b,c) : $\|\lambda L\| = |\lambda| \|L\|$, $\|L\| = 0 \Leftrightarrow L = 0$, and the triangle inequality.

It is important to know all the elements of the dual of $L^p(\Omega)$ (or any other vector space). The reason is that an element $f \in L^p(\Omega)$ can be uniquely identified (as we shall see in Theorem 2.10 (linear functionals separate)) if we know how *all* the elements of the dual act on f, i.e., if we know $L(f)$ for all $L \in L^p(\Omega)^*$.

Weak convergence.

If $f, f^1, f^2, f^3, \ldots$ is a sequence of functions in $L^p(\Omega)$, we say that f^i **converges weakly to** f (and write $f^i \rightharpoonup f$) if

$$\lim_{i \to \infty} L(f^i) = L(f) \tag{6}$$

for *every* $L \in L^p(\Omega)^*$.

An obvious but important remark is that strong convergence implies weak convergence, i.e., if $\|f^i - f\|_p \to 0$ as $i \to \infty$, then $\lim_{i \to \infty} L(f^i) = L(f)$ for all continuous linear functionals L. In particular, strong limits and weak limits have to agree, if they both exist (cf. Theorem 2.10).

Two questions that immediately present themselves are (a) what is $L^p(\Omega)^*$ and (b) how is it possible for f^i to converge weakly, but not strongly, to f? For the former, Hölder's inequality (Theorem 2.3) immediately implies that $L^{p'}(\Omega)$ is a subset of $L^p(\Omega)$ when $\frac{1}{p'} + \frac{1}{p} = 1$. A function $g \in L^{p'}(\Omega)$ acts on arbitrary functions $f \in L^p(\Omega)$ by

$$L_g(f) = \int_\Omega g(x) f(x) \mu(\mathrm{d}x). \tag{7}$$

It is easy to check that L_g is linear and continuous. A deeper question is whether (7) gives us *all of* $L^p(\Omega)^*$. The answer will turn out to be 'yes' for $1 \leq p < \infty$, and 'no' for $p = \infty$.

If we accept this conclusion for the moment we can answer question (b) above in the following heuristic way when $\Omega = \mathbb{R}^n$ and $1 < p < \infty$. There are three basic mechanisms by which $f^k \rightharpoonup f$ but $f^k \not\to f$ and we illustrate each for $n = 1$.

(i) f^k 'oscillates to death': An example is $f^k(x) = \sin kx$ for $0 \leq x \leq 1$ and zero otherwise.

(ii) f^k 'goes up the spout': An example is $f^k(x) = k^{1/p} g(kx)$, where g is any fixed function in $L^p(\mathbb{R}^1)$. This sequence becomes very large near $x = 0$.

(iii) f^k 'wanders off to infinity': An example is $f^k(x) = g(x+k)$ for some fixed function g in $L^p(\mathbb{R}^1)$.

In each case $f^k \rightharpoonup 0$ weakly but f^k does not converge strongly to zero (or to anything else). We leave it to the reader to prove this assertion; some of the theorems proved later in this section will be helpful.

We begin our study of weak convergence by showing that there are enough elements of $L^p(\Omega)^*$ to identify all elements of $L^p(\Omega)$. Much of what we prove here is normally proved with the **Hahn–Banach theorem**. We do not use it for several reasons. One is that the interested reader can easily find it in many texts. Another reason is that it is not necessary in the case of $L^p(\Omega)$ spaces and we prefer a direct 'hands on' approach to an abstract approach—wherever the abstract approach does not add significant enlightenment.

2.10 THEOREM (Linear functionals separate)

Suppose that $f \in L^p(\Omega)$ satisfies

$$L(f) = 0 \quad \text{for all } L \in L^p(\Omega)^*. \tag{1}$$

(In the case $p = \infty$ we also assume that our measure space is sigma-finite, but this restriction can be lifted by invoking transfinite induction.) Then

$$f = 0.$$

Consequently, if $f^i \rightharpoonup k$ and $f^i \rightharpoonup h$ weakly in $L^p(\Omega)$, then $k = h$.

PROOF. If $1 < p < \infty$ define

$$g(x) = |f(x)|^{p-2}\overline{f}(x)$$

when $f(x) \neq 0$, and set $g(x) = 0$ otherwise. The fact that $f \in L^p(\Omega)$ immediately implies that $g \in L^{p'}(\Omega)$. We also have that $\int gf = \|f\|_p^p$. But, as we said in 2.9(7), the functional $h \to \int gh$ is a continuous linear functional. Hence, $\int gf = \|f\|_p = 0$ by our hypothesis (1), which implies $f = 0$.

If $p = 1$ we take

$$g(x) = \overline{f}(x)/|f(x)|$$

if $f(x) \neq 0$, and $g(x) = 0$ otherwise. Then $g \in L^\infty(\Omega)$ and the above argument applies. If $p = \infty$ set $A = \{x : |f(x)| > 0\}$. If $f \not\equiv 0$, then $\mu(A) > 0$. Take any measurable subset $B \subset A$ such that $0 < \mu(B) < \infty$; such a set exists by sigma-finiteness. Set $g(x) = \overline{f}(x)/|f(x)|$ for $x \in B$ and zero otherwise. Clearly, $g \in L^1(\Omega)$ and the previous argument can be applied. ∎

2.11 THEOREM (Lower semicontinuity of norms)

*For $1 \leq p \leq \infty$ the L^p-norm is **weakly lower semicontinuous**, i.e., if $f^j \rightharpoonup f$ weakly in $L^p(\Omega)$, then*

$$\liminf_{j \to \infty} \|f^j\|_p \geq \|f\|_p. \tag{1}$$

If $p = \infty$ we make the extra technical assumption that the measure μ is sigma finite.

Moreover, if $1 < p < \infty$ and if $\lim_{j \to \infty} \|f^j\|_p = \|f\|_p$, then $f^j \to f$ strongly as $j \to \infty$.

REMARK. The second part of this theorem is very useful in practice because it often provides a way to identify strongly convergent sequences. For the connection with semicontinuity as in Sect. 1.5, cf. Exercise 1.2. Compare, also, Remark (2) after Theorem 1.9.

PROOF. For $1 \leq p < \infty$ consider the functional

$$L(h) = \int gh \quad \text{with} \quad g(x) = |f(x)|^{p-2}\overline{f}(x)$$

as in the proof of the separation theorem, Theorem 2.10. Since $L(f) = \|f\|_p^p$, we have, by Hölder's inequality with $1/p + 1/q = 1$,

$$\|f\|_p^p = \lim_{j \to \infty} L(f^j) \leq \|g\|_q \liminf_{j \to \infty} \|f^j\|_p$$

which, since $\|g\|_q = \|f\|_p^{p-1}$, gives (1).

For $p = \infty$ assume $\|f\|_\infty =: a > 0$ and consider the set

$$A_\varepsilon = \{x \in \Omega : |f(x)| > a - \varepsilon\}.$$

Since the space (Ω, μ) is σ-finite, there is a sequence of sets B_k of finite measure such that $A_\varepsilon \cap B_k$ increases to A_ε. Set $g_{k,\varepsilon} = \overline{f(x)}/|f(x)|$ if $x \in A_\varepsilon \cap B_k$ and zero otherwise. Now by Hölder's inequality

$$\mu(A_\varepsilon \cap B_k) \liminf_{j \to \infty} \|f^j\|_\infty \geq \lim_{j \to \infty} \int g_{k,\varepsilon} f^j = \int_{A_\varepsilon \cap B_k} |f(x)| \, d\mu,$$

where the last equation follows from the weak convergence of f^j to f. But

$$\int_{A_\varepsilon \cap B_k} |f(x)| \, d\mu \geq (a - \varepsilon) \mu(A_\varepsilon \cap B_k),$$

and hence $\liminf_{j\to\infty} \|f^j\|_\infty \geq \|f\|_\infty - \varepsilon$ for all $\varepsilon > 0$.

Thus far we have proved (1). To prove the second assertion for $1 < p < \infty$ we first note that $\lim \|f^j\|_p = \|f\|_p$ implies that $\lim \|f^j + f\|_p = 2\|f\|_p$ (clearly $f^j + f \rightharpoonup 2f$ and, by (1), $\liminf \|f^j + f\|_p \geq 2\|f\|_p$, but $\|f^j + f\|_p \leq \|f^j\|_p + \|f\|_p$ by the triangle inequality). For $p \leq 2$ we use the uniform convexity 2.5(2) (we leave $p > 2$ to the reader) with $g = f^j$. Taking limits we have (with $A_j = \|f + f^j\|_p$ and $B_j = \|f - f^j\|_p$)

$$\limsup_{j\to\infty} \left\{ (A_j + B_j)^p + |A_j - B_j|^p \right\} \leq 2^{p+1} \|f\|_p^p.$$

Since $x \mapsto |A + x|^p$ is *strictly* convex for $1 < p < \infty$, and since $A_j \to 2\|f\|_p$, B_j must tend to zero. ∎

● The next theorem shows that weakly convergent sequences are, at least, norm bounded.

2.12 THEOREM (Uniform boundedness principle)

Let $f^1, f^2, \ldots$ be a sequence in $L^p(\Omega)$ with the following property: For each functional $L \in L^p(\Omega)^*$ the sequence of numbers $L(f^1), L(f^2), \ldots$ is bounded. Then the norms $\|f^j\|_p$ are bounded, i.e., $\|f^j\|_p < C$ for some finite $C > 0$.

PROOF. We suppose the theorem is false and will derive a contradiction. We do this for $1 < p < \infty$, and leave the easy modifications for $p = 1$ and $p = \infty$ to the reader.

First, for the following reason, we can assume that $\|f^j\|_p = 4^j$. By choosing a subsequence (which we continue to denote by $j = 1, 2, 3, \ldots$) we can certainly arrange that $\|f^j\|_p \geq 4^j$. Then we replace the sequence f^j by the sequence

$$F^j = 4^j f^j / \|f^j\|_p,$$

which satisfies the hypothesis of the theorem since

$$L(F^j) = (4^j/\|f^j\|_p)L(f^j),$$

which is certainly bounded. Clearly $\|F^j\|_p = 4^j$ and our next step is to derive a contradiction from this fact by constructing an L for which the sequence $L(F^j)$ is not bounded.

Set $T_j(x) = |F^j(x)|^{p-2}\overline{F^j}(x)/\|F^j\|_p^{p-1}$ and define complex numbers σ_n

of modulus 1 as follows: pick $\sigma_1 = 1$ and choose σ_n recursively by requiring $\sigma_n \int T_n F^n$ to have the same argument as

$$\sum_{j=1}^{n-1} 3^{-j} \sigma_j \int T_j F^n.$$

Thus,

$$\left| \sum_{j=1}^{n} 3^{-j} \sigma_j \int T_j F^n \right| \geq 3^{-n} \int T_n F^n = 3^{-n} \|F^n\|_p = (4/3)^n.$$

Now define the linear functional L by setting

$$L(h) = \sum_{j=1}^{\infty} 3^{-j} \sigma_j \int T_j h,$$

which is obviously continuous by Hölder's inequality and the fact that $\|T_j\|_{p'} = 1$.

We can bound $|L(F^k)|$ from below as follows.

$$|L(F^k)| \geq \left| \sum_{j=1}^{k} 3^{-j} \sigma_j \int T_j F^k \right| - \left(\sum_{j=k+1}^{\infty} 3^{-j} \right) 4^k$$

$$\geq 3^{-k} 4^k - 3^{-k} 4^k \frac{1/3}{1 - (1/3)} = \frac{1}{2} \left(\frac{4}{3} \right)^k$$

which tends to ∞ as $k \to \infty$. This contradicts the boundedness of $L(F^k)$. ∎

● The next theorem, [Mazur], shows how to build strongly convergent sequences out of weakly convergent ones. It can be very useful for proving existence of minimizers for variational problems. In fact, we shall employ it in the capacitor problem in Chapter 11. The theorem holds in greater generality than the version we give here, e.g., it also holds for $L^1(\Omega)$ and $L^\infty(\Omega)$. In fact it holds for any normed space (see [Rudin 1991], Theorem 3.13). We prove it for $1 < p < \infty$ by using Lemma 2.8 (projection on convex sets). For full generality it is necessary to use the Hahn–Banach theorem, which involves the axiom of choice and which the reader can find in many texts. The proof here is somewhat more constructive and intuitive.

2.13 THEOREM (Strongly convergent convex combinations)

Let $1 < p < \infty$ and let $f^1, f^2, \ldots$ be a sequence in $L^p(\Omega)$ that converges weakly to $F \in L^p(\Omega)$. Then we can form a sequence $F^1, F^2, \ldots$ in $L^p(\Omega)$ that converges strongly to F, and such that each F^j is a convex combination of the functions $f^1, \ldots, f^j$. I.e., for each j there are nonnegative numbers $c_1^j, \ldots, c_j^j$ such that $\sum_{k=1}^{j} c_k^j = 1$ and such that the functions

$$F^j := \sum_{k=1}^{j} c_k^j f^k$$

converge strongly to F.

PROOF. First, consider the set $\widetilde{K} \subset L^p(\Omega)$ which consists of all the f^j's together with all finite convex combinations of them, i.e., all functions of the form $\sum_{k=1}^{m} d_k f^k$ with m arbitrary and with $\sum_{k=1}^{m} d_k = 1$ where $d_k \geq 0$. This set $\widetilde{K}$ is clearly convex, i.e. $f, g \in \widetilde{K} \Rightarrow \lambda f + (1-\lambda)g \in \widetilde{K}$ for all $0 \leq \lambda \leq 1$.

Next, let K denote the union of $\widetilde{K}$ and all its limit points, i.e. we add to $\widetilde{K}$ all functions in $L^p(\Omega)$ that are limits of Cauchy sequences of elements of $\widetilde{K}$. We claim that (a) K is convex and (b) K is closed. To prove (a) we note that if $f^j \to f$ and $g^j \to g$ (with $f^j, g^j \in \widetilde{K}$) then $\lambda f^j + (1-\lambda)g^j \in \widetilde{K}$ and converges to $\lambda f + (1-\lambda)g$. To prove (b), the reader can use the triangle inequality to prove that 'Cauchy sequences of Cauchy sequences are Cauchy sequences'. (Our construction here imitates the construction of the reals from the rationals.)

Our theorem amounts to the assertion that the weak limit F is in K. Suppose otherwise. By Lemma 2.8 (projection on convex sets) there is a function $h \in K$ such that $D = \text{dist}(F, K) = \|F - h\|_p > 0$. In 2.8(2) we considered the function

$$\ell(x) = [\overline{F}(x) - \overline{h}(x)]|F(x) - h(x)|^{p-2}$$

which is in $L^{p'}(\Omega)$ and showed that the continuous linear function $L(g) := \int \ell g$ satisfies

$$\operatorname{Re} L(g) - \operatorname{Re} L(h) \leq 0 \qquad (1)$$

for all $g \in K$. However, $L(F - h) = \|F - h\|_p^p$, and hence

$$\operatorname{Re} L(F) - \operatorname{Re} L(h) > 0 \qquad (2)$$

because $F - h$ is not the zero function. (2) contradicts (1) because $L(f^j) \to L(F)$ by assumption, and the f^j's are in K. ∎

• At last we come to the identification of $L^p(\Omega)^*$, the dual of $L^p(\Omega)$, for $1 \leq p < \infty$. This is **F. Riesz's representation theorem**. The dual of $L^\infty(\Omega)$ is not given because it is a huge, less useful space that requires the axiom of choice for its construction.

2.14 THEOREM (The dual of $L^p(\Omega)$)

When $1 \leq p < \infty$ the dual of $L^p(\Omega)$ is $L^q(\Omega)$, with $1/p + 1/q = 1$, in the sense that every $L \in L^p(\Omega)^*$ has the form

$$L(g) = \int_\Omega v(x) g(x) \mu(\mathrm{d}x) \tag{1}$$

for some unique $v \in L^q(\Omega)$. (In case $p = 1$ we make the additional technical assumption that (Ω, μ) is sigma-finite.) In each case the norm of L (defined in 2.9(5)) is

$$\|L\| = \|v\|_q. \tag{2}$$

PROOF. $\boxed{1 < p < \infty:}$ With $L \in L^p(\Omega)^*$ given, define the set $K = \{g \in L^p(\Omega) : L(g) = 0\} \subset L^p(\Omega)$. Clearly K is convex and K is closed (here is where the continuity of L enters). Assume $L \neq 0$, whence there is $f \in L^p(\Omega)$ such that $L(f) \neq 0$, i.e., $f \notin K$. By Lemma 2.8 (projection on convex sets) there is an $h \in K$ such that

$$\operatorname{Re} \int uk \leq 0 \tag{3}$$

for all $k \in K$. Here $u(x) = |f(x) - h(x)|^{p-2}[\overline{f}(x) - \overline{h}(x)]$, which is evidently in $L^q(\Omega)$. However, K is a linear space and hence $-k \in K$ and $ik \in K$ whenever $k \in K$. The first fact tells us that $\operatorname{Re} \int uk = 0$ and the second fact implies $\int uk = 0$ for all $k \in K$.

Now let g be an arbitrary element of $L^p(\Omega)$ and write $g = g_1 + g_2$ with

$$g_1 = \frac{L(g)}{L(f-h)}(f-h) \quad \text{and} \quad g_2 = g - g_1.$$

(Note that $L(f - h) = L(f) \neq 0$.) One easily checks that $L(g_2) = 0$, i.e., $g_2 \in K$, whence

$$\int ug = \int ug_1 + \int ug_2 = \int ug_1 = L(g)A,$$

where $A = \int u(f-h)/L(f-h) \neq 0$, since $\int u(f-h) = \int |f-h|^p$. Thus, the v in (1) equals u/A. The uniqueness of v follows from the fact that if

$\int (v-w)g = 0$ for all $g \in L^p(\Omega)$, and with $w \in L^q(\Omega)$, then we could obtain a contradiction by choosing $g = \overline{(v-w)}|v-w|^{q-2} \in L^p(\Omega)$. The easy proof of (2) is left to the reader.

$\boxed{p=1:}$ Let us assume for the moment that Ω has finite measure. In this case, Hölder's inequality implies that a continuous linear functional L on $L^1(\Omega)$ has a restriction to $L^p(\Omega)$ which is again continuous since

$$|L(f)| \leq C\|f\|_1 \leq C\mu(\Omega)^{1/q}\|f\|_p \tag{4}$$

for all $p \geq 1$. By the previous proof for $p > 1$, we have the existence of a unique $v_p \in L^q(\Omega)$ such that $L(f) = \int v_p(x) f(x) \mu(dx)$ for all $f \in L^p(\Omega)$. Moreover, since $L^r(\Omega) \subset L^p(\Omega)$ for $r \geq p$ (by Hölder's inequality) the uniqueness of v_p for each p implies that v_p is, in fact, independent of p, i.e., this function (which we now call v) is in every $L^r(\Omega)$-space for $1 < r < \infty$.

If we now pick some dual pair q and p with $p > 1$ and choose $f = |v|^{q-2}\overline{v}$ in (4) we obtain

$$\int |v|^q = L(f) \leq C(\mu(\Omega))^{1/q} \left(\int |v|^{(q-1)p} \right)^{1/p} = C(\mu(\Omega))^{1/q} \|v\|_q^{q-1},$$

and hence $\|v\|_q \leq C(\mu(\Omega))^{1/q}$ for all $q < \infty$. We claim that $v \in L^\infty(\Omega)$; in fact $\|v\|_\infty \leq C$. Suppose that $\mu(\{x \in \Omega : |v(x)| > C + \varepsilon\}) = M > 0$. Then $\|v\|_q \geq (C+\varepsilon)M^{1/q}$, which exceeds $C\mu(\Omega)^{1/q}$ if q is big enough. Thus $v \in L^\infty(\Omega)$ and $L(f) = \int v(x)f(x)\,d\mu$ for all $f \in L^p(\Omega)$ for any $p > 1$. If $f \in L^1(\Omega)$ is given, then $\int |v(x)||f(x)|\,d\mu < \infty$. Replacing $f(x)$ by $f^k(x) = f(x)$ whenever $|f(x)| \leq k$ and by zero otherwise, we note that $|f^k(x)| \leq |f(x)|$ and $f^k(x) \to f(x)$ pointwise as $k \to \infty$; hence, by dominated convergence, $f^k \to f$ in $L^1(\Omega)$ and $vf^k \to vf$ in $L^1(\Omega)$. Thus

$$L(f) = \lim_{k \to \infty} L(f^k) = \lim_{k \to \infty} \int v f^k \, d\mu = \int v f \, d\mu.$$

The previous conclusion can be extended to the case that $\mu(\Omega) \doteq \infty$ but Ω is sigma-finite. Then

$$\Omega = \bigcup_{j=1}^{\infty} \Omega_j$$

with $\mu(\Omega_j)$ finite and with $\Omega_j \cap \Omega_k$ empty whenever $j \neq k$. Any $L^1(\Omega)$ function f can be written as

$$f(x) = \sum_{j=1}^{\infty} f_j(x)$$

where $f_j = \chi_j f$ and χ_j is the characteristic function of Ω_j. $f_j \mapsto L(f_j)$ is then an element of $L^1(\Omega_j)^*$, and hence there is a function $v_j \in L^\infty(\Omega_j)$ such that $L(f_j) = \int_{\Omega_j} v_j f_j = \int_{\Omega_j} v_j f$. The important point is that each v_j is bounded in $L^\infty(\Omega_j)$ by the *same* $C = \|L\|$. Moreover, the function v, defined on all of Ω by $v(x) = v_j(x)$ for $x \in \Omega_j$, is clearly measurable and bounded by C. Thus, we have $L(f) = \int_\Omega v f$ by the countable additivity of the measure μ. Uniqueness is left to the reader. ∎

• Our next goal is the Banach–Alaoglu Theorem, 2.18, and, although it can be presented in a much more general setting, we restrict ourselves to the particular case in which Ω is a subset of $\mathbb{R}^n$ and $\mu(\mathrm{d}x)$ is Lebesgue measure. To reach it we need the separability of $L^p(\Omega)$ for $1 < p < \infty$ and to achieve that we need the density of continuous functions in $L^p(\Omega)$. The next theorem establishes this fact, and it is one of the most fundamental; its importance cannot be overstressed. It permits us to approximate $L^p(\Omega)$, functions by C_c^∞-functions (Lemma 2.19). Why then, the reader might ask, did we introduce the L^p-spaces? Why not restrict ourselves to the C^∞-functions from the outset? The answer is that the set of continuous functions is not complete in $L^p(\Omega)$, i.e., the analogue of Theorem 2.7 does not hold for them because limits of continuous functions are not necessarily continuous. As preparation we need 2.15–2.17.

2.15 CONVOLUTION

When f and g are two (complex-valued) functions on $\mathbb{R}^n$ we define their *convolution* to be the function $f * g$ given by

$$f * g(x) = \int_{\mathbb{R}^n} f(x-y) g(y) \, \mathrm{d}y. \tag{1}$$

Note that $f * g = g * f$ by changing variables. One has to be careful to make sure that (1) makes sense. One way is to require $f \in L^p(\mathbb{R}^n)$ and $g \in L^{p'}(\mathbb{R}^n)$, in which case the integral in (1) is well defined for *all* x by Hölder's inequality. More is true, as Lemma 2.20 and Theorem 4.2 (Young's inequality) show. In case f and g are in $L^1(\mathbb{R}^n)$, (1) makes sense for almost every $x \in \mathbb{R}^n$ and defines a measurable function that is in $L^1(\mathbb{R}^n)$ (see Exercise 7). Indeed, Theorem 4.2 shows that when $f \in L^p(\mathbb{R}^n)$ and $g \in L^q(\mathbb{R}^n)$ with $1/p + 1/q \geq 1$, then (1) is finite a.e. and defines a measurable function that is in $L^r(\mathbb{R}^n)$ with $1 + 1/r = 1/p + 1/q$. In the following theorem we prove this for $q = 1$.

2.16 THEOREM (Approximation by C^∞-functions)

Let j be in $L^1(\mathbb{R}^n)$ with $\int_{\mathbb{R}^n} j = 1$. For $\varepsilon > 0$, define $j_\varepsilon(x) := \varepsilon^{-n} j(x/\varepsilon)$, so that $\int_{\mathbb{R}^n} j_\varepsilon = 1$ and $\|j_\varepsilon\|_1 = \|j\|_1$. Let $f \in L^p(\mathbb{R}^n)$ for some $1 \leq p < \infty$ and define the convolution
$$f_\varepsilon := j_\varepsilon * f.$$
Then
$$f_\varepsilon \in L^p(\mathbb{R}^n) \quad \text{and} \quad \|f_\varepsilon\|_p \leq \|j\|_1 \|f\|_p. \tag{1}$$
$$f_\varepsilon \to f \quad \text{strongly in} \quad L^p(\mathbb{R}^n) \text{ as } \varepsilon \to 0. \tag{2}$$
If $j \in C_c^\infty(\mathbb{R}^n)$, then $f_\varepsilon \in C^\infty(\mathbb{R}^n)$ and (see Remark (3) below)
$$D^\alpha f_\varepsilon = (D^\alpha j_\varepsilon) * f. \tag{3}$$

REMARKS. (1) The above theorem is stated for $\mathbb{R}^n$ but it applies equally well to any measurable set $\Omega \subset \mathbb{R}^n$. Given $f \in L^p(\Omega)$ we can define $\widetilde{f} \in L^p(\mathbb{R}^n)$ by $\widetilde{f}(x) = f(x)$ for $x \in \Omega$ and $\widetilde{f}(x) = 0$ for $x \notin \Omega$. Then define
$$f_\varepsilon(x) = (j_\varepsilon * \widetilde{f})(x) \quad \text{for } x \in \Omega.$$
Equation (1) holds in $L^p(\Omega)$ since
$$\|f_\varepsilon\|_{L^p(\Omega)} \leq \|f_\varepsilon\|_{L^p(\mathbb{R}^n)} \leq \|j\|_1 \|\widetilde{f}\|_{L^p(\mathbb{R}^n)} = \|j\|_1 \|f\|_{L^p(\Omega)}.$$
Likewise, (2) is correct in $L^p(\Omega)$. If Ω is open (so that $C^\infty(\Omega)$ can be defined), then the third statement obviously holds as well with $C^\infty(\mathbb{R}^n)$ replaced by $C^\infty(\Omega)$ and f replaced by $\widetilde{f}$.

(2) We shall see in Lemma 2.19 that Theorem 2.16 can be extended in another way: The $C^\infty(\mathbb{R}^n)$ approximants, $j_\varepsilon * f$, can be modified so that they are in $C_c^\infty(\mathbb{R}^n)$ without spoiling conclusions (1) and (2). The proof of Lemma 2.19 is an easy exercise, but the lemma is stated separately because of its importance.

(3) In Chapter 6 we shall define the distributional derivative of an L^p function, f, denoted by $D^\alpha f$. It is then true that $(D^\alpha j_\varepsilon) * f = j_\varepsilon * D^\alpha f$.

PROOF. Let $J_\varepsilon(x) := |j_\varepsilon(x)|$ and $F(x) := |f(x)|$. Then $|f_\varepsilon(x)| \leq J_\varepsilon * F$ and, writing $J_\varepsilon = J_\varepsilon^{1/q} J_\varepsilon^{1/p}$ with $1/p + 1/q = 1$, we have, for $p > 1$,

$$\int_{\mathbb{R}^n} |f_\varepsilon|^p \leq \int_{\mathbb{R}^n} \left[\int_{\mathbb{R}^n} J_\varepsilon(y) F(x-y) \, dy \right]^p dx$$
$$\leq \int_{\mathbb{R}^n} \left\{ \left(\int_{\mathbb{R}^n} J_\varepsilon \right)^{p/q} \int_{\mathbb{R}^n} J_\varepsilon(y) F(x-y)^p \, dy \right\} dx \tag{4}$$
$$= \left(\int_{\mathbb{R}^n} J_\varepsilon \right)^{1+p/q} \int_{\mathbb{R}^n} F^p = \|j\|_1^p \int_{\mathbb{R}^n} F^p.$$

The first inequality is Hölder's (Theorem 2.3) applied to the dy integration. The second equality uses Fubini's theorem and

$$\int_{\mathbb{R}^n} F(x-y)^p \, dx = \int_{\mathbb{R}^n} F^p,$$

independent of y. This proves (1) for $p > 1$. The proof for $p = 1$ is the same, but without Hölder's inequality.

The bound (1) clearly shows that it suffices to prove (2) for the real and imaginary parts of f separately. Rewriting a real function as the difference of its positive and negative parts, $f = f_+ - f_-$ (see Sect. 1.4), (1) again shows that it suffices to prove (2) for nonnegative functions. Thus, let f be a nonnegative function in $L^p(\mathbb{R}^n)$ and set $f^k(x) := f(x)$ on the set $\{x \in \mathbb{R}^n : f(x) < k, |x| < k\}$ and $f^k(x) = 0$ otherwise. By dominated convergence,

$$\lim_{k \to \infty} \|f^k - f\|_p = 0$$

and, by (1),

$$\|f_\varepsilon^k - f_\varepsilon\|_p \leq \|j\|_1 \|f^k - f\|_p.$$

Hence it suffices to show (2) for those nonnegative functions f that are bounded by some constant C and vanish outside some ball B_R of radius R. Since $|f_\varepsilon(x)| \leq C \|j\|_1$ we have for $p \geq 2$ that

$$|f_\varepsilon(x) - f(x)|^p \leq [C(1 + \|j\|_1)]^{p-2} |f_\varepsilon(x) - f(x)|^2.$$

If $p \leq 2$, Hölder's inequality shows that

$$\|f_\varepsilon - f\|_p \leq |B_R|^{1/p - 1/2} \|f_\varepsilon - f\|_2$$

where $|B_R|$ denotes the volume of the ball B_R. These two inequalities reduce the problem of proving (2) for all p to the case $p = 2$.

Now,

$$\int_{\mathbb{R}^n} j_\varepsilon = 1$$

implies

$$f_\varepsilon(x) - f(x) = \int_{\mathbb{R}^n} j_\varepsilon(y) [f(x-y) - f(x)] \, dy.$$

Using the strategy of (4) with $p = 2$,

$$\|f_\varepsilon - f\|_2^2 \leq 2 \|j\|_1 \int_{\mathbb{R}^n} J_\varepsilon(y) G(y) \, dy$$

with

$$G(y) = \int_{\mathbb{R}^n} f(x) [f(x) - f(x-y)] \, dx.$$

Obviously, $G(y)$ is bounded by $2\int_{\mathbb{R}^n} f^2$. Moreover, we claim that G is continuous at $y = 0$ and $G(0) = 0$. If so, we are done because $\int_{\mathbb{R}^n} J_\varepsilon G = \int_{\mathbb{R}^n} J(y)G(\varepsilon y)\,\mathrm{d}y$, and this converges to zero as $\varepsilon \to 0$ by dominated convergence (note that $J(y)G(\varepsilon y)$ is bounded by the $L^1(\mathbb{R}^n)$ function $2(\int_{\mathbb{R}^n} f^2)J(y)$ and $G(\varepsilon y) \to 0$ as $\varepsilon \to 0$ for all y).

In short, *the essence of* (2) *is that*
$$I(y) := \int_{\mathbb{R}^n} f(x)f(x-y)\,\mathrm{d}x \quad \text{converges to} \quad \int_{\mathbb{R}^n} f^2 \text{ as } y \to 0.$$
Clearly $I(y) \leq \int_{\mathbb{R}^n} f^2$ (by Schwarz's inequality), so it suffices to prove
$$\liminf_{y \to 0} I(y) \geq \int_{\mathbb{R}^n} f^2.$$
By the layer cake representation, 1.13, and Fubini's theorem we see that it suffices to prove the key relation:
$$\liminf_{y \to 0} \int_{\mathbb{R}^n} \chi_A(x)\chi_B(x+y)\,\mathrm{d}x \geq \int_{\mathbb{R}^n} \chi_A(x)\chi_B(x)\,\mathrm{d}x \tag{5}$$
whenever χ_A and χ_B are characteristic functions of measurable sets A and B of finite measure.

Lebesgue measure has the outer regularity property, 1.2(9): Given $\delta > 0$ there is an open set $O \subset \mathbb{R}^n$ such that $B \subset O$ and $\mathcal{L}^n(O \sim B) \leq \delta$. Thus,
$$\liminf_{y \to 0} \int_{\mathbb{R}^n} \chi_A(x)\chi_B(x+y)\,\mathrm{d}x$$
$$\geq \liminf_{y \to 0} \int_{\mathbb{R}^n} \chi_A(x)\chi_O(x+y)\chi_B(x)\,\mathrm{d}x - \delta$$
$$= \int_{\mathbb{R}^n} \chi_A(x)\chi_B(x) - \delta$$
by dominated convergence and the fact that
$$\lim_{y \to 0} \chi_O(x+y) = \chi_O(x) = 1$$
if $x \in O$. Since δ was arbitrary, (5) is true and (2) is completed.

To prove (3) we shall prove that
$$\partial f_\varepsilon / \partial x_i = (\partial j_\varepsilon / \partial x_i) * f, \tag{6}$$
and that this function is continuous. This will imply that $f_\varepsilon \in C^1(\mathbb{R}^n)$ and, by induction (since $\partial j_\varepsilon / \partial x_i \in C^\infty(\mathbb{R}^n)$), that $f_\varepsilon \in C^\infty(\mathbb{R}^n)$. The continuity is an elementary consequence of the dominated convergence theorem. Since the support of j_ε is compact, the difference quotient
$$\Delta_{\varepsilon,\delta}(x) := [j_\varepsilon(\ldots, x_i + \delta, \ldots) - j_\varepsilon(\ldots, x_i, \ldots)]/\delta$$
is uniformly bounded in δ and of compact support and it is obviously bounded by some fixed $L^{p'}$-function. The desired conclusion follows again by dominated convergence. ∎

2.17 LEMMA (Separability of $L^p(\mathbb{R}^n)$)

There exists a fixed, countable set of functions $\mathcal{F} = \{\phi_1, \phi_2, \ldots\}$ (which will be constructed explicitly) with the following property: For each $1 \leq p < \infty$ and for each measurable set $\Omega \subset \mathbb{R}^n$, for each function $f \in L^p(\Omega)$ and for each $\varepsilon > 0$ we have $\|f - \phi_j\|_p < \varepsilon$ for some function ϕ_j in $\mathcal{F}$.

PROOF. It suffices to prove this for $\Omega = \mathbb{R}^n$ since we always can extend $f \in L^p(\Omega)$ to a function in $L^p(\mathbb{R}^n)$ by setting $f(x) = 0$ for $x \notin \Omega$.

To define $\mathcal{F}$ we first define a countable family, Γ, of sets in $\mathbb{R}^n$ as the collection of cubes $\Gamma_{j,m}$, for $j = 1, 2, 3, \ldots$ and for $m \in \mathbb{Z}^n$, given by

$$\Gamma_{j,m} = \{x \in \mathbb{R}^n : 2^{-j} m_i < x_i \leq 2^{-j}(m_i + 1),\ i = 1, \ldots, n\}.$$

For each j, the $\Gamma_{j,m}$'s obviously cover the whole of $\mathbb{R}^n$ as m ranges over $\mathbb{Z}^n$, the points in $\mathbb{R}^n$ with integer coordinates. The family Γ is a countable family (here we use the fact that a countable family of countable families is countable).

Next, we define the family of functions $\mathcal{F}_j$ to consist of all functions f on $\mathbb{R}^n$ with the property that $f(x) = c_{j,m} =$ constant for $x \in \Gamma_{j,m}$ and, moreover, the numbers $c_{j,m}$ are restricted to be rational complex numbers. Again this family $\mathcal{F}_j$ is countable. $\mathcal{F}$ is defined to be $\bigcup_{j=1}^\infty \mathcal{F}_j$, which is again countable.

Given $f \in L^p(\mathbb{R}^n)$, we first use Theorem 2.16 to replace f by a continuous function $\widetilde{f} \in L^p(\mathbb{R}^n)$ such that $\int |f - \widetilde{f}|^p < \varepsilon/3$. Thus, it suffices to find $f_j \in \mathcal{F}$ such that $\int |\widetilde{f} - f_j|^p < 2\varepsilon/3$. We can also assume (as in the proof of 2.16) that $\widetilde{f}(x) = 0$ for x outside some large cube γ of the form $\{x : -2^J \leq x_i < 2^J\}$ for some integer J.

For each integer j we define

$$\widetilde{f}_j(x) = 2^{-nj} \int_{\Gamma_{j,m}} \widetilde{f}(y)\,dy \quad \text{for } x \in \Gamma_{j,m},$$

i.e., $\widetilde{f}_j$ is the average of $\widetilde{f}$ over $\Gamma_{j,m}$. Since $\widetilde{f}$ is continuous, it is uniformly continuous on γ. This means that for each $\varepsilon' > 0$ there is a $\delta > 0$ such that $|\widetilde{f}(y) - \widetilde{f}(x)| < \varepsilon'$ whenever $|x - y| < \delta$. Therefore, if j is large enough so that $\delta \geq \sqrt{n} 2^{-j}$, we have

$$\int_{\mathbb{R}^n} |\widetilde{f}(x) - \widetilde{f}_j(x)|^p\,dx \leq \text{volume}(\gamma)(2\varepsilon')^p.$$

We can choose ε' to satisfy $(2\varepsilon')^p \text{volume}(\gamma) < \varepsilon/3$. Thus, $\int |f - \widetilde{f}_j|^p < \varepsilon/3$.

The final step is to replace $\widetilde{f}_j$ by a function $\widehat{f}_j$ that assumes only rational complex values in such a way that $\int |\widetilde{f}_j - \widehat{f}_j|^p < \varepsilon/3$. This is easy to do since only finitely many cubes (and hence only finitely many values of $\widetilde{f}_j$) are involved. Since $\widehat{f}_j \in \mathcal{F}$, our goal has been accomplished. ∎

● The next theorem is the **Banach–Alaoglu theorem**, but for the special case of L^p-spaces. As such, it predates Banach–Alaoglu (although we shall continue to use that appellation). For the case at hand, i.e., L^p-spaces, the axiom of choice in the realm of the uncountable is *not* needed in the proof.

2.18 THEOREM (Bounded sequences have weak limits)

Let $\Omega \in \mathbb{R}^n$ be a measurable set and consider $L^p(\Omega)$ with $1 < p < \infty$. Let $f^1, f^2, \ldots$ be a sequence of functions, bounded in $L^p(\Omega)$. Then there exist a subsequence $f^{n_1}, f^{n_2}, \ldots$ (with $n_1 < n_2 < \cdots$) and an $f \in L^p(\Omega)$ such that $f^{n_i} \rightharpoonup f$ weakly in $L^p(\Omega)$ as $i \to \infty$, i.e., for every bounded linear functional $L \in L^p(\Omega)^*$

$$L(f^{n_i}) \to L(f) \quad as \quad i \to \infty.$$

PROOF. We know from Riesz's representation theorem, Theorem 2.14, that the dual of $L^p(\Omega)$ is $L^q(\Omega)$ with $1/p + 1/q = 1$. Therefore, our first task is to find a subsequence f^{n_j} such that $\int f^{n_j}(x)g(x)\,\mathrm{d}\mu$ is a convergent sequence of numbers for every $g \in L^q(\Omega)$. In view of Lemma 2.17 (separability of $L^p(\mathbb{R}^n)$), it suffices to show this convergence only for the special countable sequence of functions ϕ^j given there.

Cantor's diagonal argument will be used. First, consider the sequence of numbers $C_1^j = \int f^j \phi_1$, which is bounded (by Hölder's inequality and the boundedness of $\|f^j\|_p$). There is then a subsequence (which we denote by f_1^j) such that C_1^j converges to some number C_1 as $j \to \infty$. Second, starting with this new sequence $f_1^1, f_1^2, \ldots$, a parallel argument shows that we can pass to a further subsequence such that $C_2^j = \int f^j \phi_2$ also converges to some number C_2. This second subsequence is denoted by $f_2^1, f_2^2, f_2^3, \ldots$. Proceeding inductively we generate a countable family of subsequences of subsequences so that for the kth subsequence (and all further subsequences) $\int f_k^j \phi_k$ converges as $j \to \infty$. Moreover, f_ℓ^j is somewhere in the sequence $f_k^1, f_k^2, \ldots$ if $k \leq \ell$.

Cantor told us how to construct one convergent subsequence from all these. The kth function in this new sequence f^{n_k} (which will henceforth be called F^k) is defined to be the kth function in the kth sequence, i.e., $F^k := f_k^k$. It is a simple exercise to show that $\int F^k \phi_\ell \to C_\ell$ as $j \to \infty$.

Our second and final task is to use the knowledge that $\int F^j g$ converges to some number (call it $L(g)$) as $j \to \infty$ for all $g \in L^q(\mathbb{R}^n)$ in order to show the existence of an $f \in L^p$ to which F^j converges weakly. To do so we note that $L(g)$ is clearly a linear functional on $L^q(\mathbb{R}^n)$ and it is also bounded (and hence continuous) since $\|F^j\|_p$ is bounded. But Theorem 2.14 tells us that the dual of $L^q(\mathbb{R}^n)$ is precisely $L^p(\mathbb{R}^n)$, and hence there is some $f \in L^p(\mathbb{R}^n)$ such that $\int F^j g \to L(g) = \int fg$. ∎

REMARK. What was really used here was the fact that the 'double dual' (or the 'dual of the dual') of $L^p(\mathbb{R}^n)$ is $L^p(\mathbb{R}^n)$. For other spaces, such as $L^1(\mathbb{R}^n)$ or $L^\infty(\mathbb{R}^n)$, the double dual is larger than the starting space, and then the analogue of Theorem 2.18 fails. Here is a counterexample in $L^1(\mathbb{R}^1)$. Let $f^j(x) = j$ for $0 \leq x \leq 1/j$ and zero otherwise. This sequence is certainly bounded: $\int |f^j| = 1$. If some subsequence had a weak limit, f, then f would have to be zero (because f would have to be zero on all intervals of the form $(-\infty, 0)$ or $(1/n, \infty)$ for any n. But $\int f^j \cdot 1 = 1 \not\to 0$, which is a contradiction since the function $f(x) \equiv 1$ is in the dual space $L^\infty(\mathbb{R}^1)$.

2.19 LEMMA (Approximation by C_c^∞-functions)

Let $\Omega \subset \mathbb{R}^n$ be an open set and let $K \subset \Omega$ be compact. Then there is a function $J_K \in C_c^\infty(\Omega)$ such that $0 \leq J_K(x) \leq 1$ for all $x \in \Omega$ and $J_K(x) = 1$ for $x \in K$.

As a consequence, there is a sequence of functions $g_1, g_2, \ldots$ in $C_c^\infty(\Omega)$ that take values in $[0, 1]$ and such that $\lim_{j \to \infty} g_j(x) = 1$ for every $x \in \Omega$.

As a second consequence, given any sequence of functions $f_1, f_2, \ldots$ in $C^\infty(\Omega)$ that converges strongly to some f in $L^p(\Omega)$ with $1 \leq p < \infty$, the sequence given by $h_i(x) = g_i(x) f_i(x)$ is in $C_c^\infty(\Omega)$ and also converges to f in the same strong sense. If, on the other hand, $f_i \rightharpoonup f$ weakly in $L^p(\mathbb{R}^n)$ for some $1 < p < \infty$, then $h_i \rightharpoonup f$ weakly in $L^p(\mathbb{R}^n)$.

PROOF. The first part of Lemma 2.19 is Urysohn's Lemma (Exercise 1.15) but we shall give a short proof using the *Lebesgue* integral instead of the Riemann integral. Since K is compact, there is a $d > 0$ such that $\{x : |x - y| \leq 2d$ for some $y \in K\} \subset \Omega$. Define $K_+ = \{x : |x - y| \leq d$ for some $y \in K\} \supset K$ and note that $K_+ \subset \Omega$ is also compact. Fix some $j \in C_c^\infty(\mathbb{R}^n)$ with support in $\{x : |x| \leq 1\}$ and such that $0 \leq j(x) \leq 1$ for all x and $\int j = 1$ (see 1.1(2) for an example). Then, with $\varepsilon = d$, we set $J_K = j_\varepsilon * \chi$, where χ is the characteristic function of K_+. It is evident that J_K has the correct properties.

It is an easy exercise to show that there is an increasing sequence of compact sets $K_1 \subset K_2 \subset \cdots \subset \Omega$ such that each $x \in \Omega$ is in $K_{m(x)}$ for some integer $m(x)$. Define $g_i := J_{K_i}$.

The strong convergence of h_i to f is a consequence of dominated convergence. The weak convergence is also a consequence of dominated convergence provided we recall that the dual of $L^p(\Omega)$ is $L^{p'}(\Omega)$, with $1 < p' < \infty$, and that the functions of compact support are dense in $L^{p'}(\Omega)$. ∎

2.20 LEMMA (Convolutions of functions in dual $L^p(\mathbb{R}^n)$-spaces are continuous)

Let f be a function in $L^p(\mathbb{R}^n)$ and let g be in $L^{p'}(\mathbb{R}^n)$ with p and $p' > 1$ and $1/p + 1/p' = 1$. Then the convolution $f * g$ is a continuous function on $\mathbb{R}^n$ that tends to zero at infinity in the strong sense that for any $\varepsilon > 0$ there is $\mathcal{R}_\varepsilon$ such that
$$\sup_{|x| > \mathcal{R}_\varepsilon} |(f * g)(x)| < \varepsilon.$$

PROOF. Note that $(f * g)(x)$ is finite and defined by $\int f(x-y)g(y)\,dy$ for *every* x. This follows from Hölder's inequality since $f \in L^p(\mathbb{R}^n)$ and $g \in L^{p'}(\mathbb{R}^n)$. For any $\delta > 0$ we can find, by Lemma 2.19 (approximation by $C_c^\infty(\Omega)$-functions), f_δ and g_δ, both in $C_c^\infty(\mathbb{R}^n)$, such that $\|f_\delta - f\|_p < \delta$ and $\|g_\delta - g\|_{p'} \leq \delta$. If we write
$$f * g - f_\delta * g_\delta = (f - f_\delta) * g + f_\delta * (g - g_\delta),$$
we see, by the triangle and Hölder's inequalities, that
$$\|f * g - f_\delta * g_\delta\|_\infty \leq \|f - f_\delta\|_p \|g\|_{p'} + \|f_\delta\|_p \|g - g_\delta\|_{p'},$$
which is bounded by $(\|g\|_{p'} + \|f\|_p)\delta$. Since $f_\delta * g_\delta$ is in $C_c^\infty(\mathbb{R}^n)$, $f * g$ is uniformly approximated by smooth functions. Hence $f * g$ is continuous and the last statement is a trivial consequence of the fact that $f_\delta * g_\delta$ has compact support. ∎

2.21 HILBERT-SPACES

The space $L^2(\Omega)$ has the special property, not shared by the other L^p-spaces, that its norm is given by an inner product—a concept familiar from elementary linear algebra. The inner product of two $L^2(\Omega)$ functions is
$$(f, g) := \int_\Omega \overline{f}(x) g(x) \mu(dx)$$

in terms of which the norm is given by $\|f\|^2 = \sqrt{(f,f)}$. Note that the complex conjugate is on the left; often it is on the right, especially in mathematical writing. Note also that the function $\overline{f}g$ is integrable, by Schwarz's inequality.

Hilbert-spaces can be defined abstractly in terms of the inner product, without mentioning functions, similar to the way a vector space can be defined without any specific representation of the vectors. In this section we shall outline the beginning of that theory.

Generally speaking, an **inner product space** V is a vector space that carries an **inner product** $(\cdot, \cdot) : V \times V \to \mathbb{C}$ having the properties
 (i) $(x, y + z) = (x, y) + (x, z)$ for all $x, y, z \in V$;
 (ii) $(x, \alpha y) = \alpha(x, y)$ for all $x, y \in V, \alpha \in \mathbb{C}$;
 (iii) $(y, x) = \overline{(x, y)}$;
 (iv) $(x, x) \geq 0$ for all x, and $(x, x) = 0$ only if $x = 0$.

Clearly, $\int \overline{f}g \, d\mu$ satisfies all these conditions.

The Schwarz inequality $|(x, y)| \leq \sqrt{(x,x)}\sqrt{(y,y)}$ can now be deduced from (i)–(iv) alone. If one of the vectors, say y, is not the zero vector, then there is equality if and only if $x = \lambda y$ for some $\lambda \in \mathbb{C}$. As an exercise the reader is asked to prove this. If we set $\|x\| = \sqrt{(x,x)}$, then, by the Schwarz inequality,

$$\|x + y\|^2 \leq \|x\|^2 + \|y\|^2 + 2\|x\|\|y\| = (\|x\| + \|y\|)^2,$$

and hence the triangle inequality $\|x + y\| \leq \|x\| + \|y\|$ holds. With the help of (ii) and (iv) the function $x \mapsto \|x\|$ is seen to be a norm.

We say that $x, y \in V$ are **orthogonal** if $(x, y) = 0$. Keeping with the tradition that every deep theorem becomes trivial with the right definition, we can state **Pythagoras' theorem** in the following way: When x and y are orthogonal, $\|x + y\|^2 = \|x\|^2 + \|y\|^2$.

An important property of $L^2(\Omega)$ is its completeness. A **Hilbert-space** $\mathcal{H}$ is *by definition* a complete inner product space, i.e., for every Cauchy sequence $x^j \in \mathcal{H}$ (meaning that $\|x^j - x^k\| \to 0$ as $j, k \to \infty$) there is some $x \in \mathcal{H}$ such that $\|x - x^j\| \to 0$ as $j \to \infty$.

With these preparations, we invite the reader to prove, as an exercise, the analogue of Lemma 2.8 (projection on convex sets) for Hilbert-spaces: Let $\mathcal{C}$ be a closed convex set in $\mathcal{H}$. Then there exists an element y of smallest norm in $\mathcal{C}$, i.e., such that $\|y\| = \inf\{\|x\| : x \in \mathcal{C}\}$.

The uniform convexity, which is needed for the projection lemma, is provided by the parallelogram identity

$$\|x + y\|^2 + \|x - y\|^2 = 2\|x\|^2 + 2\|y\|^2 \ .$$

As in Theorem 2.14, the projection lemma implies that the dual of $\mathcal{H}$, i.e., the continuous linear functionals on $\mathcal{H}$, is $\mathcal{H}$ itself.

A special case of a convex set is a **subspace** of a Hilbert-space $\mathcal{H}$, i.e., a set $M \subset \mathcal{H}$ that is closed under finite linear combinations. Let $M^\perp$ be the **orthogonal complement** of M, i.e.,

$$M^\perp := \{x \in \mathcal{H} : (x,y) = 0, y \in M\}.$$

It is easy to see that $M^\perp$ is a **closed subspace**, i.e., if $x^j \in M^\perp$ and $x^j \to x \in \mathcal{H}$, then $x \in M^\perp$. If $\overline{M}$ denotes the smallest closed subspace that contains M, then we have from the projection lemma that

$$\mathcal{H} = \overline{M} \oplus M^\perp. \tag{1}$$

This notation, $\oplus$ (called the **orthogonal sum**), means that for every $x \in \mathcal{H}$ there exist $y_1 \in \overline{M}$ and $y_2 \in M^\perp$ such that $x = y_1 + y_2$. Obviously, y_1 and y_2 are unique. y_2 is called the **normal vector** to M through x. The geometric intuition behind (1) is that if $x \in \mathcal{H}$ and M is a closed subspace, then the best least squares fit to x in M is given by $x - y_2$.

To prove (1), pick any $x \in \mathcal{H}$ and consider $\mathcal{C} = \{z \in \mathcal{H} : z = x - y, y \in \overline{M}\}$. Clearly, $\mathcal{C}$ is a closed convex set and hence there is $z_0 \in \mathcal{C}$ such that $\|z_0\| = \inf\{\|z\| : z \in \mathcal{C}\}$. Similar to the proof in Sect. 2.8, we find that z_0 is orthogonal to $\overline{M}$, $y_0 := x - z_0 \in \overline{M}$ and thus (1) is proved. It is easy to see that $\overline{M}^\perp = M^\perp$.

The reader is invited to prove the principle of uniform boundedness. That is, whenever $\{l^i\}$ is a collection of bounded linear functionals on $\mathcal{H}$ such that for every $x \in \mathcal{H}$ $\sup_i |l^i(x)| < \infty$, then $\sup_i \|l^i\| < \infty$.

Up to this point our comments concerned analogies with L^p-spaces; with the exception of (1), Hilbert-spaces have not seemed to be much different from L^p-spaces. The essential differences will be discussed next.

An orthonormal basis is a key notion in Euclidean spaces (which themselves are special examples of Hilbert-spaces) and this can be carried over to all Hilbert-spaces. Call a set $\mathcal{S} = \{w_1, w_2, \ldots\}$ of vectors in $\mathcal{H}$ an **orthonormal set** if $(w_i, w_j) = \delta_{i,j}$ for all $w_i, w_j \in \mathcal{S}$. Here $\delta_{i,j} = 1$ if $i = j$ and $\delta_{i,j} = 0$ if $i \neq j$. If $x \in \mathcal{H}$ is given, one may ask for the best quadratic fit to x by linear combinations of vectors in $\mathcal{S}$. If $\mathcal{S}$ is a finite set, then the answer is $x_N = \sum_{j=1}^N (w_j, x) w_j$ as is easily shown. Clearly,

$$0 \leq \|x - x_N\|^2 = \|x\|^2 - 2\operatorname{Re}(x, x_N) + \|x_N\|^2 = \|x\|^2 - \sum_{j=1}^N |(w_j, x)|^2$$

and we obtain the important **inequality of Bessel**

$$\sum_{j=1}^{N} |(w_j, x)|^2 \leq \|x\|^2.$$

From now on we shall assume that $\mathcal{H}$ is a **separable Hilbert-space**, i.e., there exists a countable, dense set $\mathcal{C} = \{u_1, u_2, \dots\} \subset \mathcal{H}$. (Nonseparable Hilbert-spaces are unpleasant, used rarely and best avoided.) Thus, for every element $x \in \mathcal{H}$ and for $\varepsilon > 0$, there exists N such that $\|x - u_N\| < \varepsilon$. From $\mathcal{C}$ we can construct a countable set $\mathcal{B} = \{w_1, w_2, \dots\}$ as follows. Define $w_1 := u_1/\|u_1\|$, and then recursively define $w_k := v_k/\|v_k\|$, where

$$v_k := u_k - \sum_{j=1}^{k-1} (w_j, u_k) w_j.$$

If $v_k = 0$, then throw out u_k from $\mathcal{C}$ and continue on. The set $\mathcal{B}$ is easily seen to be orthonormal and this constructive procedure for obtaining orthonormal sets is called the **Gram–Schmidt procedure**.

Suppose there is an $x \in \mathcal{H}$ such that $(x, w_k) = 0$ for all k. We claim that then $x = 0$. Recalling that $\mathcal{C} \subset \mathcal{H}$ is dense, pick $\varepsilon > 0$ and then find $u_N \in \mathcal{C}$ such that $\|x - u_N\| < \varepsilon$. By the Gram–Schmidt procedure we know that

$$u_N = v_N + \sum_{j=1}^{N-1} (w_j, u_N) w_j \quad \text{for any } N.$$

Since v_N is proportional to w_N, the condition $(x, w_k) = 0$ for all k implies that $(x, u_N) = 0$. Since $\varepsilon^2 > \|x - u_N\|^2 = \|x\|^2 + \|u_N\|^2$, we find that $\|x\| < \varepsilon$. But ε is arbitrary, so $x = 0$, as claimed.

By Bessel's inequality, the sequence

$$x_M := \sum_{j=1}^{M} (w_j, x) w_j$$

is a Cauchy sequence and hence there is an element $y \in \mathcal{H}$ such that $\|y - x_M\| \to 0$ as $M \to \infty$. Clearly, $(x - y, w_j) = 0$ for all j, and hence $x = y$. Thus we have arrived at the important fact that the set $\mathcal{B}$ is an **orthonormal basis** for our Hilbert-space, i.e., every element $x \in \mathcal{H}$ can be expanded as a **Fourier series**

$$x = \sum_{j=1}^{D} (w_j, x) w_j,$$

where D, the **dimension** of $\mathcal{H}$, is finite or infinite (we shall always write ∞ for brevity). The numbers (w_j, x) are called the **Fourier coefficients** of the element x (with respect to the basis $\mathcal{B}$, of course). It is important to note that

$$\sum_{j=1}^{\infty}(w_j, x)w_j$$

stands for the limit of the sequence

$$x_M = \sum_{j=1}^{M}(w_j, x)w_j$$

in $\mathcal{H}$ as $M \to \infty$.

It is now very simple to show the analogue of Theorem 2.18, that every ball in a separable Hilbert-space is weakly sequentially compact. To be precise, let x_i be a bounded sequence in $\mathcal{H}$. Then there exists a subsequence x_{i_k} and a point $x \in \mathcal{H}$ such that

$$\lim_{k \to \infty} (x_k, y) = (x, y)$$

for every $y \in \mathcal{H}$. Again, we leave the easy details to the reader.

There are many more fundamental points to be made about Hilbert-spaces, such as linear operators, self-adjoint operators and the spectral theorem. All these notions are not only fairly deep mathematically, but they are also the key to the interpretation of quantum mechanics; indeed, many concepts in Hilbert-space theory were developed under the stimulus of quantum mechanics in the first half of the twentieth century. There are many excellent texts that cover these topics.

Exercises for Chapter 2

1. Show that for any two nonnegative numbers a and b
$$ab \leq \frac{1}{p}a^p + \frac{1}{q}b^q$$
   where $1 \leq p, q \leq \infty$ and $\frac{1}{p} + \frac{1}{q} = 1$. Use this to derive another proof of Theorem 2.3 (Hölder's inequality).

2. Prove 2.1(6) and the statement that when $\infty \geq r \geq q \geq 1$, $f \in L^r(\Omega) \cap L^q(\Omega) \Rightarrow f \in L^p(\Omega)$ for all $r \geq p \geq q$.

3. Let $U \subset \mathbb{R}^n$ be an open set and let $K \subset U$ be compact. Show without using Theorem 2.16, i.e., without using measure theory, that there exists a C^∞-function ψ such that $\psi \equiv 1$ on K and $\operatorname{supp}\{\psi\} \subset U$.

   ▶ *Hint.* Construct first a *continuous* function ψ with $\psi \equiv 1$ on K and $\operatorname{supp}\{\psi\} \subset U$ using the inner distance function $d_i(x) = \inf\{|x - y| : y \in K\}$ and the outer distance function $d_o(x) = \inf\{|x - y| : y \in U^c\}$. (Why are these functions continuous?) Then apply the j_ε-trick using the *Riemann* integral.

4. The penultimate sentence in the remark in Sect. 2.5 is really a statement about nonnegative numbers. Prove it, i.e., for $1 \leq p \leq 2$ and for $0 < b < a$
$$(a+b)^p + (a-b)^p \geq 2a^p + p(p-1)a^{p-2}b^2.$$

5. Referring to Theorem 2.5, assume that $1 < p \leq 2$ and that f and g lie on the unit sphere in L^p, i.e., $\|f\|_p = \|g\|_p = 1$. Assume also that $\|f - g\|_p$ is small. Draw a picture of this situation. Then, using Exercise 4, explain why 2.5(2) shows that the unit sphere is 'uniformly convex'. Explain also why 2.5(1) shows that the unit sphere is 'uniformly smooth', i.e., it has no corners.

6. As needed in the proof of Theorem 2.13 (strongly convergent convex combinations), prove that 'Cauchy sequences of Cauchy sequences are Cauchy sequences'. (In particular, state clearly what this means.)

7. Assume that f and g are in $L^1(\mathbb{R}^n)$. Prove that the convolution $f * g$ in 2.15(1) is a measurable function and that this function is in $L^1(\mathbb{R}^n)$.

8. Prove that a strongly convergent sequence in $L^p(\mathbb{R}^n)$ is also a Cauchy sequence.

9. In Sect. 2.9 three ways are shown for which an $L^p(\mathbb{R}^n)$ sequence f^k can converge weakly to zero but f^k does not convergence to anything strongly. Verify this for the three examples given in 2.9.

10. Let f be a real-valued, measurable function on $\mathbb{R}$ that satisfies the equation
$$f(x+y) = f(x) + f(y)$$
for all x, y in $\mathbb{R}$. Prove that $f(x) = Ax$ for some number A.

    ▶ *Hint.* Prove this when f is continuous by examining f on the rationals. Next, convolve $\exp[if(x)]$ with a j_ε of compact support. The convolution is continuous!

11. With the usual $j_\varepsilon \in C_c^\infty$, show that if f is continuous then $j_\varepsilon * f(x)$ converges to $f(x)$ for all x, and it does so uniformly on each compact subset of $\mathbb{R}^n$.

12. Deduce Schwarz's inequality $|(x,y)| \leq \sqrt{(x,x)}\sqrt{(y,y)}$ from 2.21 (i)–(iv) *alone*. Determine all the cases of equality.

13. Prove the analogue of Lemma 2.8 (Projection on convex sets) for Hilbert-spaces.

14. For any (not necessarily closed) subspace M show that $M^\perp$ is closed and that $\overline{M}^\perp = M^\perp$.

15. Prove Riesz's representation theorem, Theorem 2.14, for Hilbert-spaces.

16. Prove the principle of uniform boundedness for Hilbert-spaces by imitating the proof in Sect. 2.12.

17. Prove that every bounded sequence in a separable Hilbert-space has a weakly convergent subsequence.

18. Prove that every convex function has a support plane at every x in the interior of its domain, as claimed in Sect. 2.1. See also Exercise 3.1.

19. Prove 2.9(4).

Chapter 3

Rearrangement Inequalities

3.1 INTRODUCTION

In Chapters 1 and 2 we laid down the general principles of measure theory and integration. That theory is quite general, for much of it holds on any abstract measure space; the geometry of $\mathbb{R}^n$ did not play a crucial role. The subject treated in this chapter — rearrangements of functions — mixes geometry and integration theory in an essential way. From the pedagogic point of view it provides a good exercise (as in the proof of Riesz's rearrangement inequality) in manipulating measurable sets. More than that, however, these rearrangement theorems (and others not mentioned here) are extremely useful analytic tools. They lead, for example, to the statement that the minimizers for the Hardy–Littlewood–Sobolev inequality (see Sect. 4.3) are spherically symmetric functions. Another consequence is Lemma 7.17 which states that rearranging a function decreases its kinetic energy. This, in turn, leads to the fact that the optimizers of the Sobolev inequalities are spherically symmetric functions. Rearrangement inequalities lead to the well-known isoperimetric inequality (not proved here) that the ball has the smallest surface area among all bodies with a given volume. In many other examples rearrangement inequalities also tell us that spherically symmetric functions are, indeed, minimizers, e.g., we show in Sect. 11.17 that balls minimize electrostatic capacity. Many more examples are given in [Pólya–Szegő]. Thus, while this topic is usually not considered a central part of analysis, we place it here as an example of conceptually interesting and practically useful mathematics.

3.2 DEFINITION OF FUNCTIONS VANISHING AT INFINITY

The functions appropriate for the definition of rearrangements are those Borel measurable functions that go to zero at infinity in the following very weak sense. If $f : \mathbb{R}^n \to \mathbb{C}$ is a Borel measurable function, then f is said to **vanish at infinity** if $\mathcal{L}^n(\{x : |f(x)| > t\})$ is finite for all $t > 0$. (Recall that $\mathcal{L}^n$ denotes Lebesgue measure.) This notion will also be used in the definition of D^1 and $D^{1/2}$ spaces, which will be the natural function spaces for Sobolev inequalities.

3.3 REARRANGEMENTS OF SETS AND FUNCTIONS

If $A \subset \mathbb{R}^n$ is a Borel set of finite Lebesgue measure, we define A^*, the **symmetric rearrangement of the set** A, to be the open ball centered at the origin whose volume is that of A. Thus,

$$A^* = \{x : |x| < r\} \quad \text{with} \quad (|\mathbb{S}^{n-1}|/n)r^n = \mathcal{L}^n(A),$$

where $|\mathbb{S}^{n-1}|$ is the surface area of $\mathbb{S}^{n-1}$.

▶ *Note.* The use of open balls is not essential. Closed balls could have been used as well, but some choice is necessary for definiteness. With our choice, the characteristic function, $\chi_{A^*}(y)$ is lower semicontinuous (see Sect. 1.5).

This definition, together with the layer cake representation (Theorem 1.13) allows us to define the **symmetric-decreasing rearrangement**, f^*, **of a function** f as follows.

The symmetric-decreasing rearrangement of a characteristic function of a set is obvious, namely

$$\chi_A^* := \chi_{A^*}.$$

Now, if $f : \mathbb{R}^n \to \mathbb{C}$ is a Borel measurable function vanishing at infinity we define

$$f^*(x) = \int_0^\infty \chi^*_{\{|f|>t\}}(x)\,dt, \tag{1}$$

which is to be compared with (see 1.13(4))

$$|f(x)| = \int_0^\infty \chi_{\{|f|>t\}}(x)\,dt. \tag{2}$$

The rearrangement f^* has a number of obvious properties:
(i) $f^*(x)$ is nonnegative.

(ii) $f^*(x)$ is radially symmetric and nonincreasing, i.e.,

$$f^*(x) = f^*(y) \quad \text{if } |x| = |y|$$

and

$$f^*(x) \geq f^*(y) \quad \text{if } |x| \leq |y|.$$

Incidentally, we say that f^* is **strictly symmetric-decreasing** if $f^*(x) > f^*(y)$ whenever $|x| < |y|$; in particular, this implies that $f^*(x) > 0$ for all x.

(iii) $f^*(x)$ is a lower semicontinuous function since the sets $\{x : f^*(x) > t\}$ are open for all $t > 0$. In particular, f^* is measurable.

(iv) The level sets of f^* are simply the rearrangements of the level sets of $|f|$, i.e.,

$$\{x : f^*(x) > t\} = \{x : |f(x)| > t\}^*.$$

A tautological, but important, consequence of this is the **equimeasurability** of the functions $|f|$ and f^*, i.e.,

$$\mathcal{L}^n(\{x : |f(x)| > t\}) = \mathcal{L}^n(\{x : f^*(x) > t\})$$

for every $t > 0$. This, together with the layer cake representation 1.13(2) yields

$$\int_{\mathbb{R}^n} \phi(|f(x)|)\, dx = \int_{\mathbb{R}^n} \phi(f^*(x))\, dx \tag{3}$$

for any function ϕ that is the difference of two monotone functions ϕ_1 and ϕ_2 and such that either $\int_{\mathbb{R}^n} \phi_1(|f(x)|)\, dx$ or $\int_{\mathbb{R}^n} \phi_2(|f(x)|)\, dx$ is finite. In particular we have the important fact that for $f \in L^p(\mathbb{R}^n)$,

$$\|f\|_p = \|f^*\|_p \tag{4}$$

for all $1 \leq p \leq \infty$.

(v) If $\Phi : \mathbb{R}^+ \to \mathbb{R}^+$ is nondecreasing, then $(\Phi \circ |f|)^* = \Phi \circ f^*$, i.e., in a slightly imprecise notation, $\bigl(\Phi(|f(x)|)\bigr)^* = \Phi(f^*(x))$. This observation yields another proof of equation (3). Simply note that by the equimeasurability of $(\phi \circ |f|)^*$ and $(\phi \circ |f|)$ we have (3) for all monotone nondecreasing functions ϕ and hence for differences of monotone nonincreasing functions ϕ.

(vi) The rearrangement is **order preserving**, i.e., suppose f and g are two nonnegative functions on $\mathbb{R}^n$, vanishing at infinity, and suppose further that $f(x) \leq g(x)$ for all x in $\mathbb{R}^n$. Then their rearrangements satisfy $f^*(x) \leq g^*(x)$ for all x in $\mathbb{R}^n$. This follows immediately from the fact that the inequality $f(x) \leq g(x)$ for all x is equivalent to the statement that the level sets of g contain the level sets of f.

3.4 THEOREM (The simplest rearrangement inequality)

Let f and g be nonnegative functions on $\mathbb{R}^n$, vanishing at infinity, and let f^* and g^* be their symmetric-decreasing rearrangements. Then

$$\int_{\mathbb{R}^n} f(x)g(x)\,\mathrm{d}x \leq \int_{\mathbb{R}^n} f^*(x)g^*(x)\,\mathrm{d}x, \tag{1}$$

with the understanding that when the left side is infinite so is the right side.

If f is strictly symmetric-decreasing (see 3.3(ii)), then there is equality in (1) if and only if $g = g^*$.

PROOF. In the following Fubini's theorem will be used freely.

We use the layer cake representation for f, g, f^* and g^*. Inequality (1) becomes

$$\int_0^\infty \int_0^\infty \int_{\mathbb{R}^n} \chi_{\{f>t\}}(x)\chi_{\{g>s\}}(x)\,\mathrm{d}x\,\mathrm{d}s\,\mathrm{d}t$$
$$\leq \int_0^\infty \int_0^\infty \int_{\mathbb{R}^n} \chi^*_{\{f>t\}}(x)\chi^*_{\{g>s\}}(x)\,\mathrm{d}x\,\mathrm{d}s\,\mathrm{d}t.$$

The general case of (1) will then follow immediately from the special case in which f and g are characteristic functions of sets of finite Lebesgue measure. Thus, we have to show that for measurable sets A and B in $\mathbb{R}^n$, $\int \chi_A \chi_B \leq \int \chi^*_A \chi^*_B$ or, what is the same thing, $\mathcal{L}^n(A \cap B) \leq \mathcal{L}^n(A^* \cap B^*)$. Assume that $\mathcal{L}^n(A) \leq \mathcal{L}^n(B)$. Then $A^* \subset B^*$ and $\mathcal{L}^n(A^* \cap B^*) = \mathcal{L}^n(A^*) = \mathcal{L}^n(A)$. But $\mathcal{L}^n(A \cap B) \leq \mathcal{L}^n(A)$, so (1) is proved.

The proof of the second part of the theorem, in which f is strictly symmetric-decreasing, is slightly more complicated. To have equality in (1) it is necessary that for Lebesgue almost every $s > 0$

$$\int_{\mathbb{R}^n} f\chi_{\{g>s\}} = \int_{\mathbb{R}^n} f\chi^*_{\{g>s\}}. \tag{2}$$

We claim that this implies that $\chi_{\{g>s\}} = \chi^*_{\{g>s\}}$ for almost every s, and hence that $g = g^*$ (by the layer cake representation). Since f is strictly symmetric-decreasing, *every* centered ball, $B_{0,r}$, is a level set of f. In fact there is a *continuous* function $r(t)$ such that $\{x : f(x) > t\} = B_{0,r(t)}$. This implies that $F_C(t) := \int \chi_{\{f>t\}}(x)\chi_C(x)\,\mathrm{d}x$ is a continuous function of t for any measurable set C. (Why?)

Now fix some $s > 0$ for which (2) holds and take $C = \{x : g(x) > s\}$. By (1), $F_C(t) \leq F_{C^*}(t)$. From (2) we have that $\int F_C(t)\,\mathrm{d}t = \int F_{C^*}(t)\,\mathrm{d}t$ and hence $F_C(t) = F_{C^*}(t)$ for almost every $t > 0$. In fact, by the continuity of F_C and F_{C^*}, we can conclude that $F_C(t) = F_{C^*}(t)$ for *every* $t > 0$. As before, this implies that for *every* $r > 0$ either $C \subset B_{0,r}$ and $C^* \subset B_{0,r}$ or else $C \supset B_{0,r}$ and $C^* \supset B_{0,r}$ (up to sets of zero $\mathcal{L}^n$ measure). Thus, $C = C^*$, up to a set of zero $\mathcal{L}^n$ measure. Hence $g = g^*$. ∎

REMARK. There is a reverse inequality which is expressed most simply for the characteristic functions of g. It is the following (for f and g nonnegative):

$$\int_{\mathbb{R}^n} f \chi_{\{g \leq s\}} \geq \int_{\mathbb{R}^n} f^* \chi_{\{g^* \leq s\}}. \tag{3}$$

(Note the $g \leq s$ in place of the usual $g > s$.) One proof is to write $\chi_{\{g \leq s\}} = 1 - \chi_{\{g > s\}}$ and then to use (1), *provided* f is summable. However, (3) is true even if f is not summable and the proof is a direct imitation of the proof above leading to (1). Again, equality in (3) for *all* s in the case that f is strictly symmetric-decreasing implies that $g = g^*$.

● The next rearrangement inequality is a refinement of (1) and uses (3). To motivate it, suppose f and g are nonnegative functions in $L^2(\mathbb{R}^n)$. Then their $L^2(\mathbb{R}^n)$ difference satisfies

$$\|f^* - g^*\|_2 \leq \|f - g\|_2, \tag{4}$$

because the difference of the square of the two sides in (4) is twice the difference of the two sides in (1). The obvious generalization is

$$\|f^* - g^*\|_p \leq \|f - g\|_p \tag{5}$$

for all $1 \leq p \leq \infty$, which means, by definition, that *rearrangement is* **nonexpansive** *on* $L^p(\mathbb{R}^n)$. The crucial point is that $|t|^p$ is a convex function of $t \in \mathbb{R}$. The following inequality validates (5) and generalizes this to arbitrary (not necessarily symmetric) convex functions, J. It is a slight generalization of a theorem of [Chiti] and [Crandall–Tartar] who proved it when $J(t) = J(-t)$.

3.5 THEOREM (Nonexpansivity of rearrangement)

Let $J : \mathbb{R} \to \mathbb{R}$ be a nonnegative convex function such that $J(0) = 0$. Let f and g be nonnegative functions on $\mathbb{R}^n$, vanishing at infinity. Then

$$\int_{\mathbb{R}^n} J(f(x)^* - g(x)^*) \, dx \leq \int_{\mathbb{R}^n} J(f(x) - g(x)) \, dx. \tag{1}$$

If we also assume that J is strictly convex, that $f = f^*$ and that f is strictly decreasing, then equality in (1) implies that $g = g^*$.

PROOF. First, we can write
$$J = J_+ + J_-$$
where $J_+(t) = 0$ for $t \leq 0$ and $J_+(t) = J(t)$ for $t \geq 0$, and similarly for J_-. Both are convex and hence it suffices to prove the theorem for J_+ and J_- separately. Since J_+ is convex, it has a right derivative $J'_+(t)$ for all t and J_+ is the integral of J'_+, i.e., $J_+(t) = \int_0^t J'_+(s)\,\mathrm{d}s$. The convexity of J_+ implies $J'_+(t)$ is a nondecreasing function of t; the strict convexity of J_+ for $t > 0$ would imply that $J'_+(t)$ is strictly increasing for $t > 0$. Therefore we can write
$$J_+(f(x) - g(x)) = \int_{g(x)}^{f(x)} J'_+(f(x) - s)\,\mathrm{d}s = \int_0^\infty J'_+(f(x) - s)\chi_{\{g \leq s\}}(x)\,\mathrm{d}s. \tag{2}$$

Now integrate (2) over $\mathbb{R}^n$ and use Fubini's theorem to exchange the s and x integrations. By 3.4(3) and Remark 3.3(v) we see that for each fixed s the $\mathbb{R}^n$-integral is not increased when f is replaced by f^* and g by g^*. A similar argument applied to J_- will yield (1).

Now assume that $f = f^*$, f is strictly decreasing and J'_+ is strictly increasing for $t > 0$. If (1) is an equality we must have that for a.e. s
$$\int_{\mathbb{R}^n} J'_+(f(x) - s)\chi_{\{g \leq s\}}(x)\,\mathrm{d}x = \int_{\mathbb{R}^n} J'_+(f(x) - s)\chi_{\{g^* \leq s\}}(x)\,\mathrm{d}x.$$

Since J'_+ is *strictly* increasing, we have, by the same argument as in the proof of Theorem 3.4, that for a.e. $r \geq s$ either $F_r \supset G_s$ or $F_r \subset G_s$, where $F_r = \{x : f(x) > r\}$ and $G_s = \{x : g(x) > s\}$. Likewise, by considering J_-, we conclude that for a.e. $r < s$ either $F_r \supset G_s$ or $F_s \subset G_r$. Since the sets F_r are centered balls whose radii vary continuously with r (here we use that f is *strictly* decreasing), we conclude that G_s is a centered ball for a.e. s (by simply choosing r such that $|F_r| = |G_s|$). ∎

● The next two rearrangement inequalities are much deeper and go back to F. Riesz [Riesz]. They have far-reaching consequences. For other proofs see [Hardy–Littlewood–Pólya].

3.6 LEMMA (Riesz's rearrangement inequality in one-dimension)

Let f, g and h be three nonnegative functions on the real line, vanishing at infinity. Denote $\int_\mathbb{R} \int_\mathbb{R} f(x)g(x-y)h(y)\,\mathrm{d}x\,\mathrm{d}y$ by $I(f, g, h)$. Then
$$I(f, g, h) \leq I(f^*, g^*, h^*),$$

with the understanding that $I(f^*, g^*, h^*) = \infty$ if $I(f, g, h) = \infty$.

PROOF. Using the layer cake representation (and Fubini's theorem) we can restrict ourselves to the case where f, g, h are characteristic functions of measurable sets of finite measure. We denote these functions by F, G, H and shall use the same letters to denote the corresponding sets. By the outer regularity of Lebesgue measure (see 1.2(9)) there exists a sequence of open sets F_k with $F \subset F_k \subset F_{k-1}$ for all k and $\lim_{k \to \infty} \mathcal{L}^1(F_k) = \mathcal{L}^1(F)$. In particular all F_k have finite measure. Similarly, we choose sets G_k and H_k. The dominated convergence theorem shows that

$$\lim_{k \to \infty} I(F_k, G_k, H_k) = I(F, G, H).$$

Clearly

$$\lim_{k \to \infty} I(F_k^*, G_k^*, H_k^*) = I(F^*, G^*, H^*).$$

Thus it suffices to prove the lemma in the case where F, G, H are open sets of finite measure.

Now every open subset F of the real line is the disjoint union of countably many intervals. We leave the proof of this fact as an exercise for the reader. Denote these intervals by $I_1, I_2, \ldots$ where the numbering is chosen such that $\mathcal{L}^1(I_{k+1}) \leq \mathcal{L}^1(I_k)$. If we set

$$F_m = \bigcup_{k=1}^{m} I_k$$

we have that

$$\lim_{m \to \infty} \mathcal{L}^1(F_m) = \sum_{k}^{\infty} \mathcal{L}^1(I_k) = \mathcal{L}(F)$$

and, by the monotone convergence theorem, we learn that

$$\lim_{m \to \infty} I(F_m, G_m, H_m) = I(F, G, H)$$

and that

$$\lim_{m \to \infty} I(F_m^*, G_m^*, H_m^*) = I(F^*, G^*, H^*).$$

The essence of all this, is that it suffices to prove the lemma for functions F, G, H that are characteristic functions of *finite* disjoint unions of open intervals.

Thus, we can write

$$F(x) = \sum_{j=1}^{k} f_j(x - a_j)$$

where f_j is the characteristic function of an interval *centered at the origin* and the a_j's are real numbers. Similarly we write

$$G(x) = \sum_{j=1}^{l} g_j(x - b_j) \quad \text{and} \quad H(x) = \sum_{j=1}^{m} h_j(x - c_j).$$

Now $I(F, G, H)$ is a sum of terms of the form

$$\int_{\mathbb{R}} \int_{\mathbb{R}} f(x - a) g(x - y - b) h(y - c) \, dx \, dy.$$

We want to show that $I(F, G, H)$ is largest if we join each family of intervals into one, which we then center at the origin. To this end consider the family of functions $F_t(x), G_t(x), H_t(x)$ where $f_j(x - a_j)$ has been replaced by $f_j(x - ta_j)$, $0 \leq t \leq 1$, etc. Now,

$$\begin{aligned} I_{jkl}(t) &= \int_{\mathbb{R}} \int_{\mathbb{R}} f_j(x - ta) g_k(x - y - tb) h_l(y - tc) \, dx \, dy \\ &= \int_{\mathbb{R}} \int_{\mathbb{R}} f_j(x) g_k(x - y) h_l(y - (a - b - c)t) \, dx \, dy \\ &= \int_{\mathbb{R}} \int_{\mathbb{R}} u_{jk}(y) h_l(y - (a - b - c)t) \, dy. \end{aligned}$$

Here, $u_{jk}(y) = \int f_j(x) g_k(x - y) \, dx$ is a symmetric-decreasing function. It is easy to see that $I_{jkl}(t)$ is nondecreasing as t varies from 1 to 0. Hence $I(F_t, G_t, H_t)$ is nondecreasing as t varies from 1 to 0. (Essentially, this is Theorem 3.4.) As t starts decreasing, the intervals associated with F_t, G_t and H_t start moving along the line toward the origin. As soon as any two intervals associated with the same function touch we stop the process and redefine it with these two intervals joined into one. Repeating this process a finite number of times will leave us eventually with three intervals, each one centered at the origin. Clearly this process did not change the total measure of these sets and $I(F, G, H)$ has not been decreased. This proves the lemma. ∎

For later use we state the following.

REMARKS. (1) $I(f, g, h) = \int_{\mathbb{R}^n} f(x)(g * h)(x) \, dx$.

(2) Defining $h_R(x) := h(-x)$, we have

$$I(f, g, h) = I(f, h, g) = I(g, f, h_R) = I(h, g_R, f) = I(h, f, g_R) = I(g, h_R, f).$$

3.7 THEOREM (Riesz's rearrangement inequality)

Let f, g and h be three nonnegative functions on $\mathbb{R}^n$. Then, with

$$I(f, g, h) := \int_{\mathbb{R}^n} \int_{\mathbb{R}^n} f(x) g(x-y) h(y) \, \mathrm{d}x \, \mathrm{d}y,$$

we have
$$I(f, g, h) \leq I(f^*, g^*, h^*), \tag{1}$$
with the understanding that $I(f^*, g^*, h^*) = \infty$ if $I(f, g, h) = \infty$.

PROOF. We shall give two proofs of this theorem in order to illustrate some of the principles about convergence developed in Chapter 2. The first uses an argument that, for want of a better word, we call a 'compactness argument' and is related to the proof in [Brascamp–Lieb–Luttinger]. The second is related to a proof in [Sobolev] which utilizes ideas due to Lusternik and Blaschke. It also uses ideas about competing symmetries from [Carlen–Loss] that will prove useful in Sects. 4.22 et seq. on the Hardy–Littlewood–Sobolev inequality. The starting point is Lemma 3.6, the one-dimensional version of our theorem. In the following, Fubini's theorem will be used freely and, by utilizing the layer cake representation, we can restrict ourselves to the case in which f, g, h are characteristic functions of measurable sets F, G, H of finite measure. We shall denote $I(\chi_F, \chi_G, \chi_H)$ simply by $I(F, G, H)$. Our proof will be a bit sketchy at some points, but the reader should have no difficulty filling in the details.

First we define **Steiner symmetrization** of any measurable function f with respect to some direction $\mathbf{e}$ in $\mathbb{R}^n$ (with $|\mathbf{e}| = 1$). Rotate $\mathbb{R}^n$ by any rotation ρ such that $\rho \mathbf{e} = (1, 0, 0, \ldots, 0)$. Let $(\rho f)(x) := f(\rho^{-1} x)$, and then replace $(\rho f)(x_1, \ldots, x_n)$ by $(\rho f)^{*1}(x_1, \ldots, x_n)$, which is defined to be the one-dimensional symmetric-decreasing rearrangement of (ρf) with respect to x_1, keeping the variables $x_2, \ldots, x_n$ fixed. The final step is to perform the inverse rotation ρ^{-1} on $\mathbb{R}^n$. The resulting function $\rho^{-1}((\rho f)^{*1})$ is the required Steiner symmetrization, and we denote it by $f^{*\mathbf{e}}$. Equivalently, we can say that we rearrange f along every line in $\mathbb{R}^n$ that is parallel to the $\mathbf{e}$-axis. The **Steiner symmetrization of a measurable set**, $F^{*\mathbf{e}}$, is, of course, the set corresponding to the rearranged characteristic function $\chi_F^{*\mathbf{e}}$.

Any set $F^{*\mathbf{e}}$ (and hence any $f^{*\mathbf{e}}$) is measurable for the following reason: First, it suffices to show that F^{*1} can be thought of as the graph of a function, m, on $\mathbb{R}^{n-1}$ defined by

$$m(x_2, \ldots, x_n) := \frac{1}{2} \int_{\mathbb{R}} \chi_F^{*1}(x_1, x_2, \ldots, x_n) \, \mathrm{d}x_1.$$

This function, m, is measurable since

$$m(x_2, \ldots, x_n) = \widehat{m}(x_2, \ldots, x_n) := \frac{1}{2} \int_{\mathbb{R}} \chi_F(x_1, x_2, \ldots, x_n) \, dx_1$$

(by the definition of the rearrangement) and $\widehat{m}$ is measurable (by Fubini's theorem). Second, as noted in Sect. 1.5, the set under the graph of a measurable function is a measurable set.

Analogously to Steiner symmetrization, one can define the **Schwarz symmetrization** of functions and sets. Instead of replacing ρf by its one-dimensional symmetric-decreasing rearrangement, we replace ρf for each value of x_1 by its $(n-1)$-dimensional rearrangement with respect to the variables $x_2, \ldots, x_n$.

For each **e** we can consider the triple of sets $F^{*\mathbf{e}}, G^{*\mathbf{e}}$ and $H^{*\mathbf{e}}$. By Lemma 3.6 and Fubini's theorem $I(F, G, H) \leq I(F^{*\mathbf{e}}, G^{*\mathbf{e}}, H^{*\mathbf{e}})$. Our goal in the following proofs will be to find a sequence of axes $\mathbf{e}_1, \mathbf{e}_2, \ldots$ such that the repeated Steiner rearrangement (by $\mathbf{e}_1, \mathbf{e}_2$, etc.) of F converges in an appropriate sense to the ball F^*. Note that G and H get rearranged along with F. By passing to a further subsequence we can assume that the sequences of G and H converge to some sets. Having done so, we will know that the supremum of $I(F, G, H)$ over all sets with given measures $\mathcal{L}^n(F), \mathcal{L}^n(G), \mathcal{L}^n(H)$ occurs when $F = F^*$. Then, employing the argument again, we will conclude that $G = G^*$ is optimal. (Note here that when $F = F^*$, further rearrangements do not change F, i.e., $(F^*)^{*\mathbf{e}} = F^*$.) Finally, we conclude that $H = H^*$ is optimal, and (1) will be proved. The main difference between the following two proofs is that the first one merely asserts the existence of such a sequence while in the second we actually construct one.

The difficult part is the $n = 2$ case, and we do that first.

$\boxed{\text{COMPACTNESS PROOF.}}$ Assume, for simplicity, that $\mathcal{L}^2(F) = 1$. By a simple approximation argument using the monotone convergence theorem it suffices to prove the theorem for bounded sets only. If $F \neq F^*$, then $\mathcal{L}^2(F \cap F^*) = \int \chi_F \chi_F^* = P < 1$. We wish to select a rearrangement axis $\mathbf{e}_1$ such that, with $F_1 := F^{*\mathbf{e}_1}$ and $\chi_1 := \chi_{F_1}$, the integral $\int \chi_1 \chi_F^* = P + \delta$ with $\delta > 0$. To find such an $\mathbf{e}_1$ we set $A = \chi_F^*(1 - \chi_F)$ and $B = (1 - \chi_F^*)\chi_F$ and consider the convolution $C(x) = \int A(x - y) B(-y) \, dy$. Since $\int C = (1 - P)^2$, C is not the zero function. There is then some $x \neq 0$ such that $C(x) > 0$, and we set $\mathbf{e}_1 = x/|x|$. It is now an elementary exercise, using the definitions of Steiner symmetrization, to show that symmetrization in the $\mathbf{e}_1$ direction has the desired effect—with $\delta \geq C(x)$, in fact (see Exercise 8).

The supremum of all such δ's is denoted by $\overline{\delta}_1 \geq 0$ and (since we do not wish to prove that $\overline{\delta}_1$ can actually be achieved) we settle for an improvement $\delta_1 \geq \frac{1}{2}\overline{\delta}_1$, which certainly can be achieved for some choice of $\mathbf{e}_1$. Thus, $\int \chi_1 \chi_F^* \geq P + \delta_1$. Next we perform a Steiner symmetrization parallel to the x_1-axis, (1,0), followed by a symmetrization parallel to the x_2-axis, (0,1). This, of course, cannot decrease $\int \chi_1 \chi_F^*$. After these last two symmetrizations, the set F_1 now lies between a certain nonnegative symmetric-decreasing function, $x_2 = S_1(x_1)$, and its reflection, $x_2 = -S_1(x_1)$.

Having done this, we repeat the process, i.e., we seek an axis $\mathbf{e}_2$ so that, with $\chi_2 := \chi_{F_2}$, we have $\int \chi_2 \chi_F^* \geq P + \delta_1 + \delta_2$ and $\delta_2 \geq \frac{1}{2}\overline{\delta}_2$, where $\overline{\delta}_2 \geq 0$ is the supremum of all possible increases. This symmetrization is followed, as before, by the two symmetrizations in the two coordinate axes, thereby giving rise to a new symmetric-decreasing function $x_2 = S_2(x_1)$.

This process is repeated indefinitely, giving us a sequence of sets $F_1, F_2, F_3, \ldots$ and functions $S_1, S_2, S_3, \ldots$ which form the boundaries of these sets. Note that since F is bounded, it is contained in some centered ball B. By 3.3(vi) all the F_j's are contained in the same ball and hence the functions S_j are uniformly bounded and have support in a fixed interval. We claim that $\int \chi_j \chi_F^*$ converges to 1, as required.

To prove this, we suppose the contrary, i.e., $\int \chi_j \chi_F^* \to Q < 1$. From the sequence of functions S_j, we can select a subsequence, which we continue to denote by S_j, so that S_j converges pointwise to some symmetric-decreasing function S. [To see this, note that since the S_j's are uniformly bounded and have support in some fixed interval, we can find a subsequence that converges on *all* the rational points $x_1 \neq 0$, since these are countable. Because the S_j are symmetric-decreasing, they converge for irrational x_1 as well. This argument is called **Helly's selection principle**.] The subsequence necessarily converges in $L^1(\mathbb{R}^1)$ to S by dominated convergence and hence, if W denotes the set lying between S and $-S$, we have that

$$\int \chi_W \chi_F^* = \lim_{j \to \infty} \int \chi_j \chi_F^* = Q,$$

while $\int \chi_W = 1$.

To obtain a contradiction, we first note, by the 'convolution' argument given at the beginning of this proof, that there is a $\delta > 0$ and an axis $\mathbf{e}$ such that $W_* := W^{*\mathbf{e}}$, with characteristic function χ_{W_*}, satisfies $\int \chi_{W_*} \chi_F^* > Q + \delta$. On the other hand, using the stated convergences, we can find an integer J such that F_J satisfies two conditions:

(a) $\int \chi_{F_J} \chi_F^* > Q - \delta/8$.
(b) $\|\chi_{F_J} - \chi_W\|_2 < \delta/4$.

Let $F_{J*} := F_J^{*\mathbf{e}}$. By Theorem 3.5 (nonexpansivity of rearrangements) or 3.4(4) we have that $\|\chi_{F_{J*}} - \chi_{W_*}\|_2 < \delta/4$. By using the Schwarz and triangle inequalities we easily conclude (proof left to the reader) that $\int \chi_{F_{J*}} \chi_F^* > Q + 3\delta/4$. This implies that the maximum improvement at the Jth step, $\overline{\delta}_J$, is greater than $3\delta/4$. On the other hand,

$$Q > \int \chi_{F_{J+1}} \chi_F^* \geq \int \chi_{F_J} \chi_F^* + \frac{1}{2}\overline{\delta}_J > Q - \frac{1}{8}\delta + \frac{1}{2}\overline{\delta}_J,$$

which implies that $\overline{\delta}_J < \delta/4$, and which is a contradiction.

The proof of the theorem for $n > 2$ is the same. We merely use Schwarz symmetrization in place of the third Steiner symmetrization, so that our boundary functions $S_1, S_2, S_3, \ldots$ are symmetric-decreasing functions of $x_1, \ldots, x_{n-1}$. Induction on n is used to insure that $(n-1)$-dimensional Schwarz symmetrization increases the integral $I(F, G, H)$. Otherwise the proof is identical to that for $n = 2$. ∎

SYMMETRY PROOF. For given sets F, G and H we shall construct sequences of sets F_k, G_k and H_k, all of them converging to balls, and such that $I(F_k, G_k, H_k)$ is an increasing sequence. The hard part is, again, the step from one to two dimensions, as already noted in the previous proof. Nevertheless we shall indicate at the end how the higher-dimensional generalization works. For the present we concentrate on two-dimensions.

Fix a rotation R_α, α indicating the angle. We choose α to be an irrational multiple of 2π. Next, for a given set $F \subset \mathbb{R}^2$ of finite Lebesgue measure, form the set $F_1 = TSR_\alpha F$, where S is the Steiner symmetrization about the x-axis and T the one about the y-axis. F_1 is a set with the same measure as F. It is reflection symmetric about the x- and y-axes and the part of F_1 contained in the upper half-plane is below the graph of a symmetric, nonincreasing function which we are free to choose to be lower semicontinuous. Note that this function is not necessarily bounded. The sets F_k, G_k, H_k are generated by applying this operation TSR_α k times to F, G and H.

First we want to show that these sequences converge strongly in $L^2(\mathbb{R}^2)$ to balls of the same volume. We note the inequalities

$$\|T\chi_F - T\chi_G\|_2 \leq \|\chi_F - \chi_G\|_2, \quad \|S\chi_F - S\chi_G\|_2 \leq \|\chi_F - \chi_G\|_2, \quad (2)$$

and the equality

$$\|R_\alpha \chi_F - R_\alpha \chi_G\|_2 = \|\chi_F - \chi_G\|_2, \quad (3)$$

valid for any two sets of finite measure. In fact the first two follow from 3.4(4) and the last one follows from the fact that rotations are measure preserving. From this we conclude that it suffices to prove the convergence result for bounded sets. Indeed, for $\varepsilon > 0$ given, we can find $\widetilde{F}$ contained in some centered ball such that $\|\chi_F - \chi_{\widetilde{F}}\|_2 < \varepsilon$. By (2) we have that $\|\chi_{F_k} - \chi_{\widetilde{F}_k}\|_2 < \varepsilon$ for all k and hence F_k converges once we have shown that $\widetilde{F}_k$ converges. Thus we can assume F, G and H to be bounded sets contained in some ball. By 3.3(vi) we know that the sequences F_k, G_k and H_k are contained in that very same ball.

The upper half-space part of F_k is bounded by the graph of a symmetric, nonincreasing lower semicontinuous function h_k which is uniformly bounded. As in the previous proof there exists a subsequence denoted by $h_{k(l)}(x)$ that converges everywhere to a lower semicontinuous function h which bounds the upper half-space part of a set D. The problem is to show that D is a disk. Consider any function, g, that is strictly symmetric-decreasing (for example, $g(x) = e^{-|x|^2}$) and define $\Delta_k = \|g - \chi_{F_k}\|_2$. Note that $Tg = Sg = R_\alpha g = g$. Thus, by Theorem 3.4, Δ_k is nonincreasing and hence it has a limit Δ. By the previous consideration we know that $\chi_{F_{k(l)}}$ converges pointwise a.e. to the characteristic function, χ_D, of D. Since $\chi_{F_{k(l)}}$ is dominated by the characteristic function of a fixed ball, we conclude by dominated convergence that
$$\Delta = \|g - \chi_D\|_2.$$

By (2) and (3) we also know that

$$\|\chi_{F_{k(l)+1}} - TSR_\alpha \chi_D\|_2 = \|TSR_\alpha \chi_{F_{k(l)}} - TSR_\alpha \chi_D\|_2 \to 0 \quad \text{as } l \to \infty.$$

Hence, by the montonicity of Δ_k, $\Delta = \|g - TSR_\alpha \chi_D\|_2$. On the other hand, since g is rotation invariant, $\|g - R_\alpha \chi_D\|_2 = \|g - \chi_D\|_2 = \Delta$. Thus

$$\|g - TSR_\alpha \chi_D\|_2 = \|g - R_\alpha \chi_D\|_2.$$

Since g is *strictly decreasing*, we conclude, using Fubini's theorem and Theorem 3.4, that $TSR_\alpha \chi_D = R_\alpha \chi_D$ a.e. In particular $R_\alpha \chi_D$ is symmetric with respect to reflection P about the x-axis and hence $R_\alpha \chi_D = PR_\alpha \chi_D = R_{-\alpha} P\chi_D = R_{-\alpha}\chi_D$, i.e., $R_{2\alpha}\chi_D = \chi_D$, or χ_D is invariant under the rotation $R_{2\alpha}$. The angle $\beta = 2\alpha$ is, by assumption, an irrational multiple of 2π and it is well known that any number $0 \leq \theta < 2\pi$ can be approximated arbitrarily closely by multiples of β mod 2π. Hence the function $\mu(\theta) := \|\chi_D - R_\theta \chi_D\|_2$ has zeros which are dense in the interval $[0, 2\pi)$. We shall show that μ is a continuous function which implies that $\chi_D = R_\theta \chi_D$ a.e. for every θ. Thus, $D = F^*$.

It suffices to show that $\int \chi_D R_\theta \chi_D = r(\theta)$ is continuous. By Theorem 2.16 there is a sequence of differentiable functions u_k such that $\delta_k = \|\chi_D - u_k\|_2 \to 0$ as $k \to \infty$. By Schwarz's inequality

$$\left| \int (\chi_D - u_k) R_\theta \chi_D \right| \le \delta_k \|\chi_D\|_2,$$

which says that the functions $r_k(\theta) = \int u_k R_\theta \chi_D$ converge to $r(\theta)$ uniformly. But $r_k(\theta) = \int (R_{-\theta} u_k) \chi_D$, which is easily seen to be continuous, and hence $r(\theta)$ is continuous.

Recall that a subsequence of the χ_{F_k} sequence converges pointwise a.e. to χ_D, and each set F_k is contained in some fixed ball. Therefore, for this subsequence, $\|\chi_D - \chi_{F_k}\|_2$ converges to zero by dominated convergence. By Theorem 3.5 (nonexpansivity of rearrangements), the whole sequence $\|\chi_D - \chi_{F_k}\|_2$ is a decreasing sequence. Since a subsequence converges to zero, the whole sequence converges to zero.

Precisely the same arguments apply to G_k and H_k, and hence χ_{F_k}, χ_{G_k} and χ_{H_k} converge strongly in $L^2(\mathbb{R}^2)$ to χ_{F^*}, χ_{G^*} and χ_{H^*}. From this it follows easily that

$$\lim_{k \to \infty} I(F_k, G_k, H_k) = I(F^*, G^*, H^*).$$

By the one-dimensional Riesz rearrangement inequality $I(F_k, G_k, H_k)$ is a nondecreasing sequence and our theorem is proved.

The generalization to higher dimensions is proved by induction. T corresponds to Steiner symmetrization along the n-axis and S is the Schwarz symmetrization perpendicular to the n-axis. The sequence to consider is $\{(TSR)^k \chi_F\}$ where R is any rotation that rotates the n-axis by $90°$. Tracing all the steps of the two-dimensional argument one obtains a limiting set D that has the following two properties: It is rotationally symmetric about the n-axis and RD is also rotationally symmetric about the n-axis. In other words, D is rotationally symmetric about two axes that are perpendicular. To see that D is a ball, consider $\chi_\varepsilon = j_\varepsilon * \chi_D$ where $j_\varepsilon(x) = \varepsilon^{-n} j(\varepsilon/n)$ and $j(x)$ is a smooth radial function with $\int_{\mathbb{R}^n} j = 1$. We know from Theorem 2.16 that χ_ε is smooth and $\chi_\varepsilon \to \chi_D$ in $L^2(\mathbb{R}^n)$ as $\varepsilon \to 0$. Moreover χ_ε has the same symmetry properties as χ_D. Thus setting $\rho^2 = x_1^2 + \cdots + x_{n-2}^2$ we find that

$$\chi_\varepsilon(x_1, \ldots, x_n) = f(\sqrt{\rho^2 + x_{n-1}^2}, x_n) = g(\sqrt{\rho^2 + x_n^2}, x_{n-1})$$

for some continuous functions f and g. We have chosen the $n-1$-axis as the other axis of symmetry. Setting $x_n = 0$ we obtain

$$g(\rho, x_{n-1}) = f(\sqrt{\rho^2 + x_{n-1}^2}, 0) \quad \text{for all } \rho > 0$$

and hence
$$\chi_\varepsilon(x_1,\ldots,x_n) = f(\sqrt{x_1^2 + \cdots + x_n^2}, 0),$$
i.e., χ_ε is radial. Hence χ_D is radial too, since it is a limit of radial functions. ∎

• The Riesz inequality, 3.7(1), concerns three functions, f, g, h and two variables x and y in $\mathbb{R}^n$. This was generalized in [Brascamp–Lieb–Luttinger] to m functions and k variables in $\mathbb{R}^n$, as given in Theorem 3.8 (without proof). The proof there follows the same strategy as in Lemma 3.6 and Theorem 3.7, namely first do the $\mathbb{R}^1$ case and then pass to $\mathbb{R}^n$ by repeated use of the $\mathbb{R}^1$ theorem. In fact, the proof given here of Lemma 3.6 originated in that paper.

3.8 THEOREM (General rearrangement inequality)

Let $f_1, f_2, \ldots, f_m$ be nonnegative functions on $\mathbb{R}^n$, vanishing at infinity. Let $k \leq m$ and let $B = \{b_{ij}\}$ be a $k \times m$ matrix (with $1 \leq i \leq k, 1 \leq j \leq m$). Define

$$I(f_1,\ldots,f_m) := \int_{\mathbb{R}^n} \cdots \int_{\mathbb{R}^n} \prod_{j=1}^{m} f_j\left(\sum_{i=1}^{k} b_{ij} x_i\right) dx_1 \cdots dx_k. \quad (1)$$

Then $I(f_1,\ldots,f_m) \leq I(f_1^*,\ldots,f_m^*)$.

REMARK. Theorem 3.7 corresponds to $m = 3$, $k = 2$ and $b = \begin{pmatrix} 1 & 1 & 0 \\ 0 & -1 & 1 \end{pmatrix}$.

3.9 THEOREM (Strict rearrangement inequality)

Let f, g and h be three nonnegative measurable functions on $\mathbb{R}^n$ with g strictly symmetric-decreasing. Then there is equality in 3.7(1) only if $f(x) = f^*(x - y)$ and $h(x) = h^*(x - y)$ for some $y \in \mathbb{R}^n$.

REMARK. By 3.6, Remark (2), the theorem holds if any one of the three functions f, g and h is radially symmetric and strictly decreasing. The word 'strictly' is important. One could ask whether it is possible to eliminate the requirement that one of the functions be radially symmetric and/or strictly decreasing. The answer is 'yes', with some caveats, as proved in [Burchard]. For example, if f, g and h are characteristic functions of three homothetic, homocentric ellipsoids, equality can be achieved in 3.7(1); this can be easily seen merely by making a linear change of coordinates in $\mathbb{R}^n$.

PROOF. By Theorem 1.13 (layer cake representation) the result follows (why?) from the one in which f and h are characteristic functions of measurable sets A, B of finite measure—which we assume henceforth.

First we prove the theorem for characteristic functions of a single variable. The general case will follow by induction on the dimension. Since g is strictly symmetric-decreasing, it follows from the layer cake representation and the Riesz rearrangement inequality that equality in 3.7(1) demands that

$$I(f, g_r, h) = I(f^*, g_r, h^*), \tag{1}$$

where g_r is the characteristic function of the centered interval of length r. The symbol I is explained in Lemma 3.6. If

$$r > |A| + |B| = \int_\mathbb{R} f + \int_\mathbb{R} h,$$

then $I(f^*, g_r, h^*) = |A||B|$. However, $I(f, g_r, h) \leq |A||B|$ with equality only if

$$g_r(x) \int_\mathbb{R} f(x+y)h(y)\,dy = \int_\mathbb{R} f(x+y)h(y)\,dy,$$

i.e., if the support of $\int f(x+y)h(y)\,dy$ is contained in the interval given by g_r. Note that, by Lemma 2.20, $\int_\mathbb{R} f(x+y)h(y)\,dy$ is a continuous function. Let J_A be the smallest interval such that $|A \cap J_A| = |A|$, and similarly for B. It is an easy exercise to see that the length of the smallest interval that contains the support of $\int_\mathbb{R} f(x+y)h(y)\,dy$ is $|J_A| + |J_B|$. Therefore $|J_A| + |J_B| < r$ for any $r > |A| + |B|$ and hence $|J_A| = |A|$ and $|J_B| = |B|$. Thus both A and B are intervals and (1) can only hold if they are centered at the same point. This proves the theorem in one-dimension.

To prove it in $n \geq 2$-dimensions we assume its truth in $(n-1)$. Equality in 3.7(1) can be expressed as

$$\int f(x', x_n) g(x' - y', x_n - y_n) h(y', y_n)\,dx'\,dy'\,dx_n\,dy_n$$
$$= \int f^*(x', x_n) g(x' - y', x_n - y_n) h^*(y', y_n)\,dx'\,dy'\,dx_n\,dy_n. \tag{2}$$

The primed variables indicate integration over $\mathbb{R}^{n-1}$. The Riesz rearrangement inequality, together with (2), implies that

$$\int f(x', x_n) g(x' - y', x_n - y_n) h(y', y_n)\,dx'\,dy'$$
$$= \int f^*(x', x_n) g(x' - y', x_n - y_n) h^*(y', y_n)\,dx'\,dy' \tag{3}$$

for Lebesgue a.e. x_n and y_n in $\mathbb{R}$. For any fixed x_n, $g(x', x_n)$ is a strictly symmetric-decreasing function of x'. Thus, by the induction hypothesis, for a.e. x_n, y_n the sets A'_{x_n}, B'_{y_n} corresponding to the characteristic functions $f(x', x_n)$ and $h(y', y_n)$ must be balls in $\mathbb{R}^{n-1}$ centered at some common point—which is independent of x_n, y_n. (Why?) In other words, up to sets of measure zero in $\mathbb{R}^n$, the sets A and B must be rotationally invariant about some common axis, $\mathbf{e}_n$, parallel to the n-direction. Similarly the two sets must be rotationally symmetric about some other common axis, say $\mathbf{e}_{n-1}$ in the $(n-1)$ direction. In particular the two axes must intersect in some point y. (Why?) We have shown at the end of the second proof of Theorem 3.7 that any measurable set in $\mathbb{R}^n$, $n \geq 3$, with two perpendicular symmetry axes must be a ball (up to a set of measure zero). Thus, the two sets A and B must be balls centered at y. The fact that A and B must be discs in the $n = 2$ case follows from the fact that every direction is a symmetry axis (as is the case for $n \geq 3$) and they all intersect pairwise (and hence at one common point) because of the nature of two-dimensional geometry. ∎

Exercises for Chapter 3

1. Show that a convex function, J, on an open interval of the real line, $\mathbb{R}$, has a right and left derivative J'_r, J'_ℓ at every point. Also show that $J(t) - J(t_0) = \int_{t_0}^t J'_r(s)\,ds = \int_{t_0}^t J'_\ell(s)\,ds$.

2. Show that every open subset of the real line is a disjoint union of open intervals.

3. Let A and B be measurable sets in $\mathbb{R}$ and let J_A and J_B be the smallest intervals such that $|A \cap J_A| = |A|$ and $|B \cap J_B| = |B|$, respectively. Show that the smallest interval that contains the support of $\chi_A * \chi_B$ has length $|J_A| + |J_B|$.

4. In the proof of Theorem 3.4 it is asserted that $r(t)$ is continuous. Prove this.

5. In the remark after the proof of Theorem 3.4 it is asserted that (3) holds even if f is not summable. Write out a proof of this fact.

6. Show that if a set in $\mathbb{R}^n$ has strictly positive but finite measure and is rotationally symmetric about two axes, these two axes must have a point in common.

7. Construct three functions, f, g and h, none of which is a translate of a symmetric-decreasing function, such that $I(f, g, h) = I(f^*, g^*, h^*)$.

8. In the first paragraph of the 'compactness proof' of Theorem 3.7 (Riesz's rearrangement inequality) it was asserted that by choosing $\mathbf{e}_1 = x/|x|$ the overlap integral $\int \chi_1 \chi_F^*$ increased from P to $P + \delta$ with $\delta \geq C(x)$. Prove this statement.

   ▶ *Hint.* Show that along each line parallel to the $\mathbf{e}_1$-axis, the overlap increases by at least $\min\{a, b\}$, where a is the $\mathcal{L}^1$ measure of the intersection of the set $F \sim F^*$ with this line, and b is the $\mathcal{L}^1$ measure of the intersection of the set $F^* \sim F$ with this line.

Chapter 4

Integral Inequalities

4.1 INTRODUCTION

Several important integral inequalities were already mentioned: Theorem 2.2 (Jensen's inequality), Theorem 2.3 (Hölder's inequality) and Theorem 2.4 (Minkowski's inequality). These are all based essentially on convexity arguments, and we had no difficulty in giving them in their sharp forms (i.e., the inequalities in question fail to be true if the constants are increased from the specified sharp values). We could also specify the cases of equality completely. These inequalities were presented in Chapter 2 because they were needed in the development of L^p-space theory.

The inequalities to be given now are far more intricate and do not follow from simple convexity. The Euclidean structure of $\mathbb{R}^n$ plays a role here. More noticeably, the determination of the sharp (or optimal) constants and the cases of equality are formidable problems. It is not too difficult to derive these inequalities (and we do so) if sharp constants are not demanded (although it has to be said that historically even the nonsharp version of Theorem 4.3 was not easy). We give these simple proofs as well. The determination of the sharp constants in Theorem 4.3, however, requires the rearrangement inequalities of Chapter 3—and that is why these integral inequalities are presented after Chapter 3. Not all the constants mentioned in Theorem 4.3 have been determined completely, however, and they present interesting open problems.

It is not obvious, or even necessarily true in all cases, that the sharp forms of inequalities can be achieved by certain specific functions. Such functions, for which the inequality becomes an equality, are called **maximizers** or **minimizers**, as the case may be. The word **optimizer** is also often used. In the cases treated in this chapter the maximizers are all determined completely. Other examples are given in Chapter 11 on the calculus of variations.

These analyses of sharp constants are in this book for two reasons. One is that they are occasionally useful, and even important. The main reason, however, is that they provide us with good examples of 'hard analysis' problems that can be carried to completion—which is usually not the case—and with the methods at our disposal. In other words, the reader is invited to use the material presented so far actually to construct and compute the solution to a minimization problem. The sharp constants are not needed in the rest of this book, however, and disinterested readers can guiltlessly omit this discussion.

Another point is that some elementary but useful facts about Gaussian functions, conformal transformations and stereographic projections will be introduced and used. Thus, we will be led to an example of the interplay between geometry and analysis.

A **Gaussian function** $g : \mathbb{R}^n \to \mathbb{C}$ is a bounded function that is the exponential of a quadratic plus linear form. That is,

$$g(x) = \exp\{-(x, Ax) + i(x, Bx) + (J, x) + C\}, \tag{1}$$

where A and B are real, symmetric matrices with A positive-semidefinite (i.e., $(x, Ax) \geq 0$ for all $x \in \mathbb{R}^n$) and with $J \in \mathbb{C}^n$. If $g \in L^p(\mathbb{R}^n)$ for some $p < \infty$, then A must be positive-definite.

Recall that $f * g$ denotes convolution, defined in Sect. 2.15.

4.2 THEOREM (Young's inequality)

Let $p, q, r \geq 1$ and $1/p + 1/q + 1/r = 2$. Let $f \in L^p(\mathbb{R}^n)$, $g \in L^q(\mathbb{R}^n)$ and $h \in L^r(\mathbb{R}^n)$. Then

$$\left| \int_{\mathbb{R}^n} f(x)(g * h)(x) \, dx \right| = \left| \int_{\mathbb{R}^n} \int_{\mathbb{R}^n} f(x)g(x-y)h(y) \, dx \, dy \right| \tag{1}$$
$$\leq C_{p,q,r;n} \|f\|_p \|g\|_q \|h\|_r.$$

The sharp constant $C_{p,q,r;n}$ equals $(C_p C_q C_r)^n$, where (with $1/p + 1/p' = 1$),

$$C_p^2 = p^{1/p}/p'^{1/p'}. \tag{2}$$

If $p, q, r > 1$, then equality can occur in (1) if and only if f, g and h are Gaussian functions

$$f(x) = A \exp[-p'(x-a, J(x-a)) + i\, k \cdot x],$$

$$g(x) = B \exp[-q'(x-b, J(x-b)) - i\, k \cdot x], \qquad (3)$$

$$h(x) = C \exp[-r'(x-c, J(x-c)) + i\, k \cdot x],$$

where $A, B, C \in \mathbb{C}$; $a, b, c, k \in \mathbb{R}^n$ with $a = b + c$; and J is any real, symmetric, positive-definite matrix.

REMARKS. (1) $C_p = 1/C_{p'}$.

(2) Using Hölder's inequality, it is easy to see that when g and h are given, the best choice for f (up to a constant) is

$$f(x) = e^{i\theta(x)} |(g * h)(x)|^{p'/p},$$

where $\theta(x)$ is defined by $g * h = e^{i\theta}|g * h|$. Thus, Young's inequality can be rephrased as follows (in which we switch p and p'):

$$\|g * h\|_p \leq (C_q C_r / C_p)^n \|g\|_q \|h\|_r = C_{p',q,r;n} \|g\|_q \|h\|_r \qquad (4)$$

with $1/q + 1/r = 1 + 1/p$.

(3) The sharp constant was found simultaneously by [Beckner] and by [Brascamp–Lieb]. The condition for equality was given in the latter.

(4) *Symmetry:* Let us denote the integral in (1) by $I(f, g, h)$ and by f_R the function $f_R(x) = f(-x)$. Then, by a simple change of variables (using Fubini's theorem)

$$I(f, g, h) = I(g, f, h_R) = I(f, h, g) = I(h, g_R, f). \qquad (5)$$

(5) Instead of viewing Young's inequality as a statement about convolution, let us consider the second integral in (1) and view it as an integral over $\mathbb{R}^{2n}$ (instead of $\mathbb{R}^n$) of a product of three functions, each of which is a composition of a linear map from $\mathbb{R}^{2n}$ to $\mathbb{R}^n$ with a function from $\mathbb{R}^n$ to $\mathbb{C}$. The ultimate generalization of Young's inequality is the following [Lieb, 1990].

Fully generalized Young's inequality. Fix $k > 1$, integers $n_1, \ldots, n_k$ and numbers $p_1, \ldots, p_k > 1$. Let $M \geq 1$ and let B_i (for $i = 1, \ldots, k$) be a linear mapping from $\mathbb{R}^M$ to $\mathbb{R}^{n_i}$. Let $Z : \mathbb{R}^M \to \mathbb{R}^+$ be some fixed Gaussian function,

$$Z(x) = \exp\{-(x, Jx)\}$$

with J a real, positive-semidefinite $M \times M$ matrix (possibly zero).

For functions f_i in $L^{p_i}(\mathbb{R}^{n_i})$ consider the integral

$$I_Z(f_1, \ldots, f_k) = \int_{\mathbb{R}^M} Z(x) \prod_{i=1}^k f_i(B_i x)\, dx \qquad (6)$$

and define

$$C_Z := \sup\{I_Z(f_1, \ldots, f_k) : \|f_i\|_{p_i} = 1 \ \text{ for } i = 1, \ldots, k\}. \qquad (7)$$

Then C_Z is determined by restricting the f's to be Gaussian functions, i.e.,

$$\begin{aligned} C_Z = \sup\{I_Z(f_1, \ldots, f_k) : \|f_i\|_{p_i} = 1 \text{ and } f_i(x) = \exp[-(x, J_i x)] \\ \text{with } J_i \text{ a real, symmetric, positive-definite } n_i \times n_i \text{ matrix}\}. \end{aligned} \qquad (8)$$

Although the sharp constant C_Z is not given explicitly, (8) contains an algorithm for computing C_Z since integrals of Gaussian functions are computable by well-known means (see the Exercises). The proof of the generalized Young's inequality (even without the sharp constant) is much more involved than the proof of the usual one, Theorem 4.2.

PROOF OF THEOREM 4.2.

(A) SIMPLE VERSION WITHOUT THE SHARP CONSTANT. We can obviously assume that f, g and h are real and nonnegative. Write the double integral in (1) as $I := \int_{\mathbb{R}^n} \int_{\mathbb{R}^n} \alpha(x, y) \beta(x, y) \gamma(x, y) \, dx\, dy$ with

$$\begin{aligned} \alpha(x, y) &= f(x)^{p/r'} g(x - y)^{q/r'}, \\ \beta(x, y) &= g(x - y)^{q/p'} h(y)^{r/p'}, \\ \gamma(x, y) &= f(x)^{p/q'} h(y)^{r/q'}. \end{aligned} \qquad (9)$$

Noting that $1/p' + 1/q' + 1/r' = 1$, we can use Hölder's inequality for three functions to obtain $|I| \le \|\alpha\|_{r'}\|\beta\|_{p'}\|\gamma\|_{q'}$. But

$$\|\alpha\|_{r'} = \left\{ \int_{\mathbb{R}^n} \int_{\mathbb{R}^n} f(x)^p g(x-y)^q \, dx \, dy \right\}^{1/r'} = \|f\|_p^{p/r'} \|g\|_q^{q/r'}, \qquad (10)$$

and similarly for β and γ. The right equality in (10) is, of course, a consequence of changing variables from y to $y - x$ and doing the y-integration first. The final result is (1) with the sharp constant $(C_p C_q C_r)^n$ replaced by the larger value 1. ∎

(B) FULL VERSION WITH THE SHARP CONSTANT. We start with an auxiliary problem that has the virtue that we can show that maximizers f, g, h exist and that we can compute them. The next part of the proof will consist in deriving the original problem from the auxiliary problem by a limiting procedure. We will not prove that the only maximizers are the functions given in (3), and will leave that to the reader, who can consult [Brascamp–Lieb] or [Lieb, 1990].

It is more convenient to prove Young's inequality in the form (4) rather than (1). Our auxiliary problem consists in replacing g on the left side of (4) by $j_\varepsilon * g$, as in Sect. 2.16 with j_ε a Gaussian with $\int j_\varepsilon = 1$. Furthermore, we multiply g and h on the left side of (4) by Gaussians $e^{-\delta x^2}$. Thus, our auxiliary problem consists in examining the function

$$K_{g,h}^{\varepsilon,\delta}(x) = \int_{\mathbb{R}^n} \int_{\mathbb{R}^n} J_n^{\varepsilon,\delta}(x,y,z) g(y) h(z) \, dy \, dz$$

with

$$J_n^{\varepsilon,\delta}(x,y,z) = (\pi\varepsilon)^{-n/2} \exp\{-(x-y-z)^2/\varepsilon - \delta y^2 - \delta z^2 - \delta x^2\}.$$

Our goal is to compute the sharp constant $C_n^{\varepsilon,\delta}$ in the inequality

$$\|K_{g,h}^{\varepsilon,\delta}\|_p \le C_n^{\varepsilon,\delta} \|g\|_q \|h\|_r, \qquad (11)$$

with $1 + 1/p = 1/q + 1/r$ as in (4). Since p, q, r are fixed, the dependence of $C_n^{\varepsilon,\delta}$ on p, q, r is not made explicit.

Note that

$$C_n^{\varepsilon,\delta} \le C_n^{\varepsilon,0} \le C_{p',q,r;n} \le 1. \qquad (12)$$

The first inequality is obvious and the second follows from $\|j_\varepsilon * g\|_p \le \|g\|_p$, which is a consequence of the nonsharp Young's inequality proved in part (A) above.

First, we show that the sharp constant is attained, i.e., that there exist two functions g and h with $\|g\|_q = \|h\|_r = 1$ such that $\|K_{g,h}^{\varepsilon,\delta}\|_p = C_n^{\varepsilon,\delta}$. Let g_i, h_i be a maximizing sequence of function pairs, i.e., $\|K_{g_i,h_i}^{\varepsilon,\delta}\|_p \to C_n^{\varepsilon,\delta}$ under the assumption that $\|g_i\|_q = \|h_i\|_r = 1$. By Theorem 2.18 (bounded sequences have weak limits) there exist $g \in L^q(\mathbb{R}^n)$, $h \in L^r(\mathbb{R}^n)$ such that $g_i \rightharpoonup g$ and $h_i \rightharpoonup h$ weakly in $L^q(\mathbb{R}^n)$, respectively $L^r(\mathbb{R}^n)$.

By Exercise 6, $K_{g_i,h_i}^{\varepsilon,\delta}$ converges strongly in $L^p(\mathbb{R}^n)$ to the function $K_{g,h}^{\varepsilon,\delta}$. By Theorem 2.11 (lower semicontinuity of norms) we know that $\|g\|_q \leq 1$ and $\|h\|_r \leq 1$. In fact, $\|g\|_q = \|h\|_r = 1$ because if they were strictly less than 1 the ratio $\|K_{g,h}^{\varepsilon,\delta}\|_p / \|g\|_q \|h\|_r$ would be strictly bigger than $C_n^{\varepsilon,\delta}$, which is a contradiction. Thus, g and h are a maximizing pair and $C_n^{\varepsilon,\delta}$ is attained, as asserted above.

The next step is to use Theorem 2.4 (Minkowski's inequality) to show that
$$C_{n+m}^{\varepsilon,\delta} = C_n^{\varepsilon,\delta} C_m^{\varepsilon,\delta}$$
and that the optimizers g and h must be Gaussian functions. This equality may seem obvious, but it is not trivial and requires proof. We write a point $x \in \mathbb{R}^n$ as $x = (x_1, x_2)$, where $x_1 \in \mathbb{R}^n$ and $x_2 \in \mathbb{R}^m$. Now, by Minkowski's inequality,

$$\begin{aligned}
C_{n+m}^{\varepsilon,\delta} &= \left(\int_{\mathbb{R}^{n+m}} \left| \int_{\mathbb{R}^{n+m}} \int_{\mathbb{R}^{n+m}} J_{n+m}^{\varepsilon,\delta}(x,y,z) g(y) h(z) \, dy \, dz \right|^p dx \right)^{1/p} \\
&\leq \left(\int_{\mathbb{R}^m} \left(\int_{\mathbb{R}^{2m}} J_m^{\varepsilon,\delta}(x_2, y_2, z_2) \right. \right. \\
&\qquad \times \left(\int_{\mathbb{R}^n} \left| \int_{\mathbb{R}^{2n}} J_n^{\varepsilon,\delta}(x_1, y_1, z_1) g(y_1, y_2) \right. \right. \\
&\qquad \left. \left. \left. \times h(z_1, z_2) \, dy_1 \, dz_1 \right|^p dx_1 \right)^{1/p} dy_2 \, dz_2 \right)^p dx_2 \right)^{1/p} \quad (13) \\
&\leq C_n^{\varepsilon,\delta} \left(\int_{\mathbb{R}^m} \left| \int_{\mathbb{R}^{2m}} J_m^{\varepsilon,\delta}(x_2, y_2, z_2) \|g(\cdot, y_2)\|_q \right. \right. \\
&\qquad \left. \left. \times \|h(\cdot, z_2)\|_r \, dy_2 \, dz_2 \right|^p dx_2 \right)^{1/p} \\
&\leq C_n^{\varepsilon,\delta} C_m^{\varepsilon,\delta} \|g\|^q \|h\|^r,
\end{aligned}$$

and hence $C_{n+m}^{\varepsilon,\delta} \leq C_n^{\varepsilon,\delta} C_m^{\varepsilon,\delta}$. Conversely, if (g_n, h_n) and (g_m, h_m) are the optimizers for the n- and m-dimensional problems, the functions

$$g(y_1, y_2) := g_n(y_1) g_m(y_2)$$

and
$$h(z_1, z_2) := h_n(z_1) h_m(z_2)$$
are optimizers for the $m + n$-dimensional problem and hence $C_{n+m}^{\varepsilon,\delta} = C_n^{\varepsilon,\delta} C_m^{\varepsilon,\delta}$.

Next, suppose that $g(y_1, y_2)$ and $h(z_1, z_2)$ are any pair of optimizers of the $m + n$-dimensional problem. Certainly they must be of one sign, which we take to be positive, and we must have equality everywhere in the chain of equalities above. In particular there must be equality in the application of Minkowski's inequality, which means that the function on the right side of (13) must factorize in the following way: There exist two functions $A_{x_2}(x_1)$ and $B_{x_2}(y_2, z_2)$ such that

$$J_m^{\varepsilon,\delta}(x_2, y_2, z_2) \int_{\mathbb{R}^{2n}} J_n^{\varepsilon,\delta}(x_1, y_1, z_1) g(y_1, y_2) h(z_1, z_2) \, dy_1 \, dz_1$$
$$= A_{x_2}(x_1) B_{x_2}(y_2, z_2) \ .$$

From this we conclude that the function
$$J_m^{\varepsilon,\delta}(x_2, y_2, z_2)^{-1} \, A_{x_2}(x_1) B_{x_2}(y_2, z_2)$$
does not depend on x_2 and hence must be of the form $C(x_1) D(y_2, z_2)$ for some functions C and D. Thus

$$\int_{\mathbb{R}^{2m}} J_m^{\varepsilon,\delta}(x_2, y_2, z_2) \int_{\mathbb{R}^{2n}} J_n^{\varepsilon,\delta}(x_1, y_1, z_1) g(y_1, y_2) h(z_1, z_2) \, dy_1 \, dz_1 \, dy_2 \, dz_2$$
$$= \int_{\mathbb{R}^{2m}} J_m^{\varepsilon,\delta}(x_2, y_2, z_2) C(x_1) D(y_2, z_2) \, dy_2 \, dz_2$$
$$= C(x_1) E(x_2)$$

for some function E.

If we interpret our inequality in the form
$$\int J_{m+n}^{\varepsilon,\delta}(x, y, z) f(x) g(y) h(z) \, dx \, dy \, dz \leq C_{n+m}^{\varepsilon,\delta} \|f\|_{p'} \|g\|_q \|h\|_r \ ,$$

the preceding statement amounts to saying that if f, g and h are optimizers, then, by Hölder's inequality, $f(x_1, x_2) = d[C(x_1) E(x_2)]^{p-1}$ where d is some constant. Since f, g and h play a symmetric role we can conclude that *all the optimizers must factorize*. Clearly, each of these factors must be an optimizer of the corresponding n- or m-dimensional problem. An immediate consequence is that the optimizers must be products of optimizers of the one-dimensional problem.

Now, let g and h be any optimizers of the one-dimensional problem. We can get new and interesting optimizers for the two-dimensional problem by considering
$$G(x_1, x_2) = g\Big(\frac{x_1 + x_2}{\sqrt{2}}\Big) g\Big(\frac{x_1 - x_2}{\sqrt{2}}\Big)$$
and
$$H(x_1, x_2) = h\Big(\frac{x_1 + x_2}{\sqrt{2}}\Big) h\Big(\frac{x_1 - x_2}{\sqrt{2}}\Big) \ .$$
We urge the reader to check, by changing variables, that G and H are indeed optimizers of the two-dimensional problem. The formula
$$J_1^{\varepsilon, \delta}(x_1, y_1, z_1) J_1^{\varepsilon, \delta}(x_2, y_2, z_2)$$
$$= J_1^{\varepsilon, \delta}\Big(\frac{x_1 + x_2}{\sqrt{2}}, \frac{y_1 + y_2}{\sqrt{2}}, \frac{z_1 + z_2}{\sqrt{2}}\Big) J_1^{\varepsilon, \delta}\Big(\frac{x_1 - x_2}{\sqrt{2}}, \frac{y_1 - y_2}{\sqrt{2}}, \frac{z_1 - z_2}{\sqrt{2}}\Big)$$
is crucial, and it is here that we first use the fact that $J_1^{\varepsilon, \delta}(x, y, z)$ is a Gaussian function. By the previous argument, since G is an optimizer, we have that
$$g\Big(\frac{x_1 + x_2}{\sqrt{2}}\Big) g\Big(\frac{x_1 - x_2}{\sqrt{2}}\Big) = u(x_1) v(x_2) \tag{14}$$
for some functions u and v. Note that $u(x_1) v(x_2) \in L^q(\mathbb{R}^2)$. We shall prove that this relation implies that g must be a Gaussian function.

Assume for the moment that g is in C^∞ and strictly positive. Then the functions $\eta(x) := \log g(x)$, $\mu(x) := \log u(x)$ and $\nu(x) := \log v(x)$ are also in C^∞ and satisfy the relation
$$\eta\left(\frac{x_1 + x_2}{\sqrt{2}}\right) + \eta\left(\frac{x_1 - x_2}{\sqrt{2}}\right) = \mu(x_1) + \nu(x_2) \ .$$
Differentiating this equation twice with respect to x_1 and x_2 yields
$$\eta''\left(\frac{x_1 + x_2}{\sqrt{2}}\right) = \eta''\left(\frac{x_1 - x_2}{\sqrt{2}}\right)$$
for all x_1 and x_2, which implies that $\eta''(x)$ must be a constant $-a/2$ and hence $\eta(x) = -ax^2 + 2bx + c$ for some constants b and c. Thus
$$g(x) = \exp[-ax^2 + 2bx + c] \ ,$$
i.e., g is a Gaussian function.

To apply this argument to the original function g which is only in $L^q(\mathbb{R})$ we consider
$$g_\lambda(x) = (\lambda/\pi)^{1/2} \int_\mathbb{R} \exp[-\lambda(x-y)^2] g(y) \, dy$$

and note that (14) holds with $g_\lambda, u_\lambda, v_\lambda$ in place of g, u, v. (Why?) Since g is nonnegative, g_λ is strictly positive and clearly in C^∞. Hence

$$g_\lambda(x) = \exp(-a_\lambda x^2 + 2b_\lambda x + c_\lambda)$$

with $a_\lambda > 0$. By Theorem 2.16 there exists a sequence $\lambda_j \to \infty$ such that $g_{\lambda_j}(x) \to g(x)$ for a.e. $x \in \mathbb{R}$. Hence $a_{\lambda_j}, b_{\lambda_j}$ and c_{λ_j} must converge and we call the limits a, b and c with $a > 0$.

The result for h is completely analogous and we can summarize our result by saying that the optimizers of inequality (11) are given by Gaussian functions. In principle these optimizers and the constant can be explicitly computed, but this is quite difficult to do. Instead, we consider first the limit of $C_1^{\varepsilon,\delta}$ as δ tends to zero. Clearly, $C_1^{\varepsilon,\delta}$ is nonincreasing in δ and is bounded by $C_1^{\varepsilon,0}$. In fact, $\lim_{\delta \to 0} C_1^{\varepsilon,\delta} = C_1^{\varepsilon,0}$, which can be seen as follows: For any $\eta > 0$ there exist nonnegative, normalized g, h such that $\|K_{g,h}^{\varepsilon,0}\|_p \geq C_1^{\varepsilon,0} - \eta$. Clearly, $C_1^{\varepsilon,\delta} \geq \|K_{g,h}^{\varepsilon,\delta}\|_p$ and, using the monotone convergence theorem, we conclude that

$$\lim_{\delta \to 0} C_1^{\varepsilon,\delta} \geq \lim_{\delta \to 0} \|K_{g,h}^{\varepsilon,\delta}\|_p = \|K_{g,h}^{\varepsilon,0}\|_p \geq C_1^{\varepsilon,0} - \eta.$$

This proves the claim since η is arbitrarily small. Thus,

$$C_1^{\varepsilon,0} = \sup_{\delta > 0} C_1^{\varepsilon,\delta} = \sup_{\delta > 0} \sup\{\|K_{g,h}^{\varepsilon,\delta}\|_p : g \text{ and } h$$
$$\text{are nonnegative Gaussians } \|g\|_q, \|h\|_r = 1\}.$$

By interchanging the two suprema (why is this allowed?) we see that $C_1^{\varepsilon,0}$ can be computed by taking the supremum over Gaussian functions. The result of this computation, which we leave to the reader, is

$$C_1^{\varepsilon,0} = C_q C_r C_{p'} . \tag{15}$$

Note that the right side does not depend on ε.

Again, we have to show that $\lim_{\varepsilon \to 0} C_1^{\varepsilon,0} = C_{p',q,r;\, 1}$. We already know that $C_1^{\varepsilon,0} \leq C_{p',q,r;\, 1}$. Now we argue as before, i.e., for each given $\eta > 0$ there exist normalized g, h such that $\|g * h\|_p \geq C_{p',q,r;\, 1} - \eta$. Again, $C_1^{\varepsilon,0} \geq \|j_\varepsilon * g * h\|_p$. Since, by Theorem 2.16, $j_\varepsilon * g \to g$ in $L^q(\mathbb{R})$, and since the right side of the preceding inequality is continuous (by the nonsharp Young's inequality), we have that $\liminf_{\varepsilon \to 0} C_1^{\varepsilon,0} \geq C_{p',q,r;\, 1} - \eta$. This shows that $C_{p',q,r;\, 1} = C_q C_r C_{p'}$. By a direct computation, one can check that the Gaussians given in the statement of Theorem 4.2 are optimizers. ∎

4.3 THEOREM (Hardy–Littlewood–Sobolev inequality)

Let $p, r > 1$ and $0 < \lambda < n$ with $1/p + \lambda/n + 1/r = 2$. Let $f \in L^p(\mathbb{R}^n)$ and $h \in L^r(\mathbb{R}^n)$. Then there exists a sharp constant $C(n, \lambda, p)$, independent of f and g, such that

$$\left| \int_{\mathbb{R}^n} \int_{\mathbb{R}^n} f(x)|x-y|^{-\lambda} h(y) \, dx \, dy \right| \leq C(n, \lambda, p) \|f\|_p \|h\|_r. \tag{1}$$

The sharp constant satisfies

$$C(n, \lambda, p) \leq \frac{n}{(n-\lambda)} (|\mathbb{S}^{n-1}|/n)^{\lambda/n} \frac{1}{pr} \left(\left(\frac{\lambda/n}{1 - 1/p} \right)^{\lambda/n} + \left(\frac{\lambda/n}{1 - 1/r} \right)^{\lambda/n} \right).$$

If $p = r = 2n/(2n - \lambda)$, then

$$C(n, \lambda, p) = C(n, \lambda) = \pi^{\lambda/2} \frac{\Gamma(n/2 - \lambda/2)}{\Gamma(n - \lambda/2)} \left\{ \frac{\Gamma(n/2)}{\Gamma(n)} \right\}^{-1+\lambda/n}. \tag{2}$$

In this case there is equality in (1) if and only if $h \equiv (\text{const.})f$ and

$$f(x) = A(\gamma^2 + (x-a)^2)^{-(2n-\lambda)/2}$$

for some $A \in \mathbb{C}$, $0 \neq \gamma \in \mathbb{R}$ and $a \in \mathbb{R}^n$.

REMARKS. (1) Inequality (1) (not in the sharp form) was proved in [Hardy–Littlewood, 1928, 1930] and [Sobolev]. The sharp version with the constant given by (2) was proved in [Lieb, 1983]. There it was also shown that in the case $p \neq r$ there exist optimizers, i.e., functions which, when inserted into (1), give equality with the smallest constant. However, neither this constant nor the optimizers are known when $p \neq r$.

(2) The inequality (1) is sometimes referred to as the weak Young inequality. Note that (1) looks almost like Young's inequality, Theorem 4.2, with $g(x)$ replaced by $|x|^{-\lambda}$. This function is, however, not in any L^p-space, but nevertheless we have an inequality analogous to Young's inequality. The term 'weak' stands for the fact that $|x|^{-\lambda}$ is in the **weak L^q-space** $L^q_w(\mathbb{R}^n)$ with $q = n/\lambda$. This space is defined as the space of all measurable functions f such that

$$\sup_{\alpha > 0} \alpha |\{x : |f(x)| > \alpha\}|^{1/q} < \infty. \tag{3}$$

Any function in $L^q(\mathbb{R}^n)$ is in $L^q_w(\mathbb{R}^n)$. Simply note that for any α

$$\|f\|_q^q \geq \int_{|f|>\alpha} |f(x)|^q \, dx \geq \alpha^q |\{x : |f(x)| > \alpha\}|. \tag{4}$$

The expression given by (3) does *not* define a norm. There is an alternative expression, equivalent to (3), that is indeed a norm. It is given by

$$\|f\|_{q,w} = \sup_A |A|^{-1/r} \int_A |f(x)|\, dx, \tag{5}$$

where $1/q + 1/r = 1$ and A denotes an arbitrary measurable set of measure $|A| < \infty$. Using Theorem 1.14 (bathtub principle) it is not hard to see that (3) and (5) are equivalent. That (5) is a true norm is an easy exercise. In particular, we note that

$$\|f\|_{n/\lambda,w} = \frac{n}{n-\lambda}[|\mathbb{S}^{n-1}|/n]^{\lambda/n} \quad \text{when } f(x) = |x|^{-\lambda}. \tag{6}$$

Here $|\mathbb{S}^{n-1}|$ is the surface area of the unit sphere $\mathbb{S}^{n-1} \subset \mathbb{R}^n$, i.e., $\mathbb{S}^{n-1} = \{x \in \mathbb{R}^n : |x| = 1\}$.

The general **weak Young inequality** states that for $g \in L^q_w(\mathbb{R}^n)$ and $\infty > p, q, r > 1$ with $1/p + 1/q + 1/r = 2$, the following inequality holds:

$$\int_{\mathbb{R}^n} \int_{\mathbb{R}^n} f(x) g(x-y) h(y) \, dx \, dy \leq C_{p,q,r} \|f\|_p \|g\|_{q,w} \|h\|_r. \tag{7}$$

A simple exercise, using the symmetric-decreasing rearrangement of Sect. 3.3, shows that for each $x \in \mathbb{R}^n$

$$\begin{aligned}(|\mathbb{S}^{n-1}|/n)^{-1/q} |x|^{-n/q} \\ = \sup\{|g(x)| : \|g\|_{q,w} \leq 1,\ g \text{ is symmetric-decreasing}\}.\end{aligned} \tag{8}$$

Hence, by Theorem 3.7 (Riesz's rearrangement inequality), we see that (7) and (1) are equivalent provided we set $\lambda = n/q$. In particular the sharp constant in the weak Young inequality is the same as for the Hardy–Littlewood–Sobolev inequality.

As in Remark (2), 4.2, we can also view the HLS inequality as the statement that convolution is a bounded map from $L^p(\mathbb{R}^n) \times L^q_w(\mathbb{R}^n)$ to $L^r(\mathbb{R}^n)$. In other words, taking $\lambda = n/q$ and setting $p = r'$ in (1),

$$\|g * h\|_r \leq \frac{1}{q'} \left(\frac{n}{|\mathbb{S}^{n-1}|}\right)^{-1/q} C(n, n/q, r') \|g\|_{q,w} \|h\|_p \tag{9}$$

with $1/p + 1/q = 1 + 1/r$.

(3) Returning to (1) we note that when $p = r$ we are allowed to take $h = \overline{f}$ in (1) because, we claim, this quadratic form is positive-definite, i.e., when $f \in L^{2n/(2n-\lambda)}(\mathbb{R}^n)$ and $f \not\equiv 0$,

$$\int_{\mathbb{R}^n} \int_{\mathbb{R}^n} \overline{f}(x) |x-y|^{-\lambda} f(y) \, dx \, dy > 0. \tag{10}$$

The proof is easy: We can write $g_\lambda(x) := |x|^{-\lambda}$ as a convolution $g_\lambda = (\text{const.}) g_{(n+\lambda)/2} * g_{(n+\lambda)/2}$. Then, with $V := g_{(n+\lambda)/2} * f$, we see that the integral in (10) is proportional to $\int |V|^2$. (These remarks are sketchy, but the assertion (10) and the proof are essentially the same as Theorem 9.8 (positivity properties of the Coulomb energy); full details are given there.)

(4) We shall give two proofs of (1). The first one is quite elementary but it will not reveal what the sharp constant is. The second proof, which is in Sect. 4.7, works only in the case $p = r$, but it will yield the sharp constant (2).

FIRST PROOF. The idea is to write the left side of (1) in terms of the layer cake representation and then estimate the integrals that occur. We can assume that both f and h are nonnegative functions and, without loss of generality, we may assume that $\|f\|_p = \|h\|_r = 1$.

We have the following formulas:

$$|x|^{-\lambda} = \lambda \int_0^\infty c^{-\lambda-1} \chi_{\{|x|<c\}}(x) \, dc, \tag{11}$$

$$f(x) = \int_0^\infty \chi_{\{f>a\}}(x) \, da, \tag{12}$$

$$h(x) = \int_0^\infty \chi_{\{h>b\}}(x) \, db. \tag{13}$$

Inserting these on the left side of (1) we obtain

$$\begin{aligned} I &:= \int_{\mathbb{R}^n} \int_{\mathbb{R}^n} f(x) |x-y|^{-\lambda} h(y) \, dx \, dy \\ &= \lambda \int_0^\infty \int_0^\infty \int_0^\infty \int_{\mathbb{R}^n} \int_{\mathbb{R}^n} c^{-\lambda-1} \chi_{\{f>a\}}(x) \chi_{\{h>b\}}(y) \\ &\quad \times \chi_{\{|x|<c\}}(x-y) \, dx \, dy \, da \, db \, dc. \end{aligned} \tag{14}$$

The integrals over x and y in (14) can be estimated from above by replacing one of the three χ's in (14) by the number 1. Thus,

$$I \leq \lambda \int \int \int c^{-\lambda-1} I(a,b,c) \, da \, db \, dc$$

and

$$I(a,b,c) := v(a) w(b) u(c) / \max\{v(a), w(b), u(c)\}, \tag{15}$$

with

$$w(b) = \int_{\mathbb{R}^n} \chi_{\{h>b\}}, \qquad v(a) = \int_{\mathbb{R}^n} \chi_{\{f>a\}}$$

and
$$u(c) = (|\mathbb{S}^{n-1}|/n)c^n.$$

The norms of f and h can be written as

$$\|f\|_p^p = p\int_0^\infty a^{p-1}v(a)\,da = 1, \qquad \|h\|_r^r = r\int_0^\infty b^{r-1}w(b)\,db = 1. \tag{16}$$

To do the c-integration we assume first that $v(a) \geq w(b)$, the other case being similar. Using (15) we compute

$$\int_0^\infty c^{-\lambda-1}I(a,b,c)\,dc$$
$$\leq \int_{u(c)\leq v(a)} c^{-\lambda-1}w(b)u(c)\,dc + \int_{u(c)>v(a)} c^{-\lambda-1}w(b)v(a)\,dc$$
$$= w(b)\left(\frac{|\mathbb{S}^{n-1}|}{n}\right)\int_0^{(v(a)n/|\mathbb{S}^{n-1}|)^{1/n}} c^{-\lambda-1+n}\,dc$$
$$+ w(b)v(a)\int_{(v(a)n/|\mathbb{S}^{n-1}|)^{1/n}}^\infty c^{-\lambda-1}\,dc \tag{17}$$
$$= \frac{1}{n-\lambda}(|\mathbb{S}^{n-1}|/n)^{\lambda/n}w(b)v(a)^{1-\lambda/n}$$
$$+ \frac{1}{\lambda}(|\mathbb{S}^{n-1}|/n)^{\lambda/n}w(b)v(a)^{1-\lambda/n}$$
$$= \frac{n}{\lambda(n-\lambda)}(|\mathbb{S}^{n-1}|/n)^{\lambda/n}w(b)v(a)^{1-\lambda/n}.$$

By repeating the same computation over the range where $w(b) \geq v(a)$ and collecting terms, one obtains

$$I \leq \frac{n}{(n-\lambda)}(|\mathbb{S}^{n-1}|/n)^{\lambda/n}$$
$$\times \int_0^\infty\int_0^\infty \min\{w(b)v(a)^{1-\lambda/n},\ w(b)^{1-\lambda/n}v(a)\}\,da\,db. \tag{18}$$

Note that $w(b) \leq v(a)$ if and only if $w(b)v(a)^{1-\lambda/n} \leq w(b)^{1-\lambda/n}v(a)$.

Next, we split the b-integral into two integrals, one from zero to $a^{p/r}$ and the other from $a^{p/r}$ to infinity. Thus, the integral in (18) is bounded above by

$$\int_0^\infty v(a)\int_0^{a^{p/r}} w(b)^{1-\lambda/n}\,db\,da + \int_0^\infty v(a)^{1-\lambda/n}\int_{a^{p/r}}^\infty w(b)\,db\,da. \tag{19}$$

It is easy to see (Exercise 3) that the second term in (19) can be rewritten as

$$\int_0^\infty w(s)\int_0^{s^{r/p}} v(t)^{1-\lambda/n}\,dt\,ds. \tag{20}$$

By Hölder's inequality with $m = (r-1)(1-\lambda/n)$

$$\int_0^{a^{p/r}} w(b)^{1-\lambda/n} b^m b^{-m}\, db$$
$$\leq \left(\int_0^{a^{p/r}} w(b)^{(1-\lambda/n)/(1-\lambda/n)} b^{m-1}\, db\right)^{1-\lambda/n} \left(\int_0^{a^{p/r}} b^{-mn/\lambda}\, db\right)^{\lambda/n}. \tag{21}$$

It is easy to see that $mn/\lambda < 1$ and hence the first term in (19) is bounded above by

$$\left(\frac{\lambda}{n - r(n-\lambda)}\right)^{\lambda/n} \left(\int_0^\infty v(a) a^{p-1}\, da\right) \left(\int_0^\infty w(b) b^{r-1}\, db\right)^{1-\lambda/n}$$
$$= \frac{1}{pr}\left(\frac{\lambda/n}{1-1/p}\right)^{\lambda/n}. \tag{22}$$

An analogous computation using (20) shows that the second term in (19) is bounded above by

$$\frac{1}{pr}\left(\frac{\lambda/n}{1-1/r}\right)^{\lambda/n}. \tag{23}$$

The desired estimate is proved by collecting terms and returning to (18) and (19). ∎

In Sect. 4.7 we shall give the proof of (1) which yields the sharp constant (2). But first some geometric concepts have to be introduced.

4.4 CONFORMAL TRANSFORMATIONS AND STEREOGRAPHIC PROJECTION

A fundamental technique is to exploit the symmetries of 4.3(1). Some of them are obvious. If we replace $f(x)$ and $h(x)$ by $(\tau_a f)(x) := f(x-a)$ and $(\tau_a h)(x) := h(x-a)$ for $a \in \mathbb{R}^n$, we see that both sides of (1) do not change their value. We then say that the inequality 4.3(1) is **translation invariant**. Similarly for $\mathcal{R} \in O(n)$, the **orthogonal group** of rotations and reflections of $\mathbb{R}^n$, we can replace f, h by $(\mathcal{R}f)(x) := f(\mathcal{R}^{-1}x), (\mathcal{R}h)(x) := h(\mathcal{R}^{-1}x)$ and again we do not change the value. Thus our inequality is invariant under the following action of the **Euclidean group**:

$$[(\mathcal{R}, a), f(x)] \mapsto f(\mathcal{R}^{-1}x - a), \qquad \mathcal{R} \in O(n), a \in \mathbb{R}^n,$$

and similarly for h.

Another simple symmetry is the **scaling symmetry**. If we replace $f(x), h(x)$ by $s^{p/n} f(sx)$, $s^{q/n} h(sx)$ for $s > 0$, then 4.3(1) is again invariant. The reader is urged to check this. Note that stretching is not a member of the Euclidean group because geometric figures do not stay congruent under scaling. It is, however, a member of another important group of transformations, **the conformal group**, namely, the group of deformations that preserve angles. There are many more maps that preserve angles and one of them is the **inversion on the unit sphere**, $\mathcal{I}: \mathbb{R}^n \to \mathbb{R}^n$,

$$x \mapsto \frac{x}{|x|^2} =: \mathcal{I}(x). \tag{1}$$

There are some remarks to be made about the inversion map. As stated it is not defined on $\mathbb{R}^n$ but only on $\mathbb{R}^n$ without the origin. One can, however, extend $\mathcal{I}$ to $\dot{\mathbb{R}}^n$, the one-point compactification of $\mathbb{R}^n$; this is nothing but $\mathbb{R}^n \cup \{\infty\}$ where ∞ is defined to be an element which is contained in all unbounded open sets. If we define $\mathcal{I}(0) = \infty$ and $\mathcal{I}(\infty) = 0$, $\mathcal{I}$ extends to $\dot{\mathbb{R}}^n$.

Now note that

$$|\mathcal{I}(x) - \mathcal{I}(y)|^2 = \left| \frac{x}{|x|^2} - \frac{y}{|y|^2} \right|^2 = \frac{1}{|x|^2} - \frac{2x \cdot y}{|x|^2 |y|^2} + \frac{1}{|y|^2} = \frac{1}{|x|^2} \frac{1}{|y|^2} |x-y|^2. \tag{2}$$

If we pick two C^1 curves $x(t), y(t)$ in $\dot{\mathbb{R}}^n$ with $x(0) = y(0) = z \neq 0$, then $u(t) := \mathcal{I}(x(t))$ and $v(t) := \mathcal{I}(y(t))$ define two new curves in $\dot{\mathbb{R}}^n$. We have to check that the angle between the tangent vectors of $u(t)$ and $v(t)$ (which are $\dot{u}$ and $\dot{v}$) has the same value at $t = 0$ as the angle between $\dot{x}(t)$ and $\dot{y}(t)$ at $t = 0$. But, by (2),

$$|\dot{u} - \dot{v}| = \lim_{t \to 0} \frac{1}{t} |\mathcal{I}(x(t)) - \mathcal{I}(z) + \mathcal{I}(z) - \mathcal{I}(y(t))| = \frac{1}{|z|^2} |\dot{x} - \dot{y}| \tag{3}$$

and, in particular, $|\dot{u}| = |\dot{x}|/|z|^2$, $|\dot{v}| = |\dot{y}|/|z|^2$, from which we find that

$$\frac{\dot{u} \cdot \dot{v}}{|\dot{u}||\dot{v}|} = \frac{\dot{x} \cdot \dot{y}}{|\dot{x}||\dot{y}|},$$

i.e., $\mathcal{I}$ is conformal. An attentive reader will actually point out that $\mathcal{I}$ is **anticonformal** since it reverses the orientation. But this distinction does not play a role in our considerations.

There is a very nice description of $\dot{\mathbb{R}}^n$ by means of **stereographic projection**. Define $s = (s_1, s_2, \ldots, s_{n+1})$ by

$$s_i = \frac{2x_i}{1 + |x|^2} \quad \text{for } i = 1, \ldots, n \text{ and } s_{n+1} = \frac{1 - |x|^2}{1 + |x|^2}. \tag{4}$$

If $x = \infty$, then $s_i = 0$ for $i = 1, \ldots, n$ and $s_{n+1} = -1$. A simple calculation shows that $\sum_{i=1}^{n+1} s_i^2 = 1$. Thus $\mathcal{S} : x \mapsto s$ is a map from $\dot{\mathbb{R}}^n \to \mathbb{S}^n$. The inverse of $\mathcal{S}$ is given by

$$x_i = \frac{s_i}{1 + s_{n+1}}, \quad i = 1, \ldots, n. \tag{5}$$

With considerable abuse of notation we shall call $\mathcal{S}^{-1}$ **stereographic coordinates** for $\mathbb{S}^n$. Of course there is no single coordinate patch that covers $\mathbb{S}^n$ nor is there one that covers $\dot{\mathbb{R}}^n$. The topology of these two spaces is, in fact, quite different from the topology of $\mathbb{R}^n$ (e.g., they are not contractible). For our purposes we do not need a coordinate description for the whole of $\mathbb{S}^n$, and thus the introduction of '∞' is a convenient way to avoid carrying around two coordinate systems. A simple calculation shows that

$$\sum_{i=1}^{n+1} (s_i - t_i)^2 = |s - t|^2 = \frac{4}{(1 + |x|^2)(1 + |y|^2)} |x - y|^2, \tag{6}$$

where $s = \mathcal{S}(x)$ and $t = \mathcal{S}(y)$. Again, as in the case of the inversion, $\mathcal{S}$ is conformal! If we consider a tiny triangle in $\dot{\mathbb{R}}^n$ and its image on $\mathbb{S}^n$ under stereographic projection, we see from (6) that the lengths of the corresponding edges have changed, but the *ratio* of the corresponding edges are the same for all three edges. Thus, the small triangle has undergone a stretching without changing its geometric shape. The term conformal stems from this fact.

Thus from the point of view of conformal geometry, i.e., by considering figures as 'congruent' if they can be transformed into each other by a conformal map, we cannot distinguish between $\mathbb{S}^n$ and $\dot{\mathbb{R}}^n$.

It is a theorem (see e.g., [Dubrovin–Fomenko–Novikov]) that the Euclidean group together with scaling and inversion generates all conformal transformations. It is another theorem that the conformal group on $\mathbb{R}^n$ is isomorphic to the Lorentz group in $(n + 1, 1)$ dimensions, also called $O(n + 1, 1)$.

The reader can relax at this point. Most of the information given above is meant as background, and it is not necessary for what follows. What is important, however, is that certain conformal transformations are easier to visualize on $\mathbb{S}^n$ and certain others on $\dot{\mathbb{R}}^n$.

It is an easy and instructive exercise to see that the inversion $\mathcal{I}$ induces a reflection of $\mathbb{S}^n$ into itself. In fact $\mathcal{S} \circ \mathcal{I} \circ \mathcal{S}^{-1}(s) = (s_1, \ldots, s_n, -s_{n+1})$.

An **isometry** of a space, generally speaking, is a map that preserves distances between points, e.g., an isometry of $L^p(\Omega)$ is a map of functions that preserves the norm $\|f - g\|_p$. The set of isometries of $\mathbb{S}^n$ is the group

$O(n + 1)$. The elements of the conformal group that are missing from this set are the translations and scaling, all of which are easier to visualize on $\dot{\mathbb{R}}^n$. If the dimensions of these groups are added we get

$$\frac{(n+1)n}{2} + n + 1 = \frac{(n+2)(n+1)}{2} = \dim O(n+1,1),$$

i.e., the dimension of the whole conformal group, which we now denote by $\mathcal{C}$.

If $\gamma : \dot{\mathbb{R}}^n \to \dot{\mathbb{R}}^n$ is in $\mathcal{C}$, then we can define an action of γ on functions f in $L^p(\mathbb{R}^n)$ as follows. Pick a sequence $f^k \in L^p(\mathbb{R}^n)$ such that f^k vanishes outside a ball B_k for all $k = 1, 2, \ldots$ and such that $f^k \to f$ in $L^p(\mathbb{R}^n)$. Next we observe that

$$(\gamma^* f^k)(x) := |\mathcal{J}_{\gamma^{-1}}(x)|^{1/p} f^k(\gamma^{-1} x) \tag{7}$$

is well defined for all k. Here $\mathcal{J}_{\gamma^{-1}}(x)$ is the Jacobian of the transformation γ^{-1}. This map γ^* is linear and, by a change of variables, it is seen that

$$\|\gamma^* f^k\|_p = \|f^k\|_p. \tag{8}$$

Thus it follows that $\gamma^* f^k$ converges strongly in $L^p(\mathbb{R}^n)$ to a function $\gamma^* f$ and this limit is independent of the approximating sequence f^k. Thus γ^* extends to an invertible isometry on $L^p(\mathbb{R}^n)$.

In the same fashion we can lift functions in $L^p(\mathbb{R}^n)$ to the sphere $\mathbb{S}^n$. Simply define

$$F(s) = (\mathcal{S}^* f)(s) = |\mathcal{J}_{\mathcal{S}^{-1}}(s)|^{1/p} f(\mathcal{S}^{-1}(x)). \tag{9}$$

Again

$$\|F\|_{L^p(\mathbb{S}^n)} = \|f\|_{L^p(\mathbb{R}^n)}. \tag{10}$$

It is necessary to compute the Jacobian of the stereographic projection $\mathcal{J}_\mathcal{S}(x)$. To this end we derive from (6) that g_{ij}, **the standard metric on $\mathbb{S}^n$** (i.e., the one inherited from $\mathbb{R}^{n+1}$) is expressed in terms of stereographic coordinates by

$$g_{ij} = \left(\frac{2}{1+|x|^2}\right)^2 \delta_{ij}. \tag{11}$$

Hence the **standard volume element on $\mathbb{S}^n$** is

$$ds = \left(\frac{2}{1+|x|^2}\right)^n dx \tag{12}$$

and therefore

$$\mathcal{J}_\mathbb{S}(x) = \left(\frac{2}{1+|x|^2}\right)^2 \quad \text{and} \quad \mathcal{J}_{\mathbb{S}^{-1}}(s) = (1+s_{n+1})^n. \tag{13}$$

Armed with (4), (7), (11) and (12) we can state the following.

4.5 THEOREM (Conformal invariance of the Hardy–Littlewood–Sobolev inequality)

Assume that $p = r$ in 4.3(1) and that $F \in L^p(\mathbb{S}^n)$ and $f \in L^p(\mathbb{R}^n)$ are related by 4.4(9). Let H and h be another pair related in the same way. Then

$$\int_{\mathbb{R}^n} \int_{\mathbb{R}^n} f(x)|x-y|^{-\lambda} h(y) \, dx \, dy = \int_{\mathbb{S}^n} \int_{\mathbb{S}^n} F(s)|s-t|^{-\lambda} H(t) \, ds \, dt \qquad (1)$$

and

$$\|F\|_p = \|f\|_p. \qquad (2)$$

Here $|s-t|^2 = \sum_{i=1}^{n+1}(s_i - t_i)^2$ is the Euclidean distance of $\mathbb{R}^{n+1}$ (and not the geodesic distance on $\mathbb{S}^n$). Manifestly, this shows the invariance under all isometries of $\mathbb{S}^n$, i.e., invariance under the group $O(n+1)$. Moreover, the HLS inequality is conformally invariant, i.e., for $\gamma \in \mathcal{C}$

$$\int_{\mathbb{R}^n} \int_{\mathbb{R}^n} (\gamma^* f)(x)|x-y|^{-\lambda} (\gamma^* h)(y) \, dx \, dy$$
$$= \int_{\mathbb{R}^n} \int_{\mathbb{R}^n} f(x)|x-y|^{-\lambda} h(y) \, dx \, dy \qquad (3)$$

and

$$\|\gamma^* f\|_p = \|f\|_p, \quad \|\gamma^* h\|_p = \|h\|_p. \qquad (4)$$

PROOF. We can write the left side of (1) as

$$\int_{\mathbb{R}^n} \int_{\mathbb{R}^n} \left(\frac{1+|x|^2}{2}\right)^{n/p} f(x) \left(\frac{2}{1+|x|^2}|x-y|^2 \frac{2}{1+|y|^2}\right)^{-\lambda/2}$$
$$\times \left(\frac{1+|y|^2}{2}\right)^{n/p} h(y) \left(\frac{2}{1+|x|^2}\right)^n dx \left(\frac{2}{1+|y|^2}\right)^n dy \qquad (5)$$

using that $2/p + \lambda/n = 2$. By (6), (9) and (12) of Sect. 4.4 this can be rewritten as

$$\int_{\mathbb{S}^n} \int_{\mathbb{S}^n} F(s)|s-t|^{-\lambda} H(t) \, ds \, dt, \qquad (6)$$

which proves (1). Equations (2) and (4) are repetitions of 4.4(10) and 4.4(8) respectively. As explained in Sect. 4.4, invariance under the isometries of $\mathbb{S}^n$, the translations and the scaling, implies invariance under the whole conformal group $\mathcal{C}$. ∎

• We turn next to the problem of finding the sharp constant in 4.3(1) if $p = r$. As explained in Remark (3), 4.3(10), we can restrict our attention to the case $h = f$ and $f \geq 0$. In other words, we are interested in the quantity

$$C(n, \lambda) = \sup\{\mathcal{H}(f) : f \in L^p(\mathbb{R}^n), \ f \geq 0, \ f \not\equiv 0\}, \tag{7}$$

where

$$\mathcal{H}(f) := \int_{\mathbb{R}^n} \int_{\mathbb{R}^n} f(x)|x - y|^{-\lambda} f(y) \, dx \, dy \bigg/ \|f\|_p^2. \tag{8}$$

Furthermore, we are interested in whether or not the supremum is a maximum, i.e., whether there exists a function f_0 such that $C(n, \lambda) = \mathcal{H}(f_0)$.

Note that in (7) the seemingly innocuous condition that f is not identically zero is crucial. If f^k is a maximizing sequence it might happen that this sequence converges to zero. One could not, then, conclude that the supremum in (7) is attained. To show that there is a maximizing sequence whose limit is nonzero is one of the key elements in the original proof [Lieb, 1983].

Here we take an approach that exploits as fully as possible the symmetries of the problem (see [Carlen–Loss]). There are two observations to be made:

(i) If we replace f by its symmetric-decreasing rearrangement f^* (see Sect. 3.3), then, by Theorem 3.7 (Riesz's rearrangement inequality), $\mathcal{H}(f) \leq \mathcal{H}(f^*)$. Thus, in order to compute $C(n, \lambda)$ it suffices to optimize within the class of symmetric-decreasing functions.

(ii) The functional (8) is conformally invariant, by the previous theorem.

The key observation is that (i) and (ii) contradict each other in the sense that if we apply a *general conformal transformation to a radial function, the result will generally no longer be radial*. We shall only give the argument here for $n \geq 2$ and shall relegate the one-dimensional case to the exercises. Pick f radial in $L^p(\mathbb{R}^n)$. Lifting it to the sphere by the prescription 4.4(9) results in the function

$$F(s) = \left(\frac{1 + |x|^2}{2}\right)^{n/p} f(x)$$

expressed in terms of stereographic coordinates. This function is invariant under rotations of $\mathbb{S}^n$ that keep the 'north pole axis' $\mathbf{n} = (0, \ldots, 0, 1)$ fixed. Those rotations correspond to the usual rotations in $\mathbb{R}^n$. Consider now a different rotation, namely the rotation by $90°$

$$D : \mathbb{S}^n \to \mathbb{S}^n, \quad Ds = (s_1, \ldots, s_{n+1}, -s_n), \tag{9}$$

which maps the north pole $\mathbf{n}$-axis into the vector $\mathbf{e} = (0, \ldots, 0, 1, 0)$. The function $F(D^{-1}s)$ is now rotationally symmetric about the $\mathbf{e}$-axis. Should

$F(D^{-1}s)$ correspond to a symmetric-decreasing function on $\mathbb{R}^n$ via 4.4(9), then it must also be symmetric about the **n**-axis. Thus, on one hand,

$$F(s) = \phi(s_{n+1}) \quad \text{for some} \quad \phi : \mathbb{S}^n \to \mathbb{R} \tag{10}$$

and, on the other hand,

$$F(D^{-1}s) = \psi(s_{n+1}) \quad \text{for some} \quad \psi : \mathbb{S}^n \to \mathbb{R}. \tag{11}$$

Consequently,

$$\phi(s_{n+1}) = F(s) = \psi((DS)_{n+1}) = \psi(-s_n)$$

for all $s \in \mathbb{S}^n$, which is only possible if F is a constant on $\mathbb{S}^n$ and hence

$$f(x) = C(1 + |x|^2)^{-n/p}.$$

It is easy to see that the function on $\mathbb{R}^n$ corresponding to $F(D^{-1}s)$ is given by

$$(D^*f)(x) = |x + a|^{-2n/p} f\left(\frac{2x_1}{|x+a|^2}, \ldots, \frac{2x_{n-1}}{|x+a|^2}, \frac{1-|x|^2}{|x+a|^2}\right) \tag{12}$$

where $a = (0, \ldots, 0, 1) \in \mathbb{R}^n$. The representation of D^* on $\mathbb{S}^n$ is, however, more illuminating. For convenience we shall drop the $*$ in the notation and call the right side of (12) $(Df)(x)$. If F is the function on $\mathbb{S}^n$ corresponding to f via (9), we set

$$(DF)(s) = F(D^{-1}s), \tag{13}$$

and we denote the symmetric-decreasing rearrangement of f by

$$(\mathcal{R}f)(x) = f^*(x).$$

Recall that $\mathcal{R}$ is norm-preserving, i.e., $\|\mathcal{R}f\|_p = \|f\|_p$. By the previous considerations we know that $D\mathcal{R}f$ is *no longer* radially symmetric. We may thus iterate these two maps and ask for the behavior of the sequence $(D\mathcal{R})^k f$. Does it converge?

For reasons that will become clear later we shall consider the map

$$\mathcal{R}D : L^p(\mathbb{R}^n) \to L^p(\mathbb{R}^n). \tag{14}$$

The following theorem was first proved in [Carlen–Loss].

4.6 THEOREM (Competing symmetries)

Let $1 < p < \infty$ and let $f \in L^p(\mathbb{R}^n)$ be any nonnegative function. Then the sequence $f^k = (\mathcal{R}D)^k f$ converges strongly in $L^p(\mathbb{R}^n)$, as $k \to \infty$, to the function $h_f := \|f\|_p h$, with

$$h(x) = |\mathbb{S}^n|^{-1/p} \left(\frac{2}{1+|x|^2} \right)^{n/p}. \tag{1}$$

REMARKS. (1) The theorem above says that the map $\mathcal{R}D$ can be viewed first of all as a discrete dynamical system on sets of the form $\{f \in L^p(\mathbb{R}^n) : \|f\|_p = C = \text{const.}\}$, and that the 'attractor' consists of a single element, the function Ch.

(2) The name 'competing symmetries' alludes to the fact that the 'symmetrization' due to the rearrangement and the conformal symmetry strive together to produce the limiting function h_f.

PROOF. We have to show that $\|h_f - f^*\|_p \to 0$ as $k \to \infty$ for every function $f \in L^p(\mathbb{R}^n)$. Actually it suffices to show this for a dense set of functions in $L^p(\mathbb{R}^n)$. To prove this sufficiency, fix $f \in L^p(\mathbb{R}^n)$ and suppose $g \in L^p(\mathbb{R}^n)$ is such that $\|f - g\|_p < \varepsilon/2$. Obviously

$$\|h_f - h_g\|_p = \big|\|f\|_p - \|g\|_p\big| \|h\|_p < \varepsilon/2,$$

since $\|h\|_p = 1$. It follows from the definition of D that

$$\|Df - Dg\|_p = \|f - g\|_p \tag{2}$$

and by Theorem 3.5 (nonexpansivity of rearrangement) we know that

$$\|\mathcal{R}f - \mathcal{R}g\|_p \leq \|f - g\|_p. \tag{3}$$

With these two inequalities we have that

$$\|f^{k+1} - g^{k+1}\|_p \leq \|f^k - g^k\|_p \leq \cdots \leq \|f - g\|_p \tag{4}$$

and therefore by the triangle inequality

$$\|h_f - f^k\|_p \leq \|h_f - h_g\|_p + \|f^k - g^k\|_p + \|h_g - g^k\|_p < \varepsilon + \|h_g - g^k\|_p.$$

Consider now the bounded functions that vanish outside a bounded set. Obviously these functions are dense in $L^p(\mathbb{R}^n)$.

If f is such a function, then there exists a constant C such that

$$f(x) \leq Ch_f(x) \text{ for almost every } x \in \mathbb{R}^n, \tag{5}$$

since h_f is strictly positive. Trivially the map D is order-preserving, i.e.,

$$f(x) \leq g(x) \text{ almost everywhere implies}$$
$$(Df)(x) \leq (Dg)(x) \text{ almost everywhere.}$$

Furthermore, the same is true for the rearrangement (see Remark 3.3(vi)). Since h_f is *invariant* under both of these operations separately, we have that (5) holds along the whole sequence, i.e., for all $k = 0, 1, 2, \ldots$

$$f^k(x) \leq Ch_f(x) \text{ for almost every } x \in \mathbb{R}^n. \tag{6}$$

The constant is the *same* as in (5)! This relation is crucial since it says that the whole sequence is uniformly bounded by a function which is p^{th}-power summable.

Define

$$A := \inf_k \|h_f - f^k\|_p = \lim_{k \to \infty} \|h_f - f^k\|_p. \tag{7}$$

The second equality follows from (2) and (3). Each of these functions f^k is a radially symmetric-decreasing function. Therefore, by using Helly's selection principle as in the compactness proof of Theorem 3.7 (Riesz's rearrangement inequality), we can pass to a further subsequence in which $f^k(x)$ converges for almost every x.

Thus, we have a subsequence f^{k_l} such that $f^{k_l}(x) \to g(x)$ as $l \to \infty$ pointwise for almost every $x \in \mathbb{R}^n$. Moreover, by (6), $f^{k_l}(x) \leq Ch_f(x)$ and hence, by dominated convergence, we have that $g \in L^p(\mathbb{R}^n)$, radially symmetric-decreasing, and

$$A = \lim_{l \to \infty} \|h_f - f^{k_l}\|_p = \|h_f - g\|_p. \tag{8}$$

Now we show that $g = h_f$. To see this apply the operation $\mathcal{R}D$ once to g. By (2) and (3) we have that in $L^p(\mathbb{R}^n)$

$$\mathcal{R}Dg = \lim_{l \to \infty} f^{k_l+1}, \tag{9}$$

and therefore, since $Dh_f = h_f$ and $\mathcal{R}h_f = h_f$,

$$A \leq \|h_f - \mathcal{R}Dg\|_p = \|\mathcal{R}Dh_f - \mathcal{R}Dg\|_p$$
$$\leq \|Dh_f - Dg\|_p = \|h_f - g\|_p = A. \tag{10}$$

Thus the equality sign must hold everywhere in (10). This says in particular that
$$\|h_f - \mathcal{R}Dg\|_p = \|h_f - Dg\|_p.$$

Since h_f is strictly symmetric-decreasing, Theorem 3.5 tells us that
$$\mathcal{R}Dg = Dg.$$

However, as explained towards the end of Sect. 4.5 the only radial functions, g, for which Dg is also radial have the form Ch. Since
$$\|g\|_p = \lim_{l \to \infty} \|f^{k_l}\|_p = \|f\|_p,$$

we have $C = \|f\|_p$ and $g = h_f$. Thus $A = 0$ and $f^{k_l} \to h_f$ in $L^p(\mathbb{R}^n)$. By (2) and (3) we have that
$$\|h_f - f^{k_l+1}\|_p \leq \|h_f - f^{k_l}\|_p,$$

and therefore the entire sequence f^k converges to h_f in $L^p(\mathbb{R}^n)$. ∎

4.7 PROOF OF THEOREM 4.3 (Sharp version of the Hardy–Littlewood–Sobolev inequality)

By 4.3(10) we may assume $h = \overline{f}$. Theorem 4.3(1) and (2) for $p = r$ will now be shown to be a corollary of Theorem 4.6 (competing symmetries). Recall that $\mathcal{H}(f)$ denotes the HLS-functional for any function $f \in L^p(\mathbb{R}^n)$ that is not identically zero. Replace f by $f^m(x) = \min(f(x), mh_f(x))$ so that f^m converges monotonically to $f(x)$ pointwise as $m \to \infty$. If we can show that $\mathcal{H}(f^m) \leq C(n, \lambda)$, then, by monotone convergence, $\mathcal{H}(f^m) \to \mathcal{H}(f)$ and thus $\mathcal{H}(f) \leq C(n, \lambda)$. For convenience we drop the m. Since $\mathcal{H}(Df) = \mathcal{H}(f)$ and $\mathcal{H}(\mathcal{R}f) \geq \mathcal{H}(f)$ by Theorem 3.7 (Riesz's rearrangement inequality), we have that $\mathcal{H}(f^k)$ is a nondecreasing sequence where $f^k = (\mathcal{R}D)^k f$. Since, by the previous theorem, f^k converges to h_f in $L^p(\mathbb{R}^n)$ as $k \to \infty$, we can pass to a subsequence (again denoted by k) and assume that f^k converges pointwise to h_f. Since
$$f^k \leq C(1 + |x|^2)^{-n/p} \quad \text{for all } k,$$

we know, by dominated convergence, that as $k \to \infty$, $\mathcal{H}(f^k)$ converges to $\mathcal{H}(h_f)$ from below. The last expression can be explicitly computed and yields 4.3(2).

It remains to determine the case of equality. It is easy to see that $f = \text{(const.)} \times \text{(a nonnegative function)}$. In short, equality in 4.3(1) can occur only if $h = (\text{const.})f$ and if $\mathcal{H}(f) = C(\lambda, n)$. Then, by the strict rearrangement inequality, Theorem 3.9, we know that f must be a translate of a symmetric-decreasing function. Moreover the same is true for Df since it is also an optimizer by the conformal invariance of $\mathcal{H}(f)$. Thus, the operation $\mathcal{R}D$ acting on f does nothing but translate Df to the origin, and hence $\mathcal{R}Df$ is nothing but a conformal transformation of f. The same is true for the whole sequence, i.e., $f^k = (\mathcal{R}D)^k f$ is a conformal image of f and we can write $f^k = C_k f$, where C_k is a sequence of conformal transformations. Since f^k converges strongly to h_f by Theorem 4.6, and since the conformal transformations (the way we have defined them) act as isometries on $L^p(\mathbb{R}^n)$, we have that

$$\lim_{k \to \infty} \|f - C_k^{-1} h_f\|_p = 0. \tag{1}$$

In Lemma 4.8 (action of the conformal group on optimizers) below, we shall prove that, due to the special nature of the function h_f in (1),

$$(C_k^{-1} h_f)(x) = \lambda_k^{n/p} |\mathbb{S}^n|^{-1/p} \|f\|_p \left(\frac{2}{\lambda_k^2 + (x - a_k)^2} \right)^{n/p} \tag{2}$$

for sequences $\lambda_k \neq 0$ and $a_k \in \mathbb{R}^n$. Since, by (1), $C_k^{-1} h_f$ converges strongly to f, it is plain that λ_k and a_k must converge to some $\lambda \neq 0$ and some $a \in \mathbb{R}^n$. Hence

$$(x) = \lambda^{n/p} |\mathbb{S}^n|^{-1/p} \|f\|_p \left(\frac{2}{\lambda^2 + (x - a)^2} \right)^{n/p}. \qquad \blacksquare$$

4.8 LEMMA (Action of the conformal group on optimizers)

Let $C \in \mathcal{C}$ be a conformal transformation and let h be given by 4.6(1). If C acts on h, then there exist $\lambda \neq 0$ and $a \in \mathbb{R}^n$ (depending on C) such that

$$(Ch)(x) = |\mathbb{S}^n|^{-1/p} \lambda^{n/p} \left(\frac{2}{\lambda^2 + (x - a)^2} \right)^{n/p}.$$

PROOF. Every element in $\mathcal{C}$ is a product of elements of the Euclidean group, scalings and inversions. The result is obvious for scalings and for Euclidean transformations. What remains to be checked is that the inversion $\mathcal{I}$ (see 4.4(1)) maps the function $u(x) = |\mathbb{S}^n|^{-1/p} \mu^{n/p} \left(\frac{2}{\mu^2 + (x-b)^2} \right)^{n/p}$ into a function of the same type. But

$$(\mathcal{I}u)(x) = |\mathbb{S}^n|^{-1/p}\mu^{n/p}|x|^{-2n/p}\left(\frac{2}{\mu^2 + (x/|x|^2 - b)^2}\right)^{n/p}$$

$$= |\mathbb{S}^n|^{-1/p}\left(\frac{\mu}{b^2 + \mu^2}\right)^{n/p}\left(\frac{2}{[\mu/(b^2 + \mu^2)]^2 + [x - b/(b^2 + \mu^2)]^2}\right)^{n/p}. \quad \blacksquare$$

Exercises for Chapter 4

1. Prove that 4.3(5) actually defines a norm—the weak L^q-norm.

2. Prove 4.3(8).

3. Use Fubini's theorem to prove that the second integral in 4.3(19) is given by 4.3(20).

4. Gaussian integrals appear frequently and it is important to know how to compute them.
   a) Show that
   $$\int_{-\infty}^{\infty} \exp(-\lambda x^2)\,dx = \sqrt{\pi/\lambda}$$
   by evaluating the square of the integral by means of polar coordinates.
   b) For A a Hermitian, positive-definite $n \times n$ matrix, show that
   $$\int_{\mathbb{R}^n} \exp[-(x, Ax)]\,dx = \pi^{n/2}/\sqrt{\mathrm{Det}\,A}$$
   where Det denotes the determinant. In the real, symmetric case this can be done by a simple change of variables. The complex case requires either an analytic continuation argument or else the argument in Sect. 5.2.
   c) For V a vector in $\mathbb{C}^n$ show, by 'completing the square', that
   $$\int_{\mathbb{R}^n} \exp[-(x, Ax) + 2(V, x)]\,dx = \left(\pi^{n/2}/\sqrt{\mathrm{Det}\,A}\right)\exp[(V, A^{-1}V)]\,.$$

5. Use Exercise 4 to verify formula 4.2(15) for the sharp constant in inequality 4.2(11) when $\delta = 0$.

6. Show that $K^{\varepsilon,\delta}_{g_i,h_i}$ converges strongly in $L^p(\mathbb{R}^n)$ as $i \to \infty$ to the function $K^{\varepsilon,\delta}_{g,h}$, as required in proof (B) of Theorem 4.2 (Young's inequality).
   ▶ *Hint.* First show that $K^{\varepsilon,\delta}_{g_i,h_i}$ converges pointwise and that it is uniformly bounded (in x and in i). Next, show that the same is true even if we multiply $K^{\varepsilon,\delta}_{g_i,h_i}$ by $\exp(+\gamma x^2)$ for some sufficiently small $\gamma > 0$.

7. *Competing symmetries in one dimension.* Let $f \in L^p(\mathbb{R})$ and denote by $F \in L^p(\mathbb{S}^1)$ the function defined on the unit circle corresponding to f via 4.4(9). Pick an angle α which is not a rational multiple of π and denote by $U_\alpha f$ the function that corresponds to $F(\theta - \alpha)$ via 4.4(9).

   Prove that $f^j = (\mathcal{R}U_\alpha)^j f$ converges strongly to
   $$h = \|f\|_p (2\pi)^{-1/p} \left(\frac{2}{1+x^2}\right)^{1/p}.$$

   Proceed as follows:
   a) By tracing the steps in the proof of Theorem 4.6, show that f^j converges to some symmetric-decreasing function $g \in L^p(\mathbb{R})$ which has the property that $U_\alpha g$ is also symmetric-decreasing.
   b) Deduce from a) that $U_{2\alpha} g = g$ and show that this implies that the function G corresponding to g via 4.4(9) must be constant, and hence that $g = h$. It is at this point that the fact that α is not a rational multiple of π is used.

Chapter 5

The Fourier Transform

The Fourier transform is a versatile tool in analysis, much loved by analysts, scientists and engineers. (In fact, in our definition below we use the engineer's convention about the placement of 2π, which eliminates the annoyance of having to multiply integrals by 2π.) The virtue of the Fourier transform is that it converts the operations of differentiation and convolution into multiplication operations. In particular it allows us to define the relativistic operators $\sqrt{-\Delta}$ and $\sqrt{-\Delta + m^2}$ and the space $H^{1/2}(\mathbb{R}^n)$ in Chapter 7. Some references for the Fourier transform are [Hörmander], [Rudin, 1991], [Reed–Simon, vol. 2], [Schwartz] and [Stein–Weiss].

5.1 DEFINITION OF THE L^1 FOURIER TRANSFORM

Let f be a function in $L^1(\mathbb{R}^n)$. The Fourier transform of f, denoted by $\widehat{f}$, is the function on $\mathbb{R}^n$ given by

$$\widehat{f}(k) = \int_{\mathbb{R}^n} e^{-2\pi i (k,x)} f(x)\, dx \qquad (1)$$

where

$$(k,x) := \sum_{i=1}^{n} k_i x_i.$$

The following algebraic properties are the main motivation for studying the Fourier transform. They are very easy to prove.

$$\text{The map } f \mapsto \widehat{f} \text{ is linear in } f, \tag{2}$$

$$\widehat{\tau_h f}(k) = e^{-2\pi i (k,h)} \widehat{f}(k), \quad h \in \mathbb{R}^n, \tag{3}$$

$$\widehat{\delta_\lambda f}(k) = \lambda^n \widehat{f}(\lambda k), \quad \lambda > 0, \tag{4}$$

where τ_h is the translation operator, $(\tau_h f)(x) = f(x - h)$, and δ_λ is the scaling operator, $(\delta_\lambda f)(x) = f(x/\lambda)$.

Two other easy to prove facts are

$$\widehat{f} \in L^\infty(\mathbb{R}^n) \quad \text{and} \quad \|\widehat{f}\|_\infty \leq \|f\|_1, \tag{5}$$

$$\widehat{f} \text{ is a continuous (and hence measurable) function.} \tag{6}$$

The latter follows from dominated convergence. In fact it is part of the **Riemann–Lebesgue lemma**, which also states that $\widehat{f}(k) \to 0$ as $|k| \to \infty$ (see Exercise 2). Note that $\|\widehat{f}\|_\infty$ equals $\|f\|_1$ whenever f is any nonnegative function; in that case

$$\|\widehat{f}\|_\infty = \widehat{f}(0) = \int f = \|f\|_1.$$

Recall from Sect. 2.15 that the convolution of two functions f and g, both in $L^1(\mathbb{R}^n)$, is given by

$$(f * g)(x) = \int_{\mathbb{R}^n} f(x - y) g(y) \, dy. \tag{7}$$

By Fubini's theorem $f * g \in L^1(\mathbb{R}^n)$, and also by Fubini's theorem

$$\widehat{(f * g)}(k) = \int_{\mathbb{R}^n} e^{-2\pi i (k,x)} \int_{\mathbb{R}^n} f(x - y) g(y) \, dy \, dx$$

$$= \int_{\mathbb{R}^n} e^{-2\pi i (k,y)} g(y) \int_{\mathbb{R}^n} e^{-2\pi i (k,(x-y))} f(x - y) \, dx \, dy \tag{8}$$

$$= \widehat{f}(k) \widehat{g}(k).$$

The following is an important example.

5.2 THEOREM (Fourier transform of a Gaussian)

For $\lambda > 0$, denote by g_λ the Gaussian function on $\mathbb{R}^n$ given by

$$g_\lambda(x) = \exp[-\pi\lambda|x|^2] \tag{1}$$

for $x \in \mathbb{R}^n$. Then

$$\widehat{g}_\lambda(k) = \lambda^{-n/2} \exp[-\pi|k|^2/\lambda].$$

REMARK. This is a special case of Exercise 4.4.

PROOF. By 5.1(4) it suffices to consider $\lambda = 1$. Since

$$g_1(x) = \prod_{i=1}^{n} \exp[-\pi(x_i)^2],$$

it suffices to consider $n = 1$. By definition (since $g_1 \in L^1(\mathbb{R})$)

$$\widehat{g}_1(k) = \int_\mathbb{R} e^{-2\pi i (x,k)} \exp[-\pi x^2] \, \mathrm{d}x = g_1(k) f(k),$$

where

$$f(k) = \int_\mathbb{R} \exp[-\pi(x + ik)^2] \, \mathrm{d}x. \tag{2}$$

A simple limiting argument using the dominated convergence theorem allows us to differentiate (2) under the integral sign as many times as we like. Therefore $f \in C^\infty(\mathbb{R})$ and

$$\begin{aligned}
\frac{\mathrm{d}f}{\mathrm{d}k}(k) &= -2\pi i \int_\mathbb{R} (x + ik) \exp[-\pi(x + ik)^2] \, \mathrm{d}x \\
&= i \int_\mathbb{R} \frac{\mathrm{d}}{\mathrm{d}x} \exp[-\pi(x + ik)^2] \, \mathrm{d}x \\
&= i \exp[-\pi(x + ik)^2]\Big|_{-\infty}^{\infty} = 0,
\end{aligned}$$

i.e., $f(k)$ is constant. But $f(0) = \int_\mathbb{R} \exp[-\pi x^2] \, \mathrm{d}x = 1$. ∎

● The Fourier transform can be defined for functions for which 5.1(1) does not make sense. In particular, it is important for quantum mechanics to define $\widehat{f}$ for $f \in L^2(\mathbb{R}^n)$. One route to this definition goes via the Schwartz space $\mathcal{S}$ (which we will not discuss here). The method below uses only Theorem 2.16 (approximation by C^∞-functions). We begin by considering functions in $L^1(\mathbb{R}^n) \cap L^2(\mathbb{R}^n)$, which are dense in $L^2(\mathbb{R}^n)$.

5.3 THEOREM (Plancherel's theorem)

If $f \in L^1(\mathbb{R}^n) \cap L^2(\mathbb{R}^n)$, then $\widehat{f}$ is in $L^2(\mathbb{R}^n)$ and the following formula of Plancherel holds:
$$\|\widehat{f}\|_2 = \|f\|_2. \tag{1}$$

The map $f \mapsto \widehat{f}$ has a unique extension to a continuous, linear map from $L^2(\mathbb{R}^n)$ into $L^2(\mathbb{R}^n)$ which is an **isometry**, i.e., Plancherel's formula (1) holds for this extension. We continue to denote this map by $f \mapsto \widehat{f}$ (even if $f \notin L^1(\mathbb{R}^n)$).

If f and g are in $L^2(\mathbb{R}^n)$, then Parseval's formula holds,
$$(f, g) := \int_{\mathbb{R}^n} \overline{f}(x) g(x) \, dx = \int_{\mathbb{R}^n} \overline{\widehat{f}}(k) \widehat{g}(k) \, dk = (\widehat{f}, \widehat{g}). \tag{2}$$

PROOF. For $f \in L^1(\mathbb{R}^n) \cap L^2(\mathbb{R}^n)$, the function $\widehat{f}(k)$ is bounded, by 5.1(5), and hence
$$\int_{\mathbb{R}^n} |\widehat{f}(k)|^2 \exp[-\varepsilon \pi |k|^2] \, dk \tag{3}$$

is defined. Since $f \in L^1(\mathbb{R}^n)$, the function $\overline{f}(x) f(y) \exp[-\varepsilon \pi |k|^2]$ of three variables is in $L^1(\mathbb{R}^{3n})$. Using Fubini's theorem and Theorem 5.2 we can express (3) as

$$\int_{\mathbb{R}^{3n}} \overline{f}(x) f(y) e^{2\pi i (k, (x-y))} \exp[-\varepsilon \pi k^2] \, dx \, dy \, dk$$
$$= \int_{\mathbb{R}^{2n}} \varepsilon^{-n/2} \exp\left[-\frac{\pi (x-y)^2}{\varepsilon}\right] \overline{f}(x) f(y) \, dx \, dy. \tag{4}$$

Using Theorem 2.16 (approximation by C^∞-functions)
$$\varepsilon^{-n/2} \int_{\mathbb{R}^n} \exp\left[-\frac{\pi (x-y)^2}{\varepsilon}\right] f(y) \, dy \to f(x)$$

in $L^2(\mathbb{R}^n)$ as $\varepsilon \to 0$, and hence (using Fubini's theorem again) (3) tends to $\int_{\mathbb{R}^n} |f(x)|^2 \, dx$. This shows that (3) is uniformly bounded in ε and the monotone convergence theorem therefore shows that $\widehat{f} \in L^2(\mathbb{R}^n)$ with
$$\|\widehat{f}\|_2 = \|f\|_2. \tag{5}$$

Now let f be in $L^2(\mathbb{R}^n)$ but not in $L^1(\mathbb{R}^n) \cap L^2(\mathbb{R}^n)$. Since $L^1(\mathbb{R}^n) \cap L^2(\mathbb{R}^n)$ is dense in $L^2(\mathbb{R}^n)$, there exists a sequence $f^j \in L^1(\mathbb{R}^n) \cap L^2(\mathbb{R}^n)$ such that $\|f - f^j\|_2 \to 0$. By (5) $\|\widehat{f^j} - \widehat{f^m}\|_2 = \|f^j - f^m\|_2$ and hence $\widehat{f^j}$

is a Cauchy sequence in $L^2(\mathbb{R}^n)$ that converges to some function in $L^2(\mathbb{R}^n)$, which we call $\widehat{f}$. It is obvious from (5) that $\widehat{f}$ does not depend on the choice of the sequence f^j. Moreover,

$$\|\widehat{f}\|_2 = \lim_{j \to \infty} \|\widehat{f^j}\|_2 = \lim_{j \to \infty} \|f^j\|_2 = \|f\|_2.$$

The continuity (in $L^2(\mathbb{R}^n)$) and the linearity of this map is left to the reader. Relation (2) follows from (1) by **polarization**, i.e., the identity

$$(f,g) = \frac{1}{2}\{\|f+g\|_2^2 - i\|f+ig\|_2^2 - (1-i)\|f\|_2^2 - (1-i)\|g\|_2^2\}.$$

Applying (1) to each of these four norms yields (2). ∎

5.4 DEFINITION OF THE L^2 FOURIER TRANSFORM

For each f in $L^2(\mathbb{R}^n)$, the $L^2(\mathbb{R}^n)$-function $\widehat{f}$ defined by the limit given in Theorem 5.3 is called the Fourier transform of f.

Theorem 5.3 is remarkable because it states that for any given $f \in L^2(\mathbb{R}^n)$ one can compute its Fourier transform $\widehat{f}$ by using any $L^1(\mathbb{R}^n)$-approximating sequence whatsoever and one always obtains, as an $L^2(\mathbb{R}^n)$ limit, a function $\widehat{f}$ which is independent of the approximation. Here are two examples with the index $j = 1, 2, 3, \ldots$:

$$\widehat{f^j}(k) = \int_{|x|<j} e^{-2\pi i(k,x)} f(x) \, dx, \tag{1}$$

$$\widehat{h^j}(k) = \int_{\mathbb{R}^n} \cos(|x|^2/j) \exp[-|x|^2/j] e^{-2\pi i(k,x)} f(x) \, dx. \tag{2}$$

The assertion is that there is an $L^2(\mathbb{R}^n)$-function $\widehat{f}$ such that $\|\widehat{f^j} - \widehat{f}\|_2 \to 0$, $\|\widehat{h^j} - \widehat{f}\|_2 \to 0$ and $\|\widehat{f^j} - \widehat{h^j}\|_2 \to 0$ as $j \to \infty$. No assertion is made that the sequences $\widehat{f^j}(k)$ and $\widehat{h^j}(k)$ converge for any k as $j \to \infty$. However, by Theorem 2.7 (completeness of L^p-spaces), there is always a subsequence $j(l)$ with $l = 1, 2, 3, \ldots$ such $\widehat{f^{j(l)}}(h)$ and $\widehat{h^{j(l)}}(k)$ converge for almost every $k \in \mathbb{R}^n$ to $\widehat{f}(k)$.

As we show next, the map $f \mapsto \widehat{f}$ is not just an isometry but it is, in fact, a **unitary transformation**, that is, an invertible isometry. The following is an explicit formula for the inverse.

5.5 THEOREM (Inversion formula)

For $f \in L^2(\mathbb{R}^n)$, we use definition 5.4 to define
$$f^\vee(x) := \widehat{f}(-x) \qquad (1)$$
(which amounts to changing i to $-i$ in 5.1(1)). Then
$$f = (\widehat{f})^\vee. \qquad (2)$$
(Note that the right side is well defined by Theorem 5.3.)

PROOF. For $f \in L^2(\mathbb{R}^n)$ the following formula holds:
$$\int_{\mathbb{R}^n} \widehat{g}_\lambda(y-x) f(y) \, dy = \int_{\mathbb{R}^n} g_\lambda(k) \widehat{f}(k) e^{2\pi i (k,x)} \, dk, \qquad (3)$$
where $g_\lambda(k) = \exp[-\lambda \pi |k|^2]$ and hence $\widehat{g}_\lambda(y-x) = \lambda^{-1/2} \exp[-\pi |x-y|^2/\lambda]$. To verify (3), approximate f by a sequence of functions f^j in $L^1(\mathbb{R}^n) \cap L^2(\mathbb{R}^n)$. For each of these functions formula (3) follows by Fubini's theorem. By Theorem 5.3 (Plancherel's theorem) we know that $f^j \to f$ in $L^2(\mathbb{R}^n)$ implies that $\widehat{f}^j \to \widehat{f}$ in $L^2(\mathbb{R}^n)$. Because g_λ and $\widehat{g}_\lambda$ are in $L^2(\mathbb{R}^n)$ the integrals converge to those in (3), and thus (3) is established in the general case.

As $\lambda \to 0$ the left side of (3) tends to $f(x)$ in $L^2(\mathbb{R}^n)$ by Theorem 2.16 (approximation by C^∞-functions). Since $g_\lambda \widehat{f} \to \widehat{f}$ in $L^2(\mathbb{R}^n)$ as $\lambda \to 0$ (by dominated convergence), we know, on account of Theorem 5.3, that $(g_\lambda \widehat{f})^\vee \to (\widehat{f})^\vee$ in $L^2(\mathbb{R}^n)$. Equating the $\lambda \to 0$ limit of the two sides of (3) gives us (2). ∎

5.6 THE FOURIER TRANSFORM IN $L^p(\mathbb{R}^n)$

The Fourier transform has been defined for $L^1(\mathbb{R}^n)$-functions (with range in $L^\infty(\mathbb{R}^n)$) and $L^2(\mathbb{R}^n)$-functions (with range in $L^2(\mathbb{R}^n)$). Can it be extended to some other $L^p(\mathbb{R}^n)$-space so that its range is in some $L^q(\mathbb{R}^n)$-space?

Let us recall the properties that have been proved so far.
$$f \in L^1(\mathbb{R}^n) \Rightarrow \widehat{f} \in L^\infty(\mathbb{R}^n) \quad \text{with} \quad \|\widehat{f}\|_\infty \leq \|f\|_1, \qquad (A)$$
but *the L^1 Fourier transform is not an invertible mapping* (i.e., not every $L^\infty(\mathbb{R}^n)$-function is the Fourier transform of some $L^1(\mathbb{R}^n)$-function; the constant function is an example).
$$f \in L^2(\mathbb{R}^n) \Rightarrow \widehat{f} \in L^2(\mathbb{R}^n) \quad \text{with} \quad \|\widehat{f}\|_2 = \|f\|_2 \qquad (B)$$

and the Fourier transform is invertible with $f = (\widehat{f})^\vee$.

One way to extend the Fourier transform for $p < \infty$ would be to imitate the $L^2(\mathbb{R}^n)$ construction. The goal would then be to find a constant $C_{p,q}$ such that for every $f \in L^p(\mathbb{R}^n) \cap L^1(\mathbb{R}^n)$ the Fourier transform is in $L^q(\mathbb{R}^n)$ and satisfies

$$\|\widehat{f}\|_q \leq C_{p,q} \|f\|_p. \tag{1}$$

Using the continuity argument of Theorem 5.3 (and the density of $L^p(\mathbb{R}^n) \cap L^1(\mathbb{R}^n)$ in $L^p(\mathbb{R}^n)$) one can then extend the Fourier transform to all of $L^p(\mathbb{R}^n)$ and (1) will continue to hold.

The first remark is that q cannot be arbitrary, in fact q must be p' (with $1/p + 1/p' = 1$). This is a simple consequence of the scaling property 5.1(4); if $q \neq p'$, then $\|\widehat{f}\|_q / \|f\|_p$ can be made arbitrarily large—even for $f \in L^1(\mathbb{R}^n)$. The second remark is that counterexamples show that no bound of type (1) can hold when $p > 2$; see Exercise 9. When $1 \leq p \leq 2$, however, (1) is true, as the following theorem (which is usually called the **Hausdorff–Young inequality**) states.

5.7 THEOREM (The sharp Hausdorff–Young inequality)

Let $1 < p < 2$ and let $f \in L^p(\mathbb{R}^n) \cap L^1(\mathbb{R}^n)$. Then, with $1/p + 1/p' = 1$,

$$\|\widehat{f}\|_{p'} \leq C_p^n \|f\|_p \tag{1}$$

with

$$C_p^2 = [p^{1/p}(p')^{-1/p'}]. \tag{2}$$

Furthermore, equality is achieved in (1) if and only if f is a Gaussian function of the form

$$f(x) = A \exp[-(x, Mx) + (B, x)] \tag{3}$$

with $A \in \mathbb{C}$, M any symmetric, real, positive-definite matrix and B any vector in $\mathbb{C}^n$.

Using the construction in Theorem 5.3, together with (1), $\widehat{f}$ can be extended to all of $L^p(\mathbb{R}^n)$ but, in contrast to the $p = 2$ case, this map is not invertible, i.e., the map is not onto all of $L^{p'}(\mathbb{R}^n)$.

REMARK. The proof of Theorem 5.7 is lengthy and we shall not attempt to give it here. The shortest proof is probably the one in [Lieb, 1990]; the basic idea is similar to that in the proof of Theorem 4.2 (Young's inequality), but the details are more involved. Inequality (1) was first proved with $C_p = 1$ by [Hausdorff] and [W. H. Young] for Fourier *series* by using the Riesz–Thorin interpolation theorem (see [Reed–Simon, vol. 2]). It was extended to Fourier integrals by [Titchmarsh] with $C_p = 1$. [Babenko] derived (2) as the sharp constant for $p' = 4, 6, 8, \ldots$ and [Beckner] proved (2) for all $1 < p < 2$. The fact that equality holds in (1) *only* when f is a Gaussian as in (3) was proved in [Lieb, 1990]. Note that $C_p = 1$ if $p = 1$ or $p = 2$, in agreement with our earlier results, but in those two cases there are many functions that give equality in (1); indeed *all $L^2(\mathbb{R}^n)$-functions give equality when $p = 2$*.

5.8 THEOREM (Convolutions)

Let $f \in L^p(\mathbb{R}^n)$ and $g \in L^q(\mathbb{R}^n)$, and let $1 + 1/r = 1/p + 1/q$. Suppose $1 \leq p, q, r \leq 2$. Then
$$\widehat{f * g}(k) = \widehat{f}(k)\,\widehat{g}(k). \tag{1}$$

PROOF. By Young's inequality, Theorem 4.2, $f * g \in L^r(\mathbb{R}^n)$. By Theorem 5.7, $\widehat{f} \in L^{p'}(\mathbb{R}^n)$ and $\widehat{g} \in L^{q'}(\mathbb{R}^n)$, so $\widehat{f}\widehat{g} \in L^{r'}(\mathbb{R}^n)$ by Hölder's inequality. Since $h := f * g$ is in $L^r(\mathbb{R}^n)$, $\widehat{h} \in L^{r'}(\mathbb{R}^n)$ by Theorem 5.7. If both f and g are also in $L^1(\mathbb{R}^n)$, then (1) is true by 5.1(8). The theorem follows by an approximation argument that is left to the reader. ∎

• The function $|x|^{2-n}$ on $\mathbb{R}^n$ with $n \geq 3$ is very important in potential theory (Chapter 9) and as the Green's function in Sect. 6.20. Hence, it is useful to know its 'Fourier transform', even though this function is not in any $L^p(\mathbb{R}^n)$ for any p. However, its action in convolution or as a multiplier on nice functions *can* be expressed easily in terms of Fourier transforms.

5.9 THEOREM (Fourier transform of $|x|^{\alpha-n}$)

Let f be a function in $C_c^\infty(\mathbb{R}^n)$ and let $0 < \alpha < n$. Then, with
$$c_\alpha := \pi^{-\alpha/2}\Gamma(\alpha/2), \tag{1}$$
$$c_\alpha (|k|^{-\alpha}\widehat{f}(k))^\vee(x) = c_{n-\alpha} \int_{\mathbb{R}^n} |x-y|^{\alpha-n} f(y)\,\mathrm{d}y . \tag{2}$$

REMARK. Since $f \in C_c^\infty(\mathbb{R}^n)$, the Fourier transform $\widehat{f}$ is a very nice function; it is in $C^\infty(\mathbb{R}^n)$ (it is analytic, in fact) and, as $|k| \to \infty$, it, and all its derivatives, decay faster than the inverse of any polynomial in k. (The verification of these two facts is recommended as an exercise using integration by parts and dominated convergence.) Therefore, the function $|k|^{-\alpha}\widehat{f}(k)$ is in $L^1(\mathbb{R}^n)$, and thus it has a Fourier transform. The function on the right side of (2) is well defined and is also in $C^\infty(\mathbb{R}^n)$, but it decays, as $|x| \to \infty$, only as $|x|^{\alpha-n}$ (in general). Thus, generally speaking, the right side of (2) is *not* in $L^p(\mathbb{R}^n)$ for any $p \leq 2$, *unless* $\alpha < n/2$ and, therefore, it does not generally have a well-defined Fourier transform. Nevertheless, (2) is true.

PROOF. Our starting point is the elementary formula

$$c_\alpha |k|^{-\alpha} = \int_0^\infty \exp[-\pi|k|^2 \lambda]\, \lambda^{\alpha/2-1}\, d\lambda. \tag{3}$$

Since $|k|^{-\alpha}\widehat{f}(k)$ is integrable, we have, by Fubini's theorem,

$$\begin{aligned}
c_\alpha(|k|^{-\alpha}\widehat{f}(k))^\vee(x) &= \int_{\mathbb{R}^n} e^{2\pi i(k,x)} \left\{ \int_0^\infty \exp[-\pi|k|^2\lambda]\lambda^{\alpha/2-1}\, d\lambda \right\} \widehat{f}(k)\, dk \\
&= \int_0^\infty \left\{ \int_{\mathbb{R}^n} e^{2\pi i(k,x)} \exp[-\pi|k|^2\lambda]\widehat{f}(k)\, dk \right\} \lambda^{\alpha/2-1}\, d\lambda \\
&= \int_0^\infty \lambda^{-n/2}\lambda^{\alpha/2-1} \left\{ \int_{\mathbb{R}^n} \exp[-\pi|x-y|^2/\lambda] f(y)\, dy \right\} d\lambda \\
&= c_{n-\alpha} \int_{\mathbb{R}^n} |x-y|^{-n+\alpha} f(y)\, dy.
\end{aligned}$$

In the penultimate equation we have used Theorem 5.2 and the convolution theorem 5.8(1). The last equation holds by Fubini's theorem. ∎

5.10 COROLLARY (Extension of 5.9 to $L^p(\mathbb{R}^n)$)

If $0 < \alpha < n/2$ and if $f \in L^p(\mathbb{R}^n)$ with $p = 2n/(n+2\alpha)$, then $\widehat{f}$ exists (by Theorem 5.7). Moreover, with c_α defined in 5.9(1), the function

$$g := c_{n-\alpha} |x|^{\alpha-n} * f$$

is an $L^2(\mathbb{R}^n)$-function (by Theorem 4.3 (HLS inequality)) and hence has a Fourier transform $\widehat{g}$.

Our new result is that the relation between $\widehat{g}$ and $\widehat{f}$ is given by

$$c_\alpha |k|^{-\alpha}\widehat{f}(k) = \widehat{g}(k). \tag{1}$$

Moreover,

$$c_{2\alpha} \int_{\mathbb{R}^n} |k|^{-2\alpha}|\widehat{f}(k)|^2\, dk = c_{n-2\alpha} \int_{\mathbb{R}^n} \int_{\mathbb{R}^n} \overline{f}(x) f(y) |x-y|^{2\alpha-n}\, dx\, dy. \tag{2}$$

REMARK. The case $\alpha = 1$ and $n \geq 3$ is especially important for potential theory (Chapter 9) and for the Green's function of the Laplacian (before 6.20). The right side of (2), without $c_{n-2\alpha}$, is twice the Coulomb potential energy of the 'charge distribution' f, 9.1(2).

PROOF. By Theorem 2.16 (approximation by C^∞-functions) we can find a sequence $f^1, f^2, \ldots$ of functions in $C_c^\infty(\mathbb{R}^n)$ such that $f^j \to f$ strongly in $L^p(\mathbb{R}^n)$. By Theorem 4.3 (HLS inequality) the functions g and

$$g^j := |x|^{\alpha-n} * f^j$$

are in $L^2(\mathbb{R}^n)$; this follows from Fubini's theorem and the fact that, for $0 < \alpha < n$, $0 < \beta < n$ and $0 < \alpha + \beta < n$, we have

$$(|x|^{\alpha-n} * |x|^{\beta-n})(y) := \int_{\mathbb{R}^n} |z|^{\alpha-n}|y-z|^{\beta-n} \, dz$$
$$= \frac{c_{n-\alpha-\beta}\, c_\alpha\, c_\beta}{c_{\alpha+\beta}\, c_{n-\alpha}\, c_{n-\beta}} |y|^{\alpha+\beta-n}, \tag{3}$$

which can be verified by a tedious but instructive computation using 5.9(3).

Since $f^j \to f$, we have $\widehat{f^j} \to \widehat{f}$ in $L^q(\mathbb{R}^n)$ with $q = 2n/(n-2\alpha)$ (by Theorem 5.7). By the HLS inequality $g^j \to g$ in $L^2(\mathbb{R}^n)$, and hence $\widehat{g^j} \to \widehat{g}$ in $L^2(\mathbb{R}^n)$ (by Theorem 5.3 (Plancherel)). By Theorem 5.9, we also know that

$$\widehat{g^j}(k) = c_\alpha |k|^{-\alpha} \widehat{f^j}(k).$$

Our problem is to show that

$$\widehat{g}(k) = c_\alpha |k|^{-\alpha} \widehat{f}(k).$$

To do this, we pass to a subsequence so that $\widehat{g^j}(k) \to \widehat{g}(k)$ and $\widehat{f^j}(k) \to \widehat{f}(k)$ pointwise a.e. (by Theorem 2.7(ii) (completeness of L^p-spaces)). Thus,

$$\widehat{g}(k) = \lim_{j \to \infty} c_\alpha |k|^{-\alpha} \widehat{f^j}(k) = c_\alpha |k|^{-\alpha} \lim_{j \to \infty} \widehat{f^j}(k) = c_\alpha |k|^{-\alpha} \widehat{f}(k)$$

for almost every k. This proves (1).

Formula (2) is just an application of Plancherel's theorem to (1), together with Fubini's theorem and (3). ∎

Exercises for Chapter 5

1. Prove that the Fourier transform has properties 5.1(2), (3) and (4).

2. Prove the Riemann–Lebesgue lemma mentioned in Sect. 5.1, i.e., for $f \in L^1(\mathbb{R}^n)$, $\widehat{f}(k) \to 0$ as $|k| \to \infty$.

   ▶ *Hint.* 5.1(3) is useful.

3. Show that the definition of the Fourier transform for functions in $L^2(\mathbb{R}^n)$, given in Sect. 5.4, does not depend on the approximating sequence.

4. Show that the definition of the Fourier transform for functions in $L^2(\mathbb{R}^n)$ gives rise to a linear map $f \mapsto \widehat{f}$.

5. Complete the proof of Theorem 5.8, i.e., work out the approximation argument mentioned at the end of Sect. 5.8.

6. For $f \in C_c^\infty(\mathbb{R}^n)$ show that its Fourier transform $\widehat{f}$ is also in C^∞ (in fact $\widehat{f}$ is analytic). Show also that $g_a(k) := |\,|k|^a \widehat{f}(k)|$ is a bounded function for each $a > 0$.

7. Verify formula 5.10(3).

8. This concerns an example of an extension of Theorem 5.8 (convolution) to the case in which $r > 2$. Suppose that f and g are $L^2(\mathbb{R}^n)$. Then we know that $f * g \in L^\infty(\mathbb{R}^n)$ and $\widehat{f}\widehat{g} \in L^1(\mathbb{R}^n)$. Although $\widehat{f * g}$ may not be obviously well defined, show that 5.1(8) holds, nevertheless, in the sense of inverse Fourier transforms, i.e.,

$$f * g = (\widehat{f}\widehat{g})^\vee.$$

9. Verify that 5.6(1) cannot hold when $p > 2$ by considering Gaussian functions, as in 5.2(1), with $\lambda = a + ib$ and with $a > 0$.

Chapter 6

Distributions

6.1 INTRODUCTION

The notion of a weak derivative is an indispensable tool in dealing with partial differential equations. Its advantage is that it allows one to dispense with subtle questions about differentiation, such as the interchange of partial derivatives. Its main point is that every locally integrable function can be weakly differentiated indefinitely many times, just as though it were a C^∞-function. The weakening of the notion of a derivative makes it easier to find solutions to equations and, once found, these 'weak' solutions can then be analyzed to find out if they are, in fact, truly differentiable in the classical sense. An analogy in elementary algebra might be trying to solve a polynomial equation by rational numbers. It is extremely important, at the beginning of the investigation, to know that solutions always exist in the larger category of real numbers; many techniques are available for this purpose, e.g. Rolle's theorem, that are not available in the category of rationals. Later on one can try to prove that the solutions are, in fact, rational.

A theory developed around the notion that every L^1_{loc}-function is differentiable is the theory of distributions (see [Schwartz], [Hörmander], [Rudin, 1991], [Reed–Simon, vol. 1]). Although we do not present some of the deeper aspects of this theory we shall state its basic techniques. In the following, for completeness, we define distributions for an arbitrary open set Ω in $\mathbb{R}^n$, but, in fact, we shall mainly need the case $\Omega = \mathbb{R}^n$ in the rest of the book.

6.2 TEST FUNCTIONS (The space $\mathcal{D}(\Omega)$)

Let Ω be an open, nonempty set in $\mathbb{R}^n$; in particular Ω can be $\mathbb{R}^n$ itself. Recall from Sect. 1.1 that $C_c^\infty(\Omega)$ denotes the space of all infinitely differentiable, complex-valued functions whose support is compact and in Ω. Recall also that the support of a continuous function is defined to be the closure of the set on which the function does not vanish, and compactness means that the closed set is also contained in some ball of finite radius. Note that Ω is never compact.

The space of **test functions**, $\mathcal{D}(\Omega)$, consists of all the functions in $C_c^\infty(\Omega)$ supplemented by the following notion of convergence: A *sequence* $\phi^m \in C_c^\infty(\Omega)$ **converges in $\mathcal{D}(\Omega)$ to the function** $\phi \in C_c^\infty(\Omega)$ if and only if there is some *fixed, compact* set $K \subset \Omega$ such that the support of $\phi^m - \phi$ is in K for all m and, for each choice of the nonnegative integers $\alpha_1, \ldots, \alpha_n$,

$$(\partial/\partial x_1)^{\alpha_1} \cdots (\partial/\partial x_n)^{\alpha_n} \phi^m \longrightarrow (\partial/\partial x_1)^{\alpha_1} \cdots (\partial/\partial x_n)^{\alpha_n} \phi \qquad (1)$$

as $m \to \infty$, *uniformly* on K. To say that a sequence of continuous functions ψ^m converges to ψ uniformly on K means that

$$\sup_{x \in K} |\psi^m(x) - \psi(x)| \to 0 \quad \text{as} \quad m \to \infty.$$

$\mathcal{D}(\Omega)$ is a linear space, i.e., functions can be added and multiplied by (complex) scalars.

6.3 DEFINITION OF DISTRIBUTIONS AND THEIR CONVERGENCE

A **distribution** T is a continuous linear functional on $\mathcal{D}(\Omega)$, i.e., $T : \mathcal{D}(\Omega) \to \mathbb{C}$ such that for $\phi, \phi_1, \phi_2 \in \mathcal{D}(\Omega)$ and $\lambda \in \mathbb{C}$

$$T(\phi_1 + \phi_2) = T(\phi_1) + T(\phi_2) \quad \text{and} \quad T(\lambda\phi) = \lambda T(\phi), \qquad (1)$$

and continuity means that whenever $\phi^n \in \mathcal{D}(\Omega)$ and $\phi^n \to \phi$ in $\mathcal{D}(\Omega)$

$$T(\phi^n) \to T(\phi).$$

Distributions can be added and multiplied by complex scalars. This linear space is denoted by $\mathcal{D}'(\Omega)$, the **dual space of** $\mathcal{D}(\Omega)$.

There is the obvious notion of convergence of distributions: A sequence of distributions $T^j \in \mathcal{D}'(\Omega)$ **converges in** $\mathcal{D}'(\Omega)$ to $T \in \mathcal{D}'(\Omega)$ if, for every $\phi \in \mathcal{D}(\Omega)$, the numbers $T^j(\phi)$ converge to $T(\phi)$.

One might suspect that this kind of convergence is rather weak. Indeed, it is! For example, we shall see in Sect. 6.6, where we develop a notion of the derivative of a distribution, that for *any* converging sequence of distributions their derivatives converge too, i.e., differentiation is a continuous operation in $\mathcal{D}'(\Omega)$. This contrasts with ordinary pointwise convergence because the derivatives of a pointwise converging sequence of functions need not, in general, converge anywhere.

Another instance, as we shall see in Sect. 6.13, is that any distribution can be approximated in $\mathcal{D}'(\Omega)$ by functions in $C^\infty(\Omega)$. To make sense of that statement, we first have to define what it means for a function to be a distribution. This is done in the next section.

6.4 LOCALLY SUMMABLE FUNCTIONS, $L^p_{\text{loc}}(\Omega)$

The foremost example of distributions are functions themselves. We begin by defining the space of **locally p^{th}-power summable functions**, $L^p_{\text{loc}}(\Omega)$, for $1 \leq p \leq \infty$. Such functions are Borel measurable functions defined on all of Ω and with the property that

$$\|f\|_{L^p(K)} < \infty \tag{1}$$

for *every* compact set $K \subset \Omega$. Equivalently, it suffices to require (1) to hold when K is any closed ball in Ω.

A sequence of functions $f^1, f^2, \ldots$ in $L^p_{\text{loc}}(\Omega)$ is said to **converge** (or **converge strongly**) to f in $L^p_{\text{loc}}(\Omega)$ (denoted by $f^j \to f$) if $f^j \to f$ in $L^p(K)$ in the usual sense (see Theorem 2.7) for every compact $K \subset \Omega$. Likewise, f^j **converges weakly** to f if $f^j \rightharpoonup f$ weakly in every $L^p(K)$ (2.9(6)).

Note (for general $p \geq 1$) that $L^p_{\text{loc}}(\Omega)$ is a vector space but it does not have a simply defined norm. Furthermore, $f \in L^p_{\text{loc}}(\Omega)$ does not imply that $f \in L^p(\Omega)$. Clearly, $L^p_{\text{loc}}(\Omega) \supset L^p(\Omega)$ and, if $r > p$, we have the inclusion

$$L^p_{\text{loc}}(\Omega) \supset L^r_{\text{loc}}(\Omega),$$

by Hölder's inequality (but it is *false*—unless Ω has finite measure—that $L^p(\Omega) \supset L^r(\Omega)$).

As far as distributions are concerned, $L^1_{\text{loc}}(\Omega)$ is the most important space. Let f be a function in $L^1_{\text{loc}}(\Omega)$. For any ϕ in $\mathcal{D}(\Omega)$ it makes sense to consider

$$T_f(\phi) := \int_\Omega f\phi \, \mathrm{d}x, \tag{2}$$

which obviously defines a linear functional on $\mathcal{D}(\Omega)$. T_f is also continuous since

$$|T_f(\phi) - T_f(\phi^m)| = \left| \int_\Omega (\phi(x) - \phi^m(x)) f(x) \, dx \right|$$
$$\leq \sup_{x \in K} |\phi(x) - \phi^m(x)| \int_K |f(x)| \, dx,$$

which tends to zero by the uniform convergence of the ϕ^m's. Thus T_f is in $\mathcal{D}'(\Omega)$. If a distribution T is given by (2) for some $f \in L^1_{\text{loc}}(\Omega)$, we say that **the distribution T is the function f**. This terminology will be justified in the next section.

An important example of a distribution that is *not* of this form is the so-called **Dirac 'delta-function'**, which is not a function at all:

$$\delta_x(\phi) = \phi(x) \tag{3}$$

with $x \in \Omega$ fixed. It is obvious that $\delta_x \in \mathcal{D}'(\Omega)$. Thus, the delta-measure of Sect. 1.2(6), like any Borel measure, can also be considered to be a distribution. In fact, one can say that it was partly the attempt to understand the true mathematical meaning of the delta function, which had been used so successfully by physicists and engineers, that led to the theory of distributions.

Although $\mathcal{D}(\Omega)$, the space of test functions, is a very restricted class of functions it is large enough to distinguish functions in $\mathcal{D}'(\Omega)$, as we now show.

6.5 THEOREM (Functions are uniquely determined by distributions)

Let $\Omega \subset \mathbb{R}^n$ be open and let f and g be functions in $L^1_{\text{loc}}(\Omega)$. Suppose that the distributions defined by f and g are equal, i.e.,

$$\int_\Omega f\phi = \int_\Omega g\phi \tag{1}$$

for all $\phi \in \mathcal{D}(\Omega)$. Then $f(x) = g(x)$ for almost every x in Ω.

PROOF. For $m = 1, 2, \ldots$ let Ω_m be the set of points $x \in \Omega$ such that $x + y \in \Omega$ whenever $|y| \leq \frac{1}{m}$. Ω_m is open. Let j be in $C_c^\infty(\mathbb{R}^n)$ with support in the unit ball and with $\int_{\mathbb{R}^n} j = 1$. Define $j_m(x) = m^n j(mx)$. Fix M. If $m \geq M$, then, by (1) with $\phi(y) = j_m(x - y)$, we have that

$(j_m * f)(x) = (j_m * g)(x)$ for all $x \in \Omega_M$ (see Sect. 2.15 for the definition of the convolution $*$). By Theorem 2.16, $j_m * f \to f$ and $j_m * g \to g$ in $L^1_{\text{loc}}(\Omega_M)$ as $m \to \infty$. Thus $f = g$ in $L^1_{\text{loc}}(\Omega_M)$ and therefore $f(x) = g(x)$ for almost every $x \in \Omega_M$. Finally let M tend to ∞. ∎

6.6 DERIVATIVES OF DISTRIBUTIONS

We now define the notion of **distributional** or **weak derivative**. Let T be in $\mathcal{D}'(\Omega)$ and let $\alpha_1, \ldots, \alpha_n$ be nonnegative integers. We define the distribution $(\partial/\partial x_1)^{\alpha_1} \cdots (\partial/\partial x_n)^{\alpha_n} T$, denoted by $D^\alpha T$, by its action on each $\phi \in \mathcal{D}(\Omega)$ as follows:

$$(D^\alpha T)(\phi) = (-1)^{|\alpha|} T(D^\alpha \phi) \tag{1}$$

with the notation

$$|\alpha| = \sum_{i=1}^n \alpha_i. \tag{2}$$

The symbol
$$\partial_i T$$
denotes $D^\alpha T$ in the special case $\alpha_i = 1, \alpha_j = 0$ for $j \neq i$.

The symbol ∇T, called the **distributional gradient of** T, denotes the n-tuple $(\partial_1 T, \partial_2 T, \ldots, \partial_n T)$.

If f is a $C^{|\alpha|}(\Omega)$-function (not necessarily of compact support), then

$$(D^\alpha T_f)(\phi) := (-1)^{|\alpha|} \int_\Omega (D^\alpha \phi) f \, dx = \int_\Omega (D^\alpha f) \phi \, dx =: T_{D^\alpha f}(\phi),$$

where the middle equality holds by partial integration. Hence the notion of weak derivative extends the classical one and it agrees with the classical one whenever the classical derivative exists and is continuous (see Theorem 6.10 (equivalence of classical and distributional derivatives)). Obviously, in this weak sense, *every distribution is infinitely often differentiable* and this is one of the main virtues of the theory. Note however, that the distributional derivative of a nondifferentiable function (in the classical sense) is not necessarily a function.

Let us show that $D^\alpha T$ actually is a distribution. Obviously it is linear, so we only have to check its continuity on $\mathcal{D}(\Omega)$. Let $\phi^m \to \phi$ in $\mathcal{D}(\Omega)$. Then $D^\alpha \phi^m \to D^\alpha \phi$ in $\mathcal{D}(\Omega)$ since

$$\text{supp}\{D^\alpha \phi^m - D^\alpha \phi\} \subset \text{supp}\{\phi^m - \phi\} \subset K$$

and
$$D^\beta(D^\alpha \phi^m - D^\alpha \phi) = D^{\beta+\alpha}\phi^m - D^{\beta+\alpha}\phi$$

converges to zero uniformly on compact sets. [Here $\beta + \alpha$ simply denotes the multi-index given by $(\beta_1 + \alpha_1, \beta_2 + \alpha_2, \ldots, \beta_n + \alpha_n)$]. Thus, $D^\alpha \phi$ and $D^\alpha \phi^m$ are themselves functions in $\mathcal{D}(\Omega)$ with $D^\alpha \phi^m \to D^\alpha \phi$ as $m \to \infty$. Hence, as $m \to \infty$,

$$(D^\alpha T)(\phi^m) := (-1)^{|\alpha|} T(D^\alpha \phi^m) \longrightarrow (-1)^{|\alpha|} T(D^\alpha \phi) =: (D^\alpha T)(\phi).$$

We end this section by showing that differentiation of distributions is a continuous operation in $\mathcal{D}'(\Omega)$. Indeed, if $T^j(\phi) \to T(\phi)$ for all $\phi \in \mathcal{D}(\Omega)$, then, by the definition of the derivative of a distribution

$$(D^\alpha T^j)(\phi) = (-1)^{|\alpha|} T^j(D^\alpha \phi) \underset{j \to \infty}{\to} (-1)^{|\alpha|} T(D^\alpha \phi) = (D^\alpha T)(\phi)$$

since $D^\alpha \phi \in \mathcal{D}(\Omega)$.

6.7 DEFINITION OF $W^{1,p}_{\mathrm{loc}}(\Omega)$ AND $W^{1,p}(\Omega)$

$L^1_{\mathrm{loc}}(\Omega)$-functions are an important class of distributions, but we can usefully refine that class by studying functions whose distributional derivatives are also $L^1_{\mathrm{loc}}(\Omega)$-functions. This class is denoted by $W^{1,1}_{\mathrm{loc}}(\Omega)$. Furthermore, just as $L^p_{\mathrm{loc}}(\Omega)$ is related to $L^1_{\mathrm{loc}}(\Omega)$ we can also define the class of functions $W^{1,p}_{\mathrm{loc}}(\Omega)$ for each $1 \leq p \leq \infty$. Thus,

$$W^{1,p}_{\mathrm{loc}}(\Omega) = \{f : \Omega \to \mathbb{C} : f \in L^p_{\mathrm{loc}}(\Omega) \text{ and } \partial_i f, \text{ as a distribution}$$
$$\text{in } \mathcal{D}'(\Omega), \text{ is an } L^p_{\mathrm{loc}}(\Omega)\text{-function for } i = 1, \ldots, n\}.$$

We urge the reader not to use the symbol ∇f at first, since it is tempting to apply the rules of calculus which we have not established yet. One should just think of ∇f as n functions $\mathbf{g} = (g_1, \ldots, g_n)$, each of which is in $L^p_{\mathrm{loc}}(\Omega)$, such that

$$\int_\Omega f \nabla \phi = -\int_\Omega \mathbf{g} \phi \quad \text{for all } \phi \in \mathcal{D}(\Omega).$$

This set of functions, $W^{1,p}_{\mathrm{loc}}(\Omega)$, forms a vector space but not a normed one. We have the inclusion $W^{1,p}_{\mathrm{loc}}(\Omega) \supset W^{1,r}_{\mathrm{loc}}(\Omega)$ if $r > p$.

We can also define $W^{1,p}(\Omega) \subset W^{1,p}_{\mathrm{loc}}(\Omega)$ analogously:

Sections 6.6–6.7

$$W^{1,p}(\Omega) = \{f : \Omega \to \mathbb{C} : f \text{ and } \partial_i f \text{ are in } L^p(\Omega) \text{ for } i = 1,\ldots,n\}.$$

We can make $W^{1,p}(\Omega)$ into a normed space, by defining

$$\|f\|_{W^{1,p}(\Omega)} = \left\{\|f\|^p_{L^p(\Omega)} + \sum_{j=1}^n \|\partial_j f\|^p_{L^p(\Omega)}\right\}^{1/p} \tag{1}$$

and it is complete, i.e., every Cauchy sequence in this norm has a limit in $W^{1,p}(\Omega)$. This follows easily from the completeness of $L^p(\Omega)$ (Theorem 2.7) together with the definition 6.6 of the distributional derivative, i.e., if $f^j \to f$ and $\partial_i f^j \to g_i$, then it follows that $g_i = \partial_i f$ in $\mathcal{D}'(\Omega)$. The proof is a simple adaptation of the one for $W^{1,2}(\Omega) = H^1(\Omega)$ in Theorem 7.3 (see Remark 7.5). We leave the details to the reader.

The spaces $W^{1,p}(\Omega)$ are called **Sobolev spaces**. In this chapter only $W^{1,1}_{\text{loc}}(\Omega)$ will play a role.

The superscript 1 in $W^{1,p}(\Omega)$ denotes the fact that the first derivatives of f are p^{th}-power summable functions.

As with $L^p(\Omega)$ and $L^p_{\text{loc}}(\Omega)$, we can define the notions of **strong** and **weak convergence** in the spaces $W^{1,p}_{\text{loc}}(\Omega)$ or $W^{1,p}(\Omega)$ of a sequence of functions $f^1, f^2, \ldots$ to a function f. Strong convergence simply means that the sequence converges strongly to f in $L^p(\Omega)$ and the n sequences $\{\partial_1 f^j\}, \ldots, \{\partial_n f^j\}$, formed from the derivatives of f^j, converge in $L^p(\Omega)$ to the n functions $\partial_1 f, \ldots, \partial_n f$ in $L^p(\Omega)$. In the case of $W^{1,p}_{\text{loc}}(\Omega)$ we require this convergence only on every compact subset of Ω. Weak convergence in $W^{1,p}(\Omega)$ is defined similarly, i.e., for each $L^{p'}(\Omega)$-function g we require that $\int g(f^j - f) \to 0$ and, for each i, $\int g(\partial_i f^j - \partial_i f) \to 0$ as $j \to \infty$. Here, $1/p + 1/p' = 1$. For $W^{1,p}_{\text{loc}}(\Omega)$ we require this only for each g with compact support.

Similar definitions apply to $W^{m,p}(\Omega)$ and $W^{m,p}_{\text{loc}}(\Omega)$ with $m > 1$. The first m derivatives of these functions are $L^p(\Omega)$-functions and, similarly to (1),

$$\|f\|^p_{W^{m,p}(\Omega)} := \|f\|^p_{L^p(\Omega)} + \sum_{j=1}^n \|\partial_j f\|^p_{L^p(\Omega)}$$
$$+ \cdots + \sum_{j_1=1}^n \cdots \sum_{j_m=1}^n \|\partial_{j_1} \cdots \partial_{j_m} f\|^p_{L^p(\Omega)}. \tag{2}$$

● In the following it will be convenient to denote by ϕ_z the function ϕ translated by $z \in \mathbb{R}^n$, i.e.,

$$\phi_z(x) := \phi(x - z). \tag{3}$$

6.8 LEMMA (Interchanging convolutions with distributions)

Let $\Omega \subset \mathbb{R}^n$ be open and let $\phi \in \mathcal{D}(\Omega)$. Let $\mathcal{O}_\phi \subset \mathbb{R}^n$ be the set

$$\mathcal{O}_\phi = \{y \; : \; \operatorname{supp}\{\phi_y\} \subset \Omega\}.$$

It is elementary that $\mathcal{O}_\phi$ is open and not empty. Let $T \in \mathcal{D}'(\Omega)$. Then the function $y \mapsto T(\phi_y)$ is in $C^\infty(\mathcal{O}_\phi)$. In fact, with D_y^α denoting derivatives with respect to y,

$$D_y^\alpha T(\phi_y) = (-1)^{|\alpha|} T((D^\alpha \phi)_y) = (D^\alpha T)(\phi_y). \tag{1}$$

Now let $\psi \in L^1(\mathcal{O}_\phi)$ have compact support. Then

$$\int_{\mathcal{O}_\phi} \psi(y) T(\phi_y) \, \mathrm{d}y = T(\psi * \phi). \tag{2}$$

PROOF. If $y \in \mathcal{O}_\phi$ and if $\varepsilon > 0$ is chosen so that $y + z \in \mathcal{O}_\phi$ for all $|z| < \varepsilon$, we have that for all $x \in \Omega$

$$|\phi_y(x) - \phi_{y+z}(x)| = |\phi(x - y) - \phi(x - y - z)| < C\varepsilon \tag{3}$$

for some number $C < \infty$. This is so because ϕ has continuous derivatives and (since it has compact support) these derivatives are *uniformly* continuous. For the same reason, (3) holds for all derivatives of ϕ (with C depending on the order of the derivative). This means that ϕ_{y+z} converges to ϕ_y as $z \to 0$ in $\mathcal{D}(\Omega)$ (see Sect. 6.2). Therefore, $T(\phi_{y+z}) \to T(\phi_y)$ as $z \to 0$, and thus $y \mapsto T(\phi_y)$ is continuous on $\mathcal{O}_\phi$.

Similarly, we have that

$$\{[\phi(x + \delta z) - \phi(x)]/\delta - \nabla \phi(x) \cdot z\} \leq C' \delta |z|$$

and thus, by a similar argument, $y \mapsto T(\phi_y)$ is differentiable. Continuing in this manner we find that (1) holds.

To prove (2) it suffices to assume that $\psi \in C_c^\infty(\mathcal{O}_\phi)$. To verify this, we use Theorem 2.16 to find, for each $\delta > 0$, $\psi^\delta \in C_c^\infty(\mathcal{O}_\phi)$ so that $\int_{\mathcal{O}_\phi} |\psi^\delta - \psi| < \delta$. In fact, we can assume that $\text{supp}\{\psi^\delta\}$ is contained in some fixed compact subset, K, of $\mathcal{O}_\phi$, independent of δ. Then

$$\left| \int \{\psi(y) - \psi^\delta(y)\} T(\phi_y) \, dy \right| \leq \delta \sup \{|T(\phi_y)| : y \in K\}.$$

It is also easy to see that $\psi^\delta * \phi$ converges to $\psi * \phi$ in $\mathcal{D}(\Omega)$ and therefore $T(\psi^\delta * \phi) \to T(\psi * \phi)$.

With ψ now in $C_c^\infty(\mathcal{O}_\phi)$ we note that the integrand in (2) is a product of two C_c^∞-functions. Hence the integral can be taken as a Riemann integral and thus can be approximated by finite sums of the form

$$\Delta_m \sum_{j=1}^{m} \psi(y_j) T(\phi_{y_j}) \quad \text{with } \Delta_m \to 0 \text{ as } m \to \infty.$$

Likewise, for any multi-index α, $(D^\alpha(\psi * \phi))(x)$ is uniformly approximated by $\Delta_m \sum_{j=1}^{m} \psi(y_j) D^\alpha \phi(x - y_j)$ as $m \to \infty$ (because $\phi \in C_c^\infty(\Omega)$). Note that for m sufficiently large every member of this sequence has support in a fixed compact set $K \subset \Omega$. Since T_m is continuous (by definition) and the function $\eta_m(x) = \Delta_m \sum_{j=1}^{n} \psi(y_j) \phi(x - y_j)$ converges to $(\psi * \phi)(x)$ as $m \to \infty$, we conclude that $T(\eta_m)$ converges to $T(\psi * \phi)$ as $m \to \infty$. ∎

6.9 THEOREM (Fundamental theorem of calculus for distributions)

Let $\Omega \subset \mathbb{R}^n$ be open, let $T \in \mathcal{D}'(\Omega)$ be a distribution and let $\phi \in \mathcal{D}(\Omega)$ be a test function. Suppose that for some $y \in \mathbb{R}^n$ the function ϕ_{ty} is also in $\mathcal{D}(\Omega)$ for all $0 \leq t \leq 1$ (see 6.7(3)). Then

$$T(\phi_y) - T(\phi) = \int_0^1 \sum_{j=1}^{n} y_j (\partial_j T)(\phi_{ty}) \, dt. \tag{1}$$

As a particular case of (1), suppose that $f \in W^{1,1}_{\text{loc}}(\mathbb{R}^n)$. Then, for each y in $\mathbb{R}^n$ and almost every $x \in \mathbb{R}^n$,

$$f(x + y) - f(x) = \int_0^1 y \cdot \nabla f(x + ty) \, dt. \tag{2}$$

PROOF. Let $\mathcal{O}_\phi = \{z \in \mathbb{R}^n : \phi_z \in \mathcal{D}(\Omega)\}$. It is clearly open and nonempty. Denote the right side of (1) by $F(y)$. Observe that by Lemma 6.8, $z \mapsto (\partial_j T)(\phi_z)$ is a C^∞-function on $\mathcal{O}_\phi$ and $\partial(\partial_j T(\phi_z))/\partial z_i = -\partial_j T(\partial_i \phi_z)$.

With this infinite differentiability in mind we can now interchange derivatives and integrals, and compute

$$\partial_i F(y) = -\sum_{j=1}^n \int_0^1 t(\partial_j T)(\partial_i \phi_{ty}) y_j \, dt + \int_0^1 (\partial_i T)(\phi_{ty}) \, dt.$$

The first term is, by the definition of the derivative of a distribution,

$$\sum_{j=1}^n \int_0^1 tT(\partial_j \partial_i \phi_{ty}) y_j \, dt = -\int_0^1 \sum_{j=1}^n t(\partial_i T)(\partial_j \phi_{ty}) y_j \, dt,$$

which can be rewritten (for the same reason as before) as

$$\int_0^1 t \frac{d}{dt}(\partial_i T)(\phi_{ty}) \, dt.$$

A simple integration by parts then yields $\partial_i F(y) = (\partial_i T)(\phi_y)$. The function $y \mapsto G(y) = T(\phi_y) - T(\phi)$ is also C^∞ in y (by Lemma 6.8) and also has $(\partial_i T)(\phi_y)$ as its partial derivatives. Since $F(0) = G(0) = 0$, the two C^∞-functions F and G must be the same. This proves (1).

To prove (2), note that since

$$(\partial_j f)(\phi_{ty}) = \int \phi(x)(\partial_j f)(x + ty) \, dx,$$

(1) implies that

$$\int_{\mathbb{R}^n} \phi(x)[f(x+y) - f(x)] \, dx = \int_0^1 \sum_{j=1}^n y_j \left\{ \int_{\mathbb{R}^n} \phi(x)(\partial_j f)(x+ty) \, dx \right\} dt.$$

Since ϕ has compact support, the integrand is (t,x) integrable (even if $\partial_j f \notin L^1(\mathbb{R}^n)$), and hence Fubini's theorem can be used to interchange the t and x integrations. Conclusion (2) then follows from Theorem 6.5. ∎

6.10 THEOREM (Equivalence of classical and distributional derivatives)

Let $\Omega \subset \mathbb{R}^n$ be open, let $T \in \mathcal{D}'(\Omega)$ and set $G_i := \partial_i T \in \mathcal{D}'(\Omega)$ for $i = 1, 2, \ldots, n$. The following are equivalent.

(i) T is a function $f \in C^1(\Omega)$.

(ii) G_i is a function $g_i \in C^0(\Omega)$ for each $i = 1, \ldots, n$.

In each case, g_i is $\partial f / \partial x_i$, the classical derivative of f.

REMARK. The assertion $f \in C^1(\Omega)$ means, of course, that there is a $C^1(\Omega)$-function in the equivalence class of f. A similar remark applies to $g_i \in C^0(\Omega)$.

PROOF. $\boxed{\text{(i)} \Rightarrow \text{(ii)}.}$ $G_i(\phi) = (\partial_i T)(\phi) = -\int_\Omega (\partial_i \phi) f$ by the definition of distributional derivative. On the other hand, the classical integration by parts formula yields

$$\int_\Omega (\partial_i \phi) f = -\int_\Omega \phi (\partial f / \partial x_i)$$

since ϕ has compact support in Ω and $f \in C^1(\Omega)$. Therefore, by the terminology of Sect. 6.4 and Theorem 6.5, G_i is the function $\partial f / \partial x_i$.

$\boxed{\text{(ii)} \Rightarrow \text{(i)}.}$ Fix $R > 0$ and let $\omega = \{x \in \Omega : |x - z| > R \text{ for all } z \notin \Omega\}$. Clearly ω is open and nonempty for R small enough, which we henceforth assume. Take $\phi \in \mathcal{D}(\omega) \subset \mathcal{D}(\Omega)$ and $|y| < R$. Then $\phi_{ty} \in \mathcal{D}(\Omega)$ for $-1 \leq t \leq 1$. By 6.9(1) and Fubini's theorem

$$\begin{aligned} T(\phi_y) - T(\phi) &= \int_0^1 \sum_{j=1}^n y_j \int_\omega g_j(x) \phi(x - ty) \, dx \, dt \\ &= \int_\omega \left\{ \int_0^1 \sum_{j=1}^n g_j(x + ty) y_j \, dt \right\} \phi(x) \, dx. \end{aligned} \quad (1)$$

Pick $\psi \in C_c^\infty(\mathbb{R}^n)$ nonnegative with $\operatorname{supp}\{\psi\} \subset B := \{y : |y| < R\}$ and $\int \psi = 1$. The convolution $\int_B \psi(y) \phi(x - y) \, dy$ with $\phi \in \mathcal{D}(\omega)$ defines a function in $\mathcal{D}(\Omega)$. Integrating (1) against ψ we obtain, using Fubini's theorem,

$$\begin{aligned} \int_B \psi(y) T(\phi_y) \, dy - T(\phi) \\ = \int_\omega \left\{ \sum_{j=1}^n \int_B \psi(y) \int_0^1 y_j g_j(x + ty) \, dt \, dy \right\} \phi(x) \, dx. \end{aligned} \quad (2)$$

The first term on the left is $\int_\omega \phi(x) T(\psi_x) \, dx$, which follows from Lemma 6.8 by noting that ψ_x for $x \in \omega$ is an element of $\mathcal{D}(\Omega)$. Hence

$$T(\phi) = \int_\omega \left\{ T(\psi_x) - \sum_{j=1}^n \int_B \psi(y) \int_0^1 y_j g_j(x + ty) \, dt \, dy \right\} \phi(x) \, dx,$$

which displays T explicitly as a function, which we denote by f.

Finally, by Theorem 6.9(2)

$$f(x+y) - f(x) = \int_0^1 \sum_{j=1}^n g_j(x+ty) y_j \, dt \qquad (3)$$

for $x \in \omega$ and $|y| < R$. The right side is

$$\sum_{j=1}^n g_j(x) y_j + o(|y|)$$

and this proves that $f \in C^1(\omega)$ with derivatives g_i. This suffices, since x can be arbitrarily chosen in Ω by choosing R to be small enough. ∎

The following is a special case of Theorem 6.10, which we state separately for emphasis.

6.11 THEOREM (Distributions with zero derivatives are constants)

Let $\Omega \subset \mathbb{R}^n$ be a connected, open set and let $T \in \mathcal{D}'(\Omega)$. Suppose that $\partial_i T = 0$ for each $i = 1, \ldots, n$. Then there is a constant C such that

$$T(\phi) = C \int_\Omega \phi$$

for all $\phi \in \mathcal{D}(\Omega)$.

PROOF. By Theorem 6.10, T is a $C^1(\Omega)$-function, f, and $\partial f / \partial x^i = 0$. Application of 6.10(3) to f shows that f is constant. ∎

6.12 MULTIPLICATION AND CONVOLUTION OF DISTRIBUTIONS BY C^∞-FUNCTIONS

A useful fact is that distributions can be multiplied by C^∞-functions. Consider T in $\mathcal{D}'(\Omega)$ and ψ in $C^\infty(\Omega)$. Define the product ψT by its action on $\phi \in \mathcal{D}(\Omega)$ as

$$(\psi T)(\phi) := T(\psi \phi) \qquad (1)$$

for all $\phi \in \mathcal{D}(\Omega)$. That ψT is a distribution follows from the fact that the product $\psi \phi \in C_c^\infty(\Omega)$ if $\phi \in C_c^\infty(\Omega)$. Moreover, if $\phi^n \to \phi$ in $\mathcal{D}(\Omega)$, then

$\psi\phi^n \to \psi\phi$ in $\mathcal{D}(\Omega)$. To differentiate ψT we simply apply the product rule, namely
$$\partial_i(\psi T)(\phi) = \psi(\partial_i T)(\phi) + (\partial_i \psi)T(\phi), \tag{2}$$
which is easily verified from the basic definition 6.6(1) and Leibniz's differentiation formula $\partial_i(\psi\phi) = \phi\partial_i\psi + \psi\partial_i\phi$ for C^∞-functions.

Observe that when $T = T_f$ for some f in $L^1_{\text{loc}}(\Omega)$, then $\psi T = T_{\psi f}$. If, moreover, $f \in W^{1,p}_{\text{loc}}(\Omega)$, then $\psi f \in W^{1,p}_{\text{loc}}(\Omega)$ and (2) reads
$$\partial_i(f\psi)(x) = f(x)\partial_i\psi(x) + \psi(x)(\partial_i f)(x) \tag{3}$$
for almost every x. The same holds for $W^{1,p}(\Omega)$ and it also clearly extends to $W^{k,p}_{\text{loc}}(\Omega)$ and $W^{k,p}(\Omega)$.

The convolution of a distribution T with a $C^\infty_c(\mathbb{R}^n)$-function j is defined by
$$(j*T)(\phi) := T(j_R * \phi) = T\left(\int_{\mathbb{R}^n} j(y)\phi_{-y}\,dy\right) \tag{4}$$
for all $\phi \in \mathcal{D}(\mathbb{R}^n)$, where $j_R(x) := j(-x)$. Since $j_R * \phi \in C^\infty_c(\mathbb{R}^n)$, $j*T$ makes sense and is in $\mathcal{D}'(\mathbb{R}^n)$. The reader can check that when T is a function, i.e., $T = T_f$, then, with this definition, $(j*T_f)(\phi) = T_{j*f}(\phi)$ where $(j*f)(x) = \int_{\mathbb{R}^n} j(x-y)f(y)\,dy$ is the usual convolution.

▶ Note the requirement that j must have compact support.

6.13 THEOREM (Approximation of distributions by C^∞-functions)

Let $T \in \mathcal{D}'(\mathbb{R}^n)$ and let $j \in C^\infty_c(\mathbb{R}^n)$. Then there exists a function $t \in C^\infty(\mathbb{R}^n)$ (depending only on T and j) such that
$$(j*T)(\phi) = \int_{\mathbb{R}^n} t(y)\phi(y)\,dy \tag{1}$$
for every $\phi \in \mathcal{D}(\mathbb{R}^n)$. If we further assume that $\int_{\mathbb{R}^n} j = 1$, and if we set $j_\varepsilon(x) = \varepsilon^{-n} j(x/\varepsilon)$ for $\varepsilon > 0$, then $j_\varepsilon * T$ converges to T in $\mathcal{D}'(\mathbb{R}^n)$ as $\varepsilon \to 0$.

PROOF. By definition we have that
$$(j*T)(\phi) := T(j_R * \phi) = T\left(\int_{\mathbb{R}^n} j(y - \cdot)\phi(y)\,dy\right)$$

which, by Lemma 6.8, equals $\int_{\mathbb{R}^n} T(j(y-\cdot))\phi(y)\,dy$. If we now define $t(y) := T(j(y-\cdot))$, then, by 6.8(1), $t \in C^\infty(\mathbb{R}^n)$. This proves (1). To verify the convergence of $j_\varepsilon * T$ to T, simply observe that

$$\begin{aligned}(j_\varepsilon * T)(\phi) &:= T\left(\int_{\mathbb{R}^n} j_\varepsilon(y)\phi_{-y}\,dy\right) \\ &= \int_{\mathbb{R}^n} j_\varepsilon(y) T(\phi_{-y})\,dy = \int_{\mathbb{R}^n} j(y) T(\phi_{-\varepsilon y})\,dy\end{aligned} \qquad (2)$$

by changing variables. It is clear that the last term in (2) tends to $T(\phi)$ since $T(\phi_{-y})$ is C^∞ as a function of y, and j has compact support. ∎

• The **kernel** or **null-space** of a distribution $T \in \mathcal{D}'(\Omega)$ is defined by $\mathcal{N}_T = \{\phi \in \mathcal{D}(\Omega) : T(\phi) = 0\}$. It forms a closed linear subspace of $\mathcal{D}(\Omega)$. The following theorem about the intersection of kernels is useful in connection with Lagrange multipliers in the calculus of variations. (See Sect. 11.6.)

6.14 THEOREM (Linear dependence of distributions)

Let $S_1, \ldots, S_N \in \mathcal{D}'(\Omega)$ be distributions. Suppose that $T \in \mathcal{D}'(\Omega)$ has the property that $T(\phi) = 0$ for all $\phi \in \bigcap_{i=1}^N \mathcal{N}_{S_i}$. Then there exist complex numbers $c_1, \ldots, c_N$ such that

$$T = \sum_{i=1}^N c_i S_i. \qquad (1)$$

PROOF. Without loss of generality it can be assumed that the S_i's are linearly independent. First, we show that there exist N fixed functions $u_1, \ldots, u_N \in \mathcal{D}(\Omega)$ such that every $\phi \in \mathcal{D}(\Omega)$ can be written as

$$\phi = v + \sum_{i=1}^N \lambda_i(\phi) u_i \qquad (2)$$

for some $\lambda_i(\phi) \in \mathbb{C}$, $i = 1, \ldots, N$, and $v \in \bigcap_{i=1}^N \mathcal{N}_{S_i}$. To see this consider the set of vectors

$$V = \{\underline{S}(\phi) : \phi \in \mathcal{D}(\Omega)\}, \qquad (3)$$

where $\underline{S}(\phi) = (S_1(\phi), \ldots, S_N(\phi))$. It is obvious that V is a vector space of dimension N since the S_i's are linearly independent. Hence there exist functions $u_1, \ldots, u_N \in \mathcal{D}(\Omega)$ such that $\underline{S}(u_1), \ldots, \underline{S}(u_N)$ span V. Thus, the $N \times N$ matrix given by $M_{ij} = S_i(u_j)$ is invertible. With

$$\lambda_i(\phi) = \sum_{j=1}^{N} (M^{-1})_{ij} S_j(\phi), \qquad (4)$$

it is easily seen that (2) holds.

Applying T to formula (2) yields (using $T(v) = 0$)

$$T(\phi) = \sum_{i,j=1}^{N} (M^{-1})_{ij} T(u_i) S_j(\phi),$$

which gives (1) with $c_i = \sum_{j=1}^{N} (M^{-1})_{ji} T(u_j)$. ∎

6.15 THEOREM ($C^\infty(\Omega)$ is 'dense' in $W^{1,p}_{\text{loc}}(\Omega)$)

Let f be in $W^{1,p}_{\text{loc}}(\Omega)$. For any open set $\mathcal{O}$ with the property that there exists a compact set $K \subset \Omega$ such that $\mathcal{O} \subset K \subset \Omega$, we can find a sequence $f^1, f^2, f^3, \ldots \in C^\infty(\mathcal{O})$ such that

$$\|f - f^k\|_{L^p(\mathcal{O})} + \sum_i \|\partial_i f - \partial_i f^k\|_{L^p(\mathcal{O})} \to 0 \quad \text{as} \quad k \to \infty. \qquad (1)$$

PROOF. For $\varepsilon > 0$ consider the function $j_\varepsilon * f$, where $j_\varepsilon(x) = \varepsilon^{-n} j(x/\varepsilon)$ and j is a C^∞-function with support in the unit ball centered at the origin with $\int_{\mathbb{R}^n} j(x)\, dx = 1$. For any open set $\mathcal{O}$ with the properties stated above we have that $j_\varepsilon * f \in C^\infty(\mathcal{O})$ if ε is sufficiently small since, on $\mathcal{O}$,

$$D^\alpha(j_\varepsilon * f)(x) = \int_{\mathbb{R}^n} (D^\alpha j_\varepsilon)(x-y) f(y)\, dy$$

for derivatives of any order α. Further, since $\mathcal{O} \subset K \subset \Omega$ with K compact, we can assume, by choosing ε small enough, that

$$\mathcal{O} + \text{supp}\{j_\varepsilon\} := \{x + z : x \in \mathcal{O}, z \in \text{supp}\{j_\varepsilon\}\} \subset K.$$

Thus, since

$$\partial_i \int_K j_\varepsilon(x-y) f(y)\, dy = \int_K j_\varepsilon(x-y) (\partial_i f)(y)\, dy,$$

and since f and $\partial_i f$ are in $L^p(K)$ for $i = 1, \ldots, n$, (1) follows from Theorem 2.16 by choosing $\varepsilon = 1/k$ with k large enough. ∎

● The reader is invited to jump ahead for the moment and compare Theorem 6.15 for $p = 2$ with the much deeper Meyers–Serrin Theorem 7.6 (density of $C^\infty(\Omega)$ in $H^1(\Omega)$). The latter easily generalizes to $p \neq 2$, i.e., to $W^{1,p}(\Omega)$ and, in each case, implies 6.15. The important point is that if $f \in H^1(\Omega)$, then $\nabla f(x)$ can go to infinity as x goes to the boundary of Ω. Thus, convergence of the smooth functions f^k to f in the $H^1(\Omega)$-norm, as in 7.6, is not easy to achieve. Theorem 6.15 only requires convergence arbitrarily close to, but not up to, the boundary of Ω. The sequence f^k in 6.15 is allowed to depend on the open subset $\mathcal{O} \subset \Omega$. In contrast, in 7.6 the fixed sequence f^k must yield convergence in $H^1(\Omega)$. On the other hand, a function in $W^{1,2}_{\text{loc}}(\Omega)$ need not be in $H^1(\Omega)$; it need not even be in $L^1(\Omega)$.

6.16 THEOREM (Chain rule)

Let $G : \mathbb{R}^N \to \mathbb{C}$ be a differentiable function with bounded and continuous derivatives. We denote it explicitly by $G(s_1, \ldots, s_N)$. If

$$u(x) = (u_1(x), \ldots, u_N(x))$$

denotes N functions in $W^{1,p}_{\text{loc}}(\Omega)$, then the function $K : \Omega \to \mathbb{C}$ given by

$$K(x) = (G \circ u)(x) = G(u(x))$$

is in $W^{1,p}_{\text{loc}}(\Omega)$ and

$$\frac{\partial}{\partial x_i} K = \sum_{k=1}^N \frac{\partial G}{\partial s_k}(u) \cdot \frac{\partial u_k}{\partial x_i} \tag{1}$$

in $\mathcal{D}'(\Omega)$.

If $u_1, \ldots, u_N$ are in $W^{1,p}(\Omega)$, then K is also in $W^{1,p}(\Omega)$ and (1) holds—provided we make the additional assumption, in case $|\Omega| = \infty$, that $G(0) = 0$.

PROOF. It suffices to prove that $K \in W^{1,p}(\mathcal{O})$ and verify formula (1) for any open set $\mathcal{O}$ with the property that $\mathcal{O} \subset C \subset \Omega$, with C compact.

By Theorem 6.15 we can find a sequence of functions $\phi^m = (\phi_1^m, \ldots, \phi_N^m)$ in $(C^\infty(\mathcal{O}))^N$ such that (with an obvious abuse of notation)

$$\|\phi^m - u\|_{W^{1,p}(\mathcal{O})} \to 0 \tag{2}$$

as $m \to \infty$. By passing to a subsequence we may assume that $\phi^m \to u$ pointwise a.e. and $\frac{\partial}{\partial x_i}\phi^m \to \frac{\partial}{\partial x_i} u$ pointwise a.e. for all $i = 1, \ldots, n$. Set $K^m(x) = G(\phi^m(x))$. Since

$$\max_i \left| \frac{\partial G}{\partial s_i} \right| \leq M,$$

a simple application of the fundamental theorem of calculus and Hölder's inequality in $\mathbb{R}^N$ shows that for $s, t \in \mathbb{R}^N$

$$|G(s) - G(t)| \leq MN^{1/p'} \left(\sum_{i=1}^{N} |s_i - t_i|^p \right)^{1/p}. \tag{3}$$

Here $1/p + 1/p' = 1$. Since $\mathcal{O} \subset C$ and G is bounded, K is in $L^p_{\text{loc}}(\mathcal{O})$.

Next, for $\psi \in \mathcal{D}(\Omega)$

$$\begin{aligned}\int_\Omega \frac{\partial \psi}{\partial x_k} K^m \, dx &= -\int_\Omega \psi \frac{\partial}{\partial x_k} K^m \, dx \\ &= -\sum_{l=1}^{N} \int_\Omega \psi \frac{\partial G}{\partial s_l}(\phi^m) \frac{\partial}{\partial x_k} \phi_l^m \, dx.\end{aligned} \tag{4}$$

In (4) the ordinary chain rule for C^1-functions has been used. Using (3) we find that

$$|K(x) - K^m(x)| \leq MN^{1/p'} \left(\sum_{i=1}^{N} |u_i(x) - \phi_i^m(x)|^p \right)^{1/p},$$

which implies that $K^m \to K$ in $L^p(\mathcal{O})$, and therefore the left side of (4) tends to $\int_\Omega \frac{\partial \psi(x)}{\partial x_k} K(x) \, dx$. Each term on the right side can be written as

$$\int_\mathcal{O} \psi \frac{\partial G}{\partial s_l}(\phi^m) \frac{\partial}{\partial x_k} u_l \, dx + \int_\mathcal{O} \psi \frac{\partial G}{\partial s_l}(\phi^m) \left(\frac{\partial}{\partial x_k} \phi_l^m - \frac{\partial}{\partial x_k} u_l \right) dx. \tag{5}$$

The first term tends to

$$\int_\mathcal{O} \psi \frac{\partial G}{\partial s_l}(u) \frac{\partial u_l}{\partial x_k} \, dx$$

by dominated convergence and the second tends to zero since $\frac{\partial G}{\partial s_l}$ is uniformly bounded and $\frac{\partial \phi_l^m}{\partial x_k} - \frac{\partial u_l}{\partial x_k} \to 0$ in $L^p(\mathcal{O})$. Clearly $\frac{\partial G}{\partial s_l}(u) \frac{\partial u_l}{\partial x_k}$, which is a bounded function times an $L^p(\mathcal{O})$-function, is itself in $L^p(\mathcal{O})$.

To verify the second statement about $W^{1,p}(\Omega)$, note that $\partial G/\partial s_k$ is bounded for all $k = 1, 2, \ldots, N$ and, since $\nabla u_k \in L^p(\Omega)$, it follows from (1) that $\nabla K \in L^p(\Omega)$ also. The only thing to check is that K itself is in $L^p(\Omega)$. It follows from (3) that

$$|K(x)|^p \leq A + B \sum_{k=1}^{N} |u_k(x)|^p, \tag{6}$$

where A and B are some constants. If $|\Omega| < \infty$, (6) implies that $K \in L^p(\Omega)$. If $|\Omega| = \infty$ we have to use the assumption $G(0) = 0$, which implies that we can take $A = 0$ in (6). Again, $K \in L^p(\Omega)$. ■

6.17 THEOREM (Derivative of the absolute value)

Let f be in $W^{1,p}(\Omega)$. Then the absolute value of f, denoted by $|f|$ and defined by $|f|(x) = |f(x)|$, is in $W^{1,p}(\Omega)$ with $\nabla|f|$ being the function

$$(\nabla|f|)(x) = \begin{cases} \frac{1}{|f|(x)}(R(x)\nabla R(x) + I(x)\nabla I(x)) & \text{if } f(x) \neq 0, \\ 0 & \text{if } f(x) = 0; \end{cases} \quad (1)$$

here $R(x)$ and $I(x)$ denote the real and imaginary parts of f. In particular, if f is real-valued,

$$(\nabla|f|)(x) = \begin{cases} \nabla f(x) & \text{if } f(x) > 0, \\ -\nabla f(x) & \text{if } f(x) < 0, \\ 0 & \text{if } f(x) = 0. \end{cases} \quad (2)$$

Thus $\big|\nabla|f|\big| \leq |\nabla f|$ a.e. if f is complex-valued and $\big|\nabla|f|\big| = |\nabla f|$ a.e. if f is real-valued.

PROOF. We follow [Gilbarg–Trudinger]. That $|f|$ is in $L^p(\Omega)$ follows from the definition of $\|f\|_p$. Further, since

$$\left|\frac{1}{|f|}(R\nabla R + I\nabla I)\right|^2 \leq (\nabla R)^2 + (\nabla I)^2 \quad (3)$$

pointwise, $\nabla|f|$ is also in $L^p(\Omega)$ once the claimed equality (1) is proved. Consider the function

$$G_\varepsilon(s_1, s_2) = \sqrt{\varepsilon^2 + s_1^2 + s_2^2} - \varepsilon . \quad (4)$$

Obviously $G_\varepsilon(0,0) = 0$ and

$$\left|\frac{\partial G_\varepsilon}{\partial s_i}\right| = \left|\frac{s_i}{\sqrt{\varepsilon^2 + s_1^2 + s_2^2}}\right| \leq 1 . \quad (5)$$

Hence, by 6.16, the function $K_\varepsilon(x) = G_\varepsilon(R(x), I(x))$ is in $W^{1,p}(\Omega)$ and for all ϕ in $\mathcal{D}(\Omega)$

$$\begin{aligned}\int_\Omega \nabla\phi(x) K_\varepsilon(x)\,dx &= -\int_\Omega \phi(x)\nabla K_\varepsilon(x)\,dx \\ &= -\int_\Omega \phi(x)\frac{R(x)\nabla R(x) + I(x)\nabla I(x)}{\sqrt{\varepsilon^2 + |f(x)|^2}}\,dx.\end{aligned} \quad (6)$$

Since $K_\varepsilon(x) \leq |f(x)|$ and
$$\left| \frac{R(x)\nabla R(x) + I(x)\nabla I(x)}{\sqrt{\varepsilon^2 + |f(x)|^2}} \right| \leq |\nabla f(x)|^2 ,$$
and since the two functions (4) and (5) converge pointwise to the claimed expressions as $\varepsilon \to 0$, the result follows by dominated convergence. ∎

6.18 COROLLARY (Min and Max of $W^{1,p}$-functions are in $W^{1,p}$)

Let f and g be two real-valued functions in $W^{1,p}(\Omega)$. Then the minimum of $(f(x), g(x))$ and the maximum of $(f(x), g(x))$ are functions in $W^{1,p}(\Omega)$ and the gradients are given by

$$\nabla \max(f(x), g(x)) = \begin{cases} \nabla f(x) & \text{when } f(x) > g(x), \\ \nabla g(x) & \text{when } f(x) < g(x), \\ \nabla f(x) = \nabla g(x) & \text{when } f(x) = g(x), \end{cases} \quad (1)$$

$$\nabla \min(f(x), g(x)) = \begin{cases} \nabla g(x) & \text{when } f(x) > g(x), \\ \nabla f(x) & \text{when } f(x) < g(x), \\ \nabla f(x) = \nabla g(x) & \text{when } f(x) = g(x). \end{cases} \quad (2)$$

PROOF. That these two functions are in $W^{1,p}(\Omega)$ follows from the formulas
$$\min(f(x), g(x)) = \frac{1}{2}\big[(f(x) + g(x)) - |f(x) - g(x)|\big]$$
and
$$\max(f(x), g(x)) = \frac{1}{2}\big[(f(x) + g(x)) + |f(x) - g(x)|\big] .$$

The formulas (1) and (2) follow immediately from Theorem 6.17 in the cases where $f(x) > g(x)$ or $f(x) < g(x)$. To understand the case $f(x) = g(x)$ consider
$$h(x) = (f(x) - g(x))_+ = \frac{1}{2}\{|f(x) - g(x)| + (f(x) - g(x))\}.$$
Obviously $|h|(x) = h(x)$, and hence by 6.17
$$\nabla h(x) = \nabla |h|(x) = 0 \quad \text{when } f(x) \leq g(x).$$
But again by 6.17 $\nabla h(x) = \frac{1}{2}(\nabla(f - g))(x)$, when $f(x) = g(x)$ and hence
$$(\nabla f)(x) = (\nabla g)(x) \quad \text{when } f(x) = g(x),$$
which yields (1) and (2) in the case $f(x) = g(x)$. ∎

It is an easy exercise to extend the above result to truncations of $W^{1,p}(\Omega)$-functions defined by

$$f_{<\alpha}(x) = \min(f(x), \alpha).$$

The gradient is then given by

$$(\nabla f_{<\alpha})(x) = \begin{cases} \nabla f(x) & \text{if } f(x) < \alpha, \\ 0 & \text{otherwise.} \end{cases}$$

Analogously, define

$$f_{>\alpha}(x) = \max(f(x), \alpha).$$

Then

$$(\nabla f_{>\alpha})(x) = \begin{cases} \nabla f(x) & \text{if } f(x) > \alpha, \\ 0 & \text{otherwise.} \end{cases}$$

Note that when Ω is unbounded $f_{<\alpha} \in W^{1,p}(\Omega)$ only if $\alpha \geq 0$, and $f_{>\alpha} \in W^{1,p}(\Omega)$ only if $\alpha \leq 0$.

The foregoing implies that if $u \in W^{1,1}_{\text{loc}}(\Omega)$, if $\alpha \in \mathbb{R}$ and if $u(x) = \alpha$ on a set of positive measure in $\mathbb{R}^n$, then $(\nabla u)(x) = 0$ for almost every x in this set. This can be derived easily from 6.18. The following theorem, to be found in [Almgren–Lieb], generalizes this fact by replacing the single point $\alpha \in \mathbb{R}$ by a Borel set A of zero measure. Such sets need not be 'small', e.g. α could be all the rational numbers, and hence α could be dense in $\mathbb{R}$. Note that if f is a Borel measurable function, then $f^{-1}(A) := \{x \in \mathbb{R}^n : f(x) \in A\}$ is a Borel set, and hence is measurable. This follows from the statement in Sect. 1.5 and Exercise 1.3 that $x \mapsto \chi_A(f(x))$ is measurable.

6.19 THEOREM (Gradients vanish on the inverse of small sets)

Let $A \subset \mathbb{R}$ be a Borel set with zero Lebesgue measure and let $f : \Omega \to \mathbb{R}$ be in $W^{1,1}_{\text{loc}}(\Omega)$. Let

$$B = f^{-1}(A) := \{x \in \Omega : f(x) \in A\} \subset \Omega.$$

Then $\nabla f(x) = 0$ for almost every $x \in B$.

PROOF. Our goal will be to establish the formula

$$\int_\Omega \phi(x)\chi_{\mathcal{O}}(f(x))\nabla f(x)\,dx = -\int_\Omega \nabla\phi(x)G_{\mathcal{O}}(f(x))\,dx \tag{1}$$

for each open set $\mathcal{O} \subset \mathbb{R}$. Here $\chi_\mathcal{O}$ is the characteristic function of $\mathcal{O}$ and $G_\mathcal{O}(t) = \int_{-\infty}^t \chi_\mathcal{O}(s)\,ds$. Equation (1) is just like the chain rule except that $G_\mathcal{O}$ is not in $C^1(\mathbb{R})$. Assuming (1) for the moment, we can conclude the proof of our theorem as follows. By the outer regularity of Borel measure we can find a decreasing sequence $\mathcal{O}^1 \supset \mathcal{O}^2 \supset \mathcal{O}^3 \supset \cdots$ of open sets such that $A \subset \mathcal{O}^j$ for each j and $\mathcal{L}^1(\mathcal{O}^j) \to 0$ as $j \to \infty$. Thus $A \subset C := \bigcap_{j=1}^\infty \mathcal{O}^j$ (but it could happen that A is strictly smaller than C) and $\mathcal{L}^1(C) = 0$. By definition, $G_j(t) := G_{\mathcal{O}_j}(t)$ satisfies $G_j(t) \leq \mathcal{L}^1(\mathcal{O}^j)$, and thus $G_j(t)$ goes uniformly to zero as $j \to \infty$. The right side of (1) (with $\mathcal{O}$ replaced by $\mathcal{O}^j$) therefore tends to zero as $j \to \infty$. On the other hand, $\chi_j := \chi_{\mathcal{O}_j}$ is bounded by 1, and $\chi_j(f(x)) \to \chi_{f^{-1}(C)}(x)$ for every $x \in \mathbb{R}^n$. By dominated convergence, the left side of (1) converges to $\int_\Omega \phi \chi_{f^{-1}(C)} \nabla f$, and this equals zero for every $\phi \in \mathcal{D}(\Omega)$. By the uniqueness of distributions, the function $\chi_{f^{-1}(C)}(x) \nabla f(x) = 0$ for almost every x, which is what we wished to prove.

It remains to prove (1). Observe that every open set $\mathcal{O} \subset \mathbb{R}$ is the union of countably many disjoint open intervals. (Why?) Thus $\mathcal{O} = \bigcup_{j=1}^\infty U_j$ with $U_j = (a_j, b_j)$. Since f is a function, $f^{-1}(U_j)$ is disjoint from $f^{-1}(U_k)$ when $j \neq k$. By the countable additivity of measure, therefore, it suffices to prove (1) when $\mathcal{O}$ is just one interval (a, b). We can easily find a sequence $\chi^1, \chi^2, \chi^3, \ldots$ of continuous functions such that $\chi^j(t) \to \chi_\mathcal{O}(t)$ for *every* $t \in \mathbb{R}$ and $0 \leq \chi^j(t) \leq 1$ for every $t \in \mathbb{R}$. The everywhere (not just almost everywhere) convergence is crucial and we leave the simple construction of $\{\chi^j\}$ to the reader. Then, with $G^j = \int_{-\infty}^t \chi^j$, equation (1) is obtained by taking the limit $j \to \infty$ on both sides and using dominated convergence. The easy verification is again left to the reader. ∎

• An amusing—and useful—exercise in the computation of distributional derivatives is the computation of Green's functions. Let $y \in \mathbb{R}^n$, $n \geq 1$, and let $G_y : \mathbb{R}^n \to \mathbb{R}$ be defined by

$$G_y(x) = -|\mathbb{S}^1|^{-1} \ln(|x-y|), \qquad n = 2,$$
$$G_y(x) = [(n-2)|\mathbb{S}^{n-1}|]^{-1} |x-y|^{2-n}, \qquad n \neq 2, \tag{2}$$

where $|\mathbb{S}^{n-1}|$ is the area of the unit sphere $\mathbb{S}^{n-1} \subset \mathbb{R}^n$.

$$|\mathbb{S}^0| = 2, \quad |\mathbb{S}^1| = 2\pi, \quad |\mathbb{S}^2| = 4\pi, \quad |\mathbb{S}^{n-1}| = 2\pi^{n/2}/\Gamma(n/2).$$

These are the **Green's functions for Poisson's equation in $\mathbb{R}^n$**. Recall that the **Laplacian**, Δ, is defined by $\Delta := \sum_{i=1}^n \partial^2/\partial x_i^2$. The notation notwithstanding, $G_y(x)$ is actually *symmetric*, i.e., $G_y(x) = G_x(y)$.

6.20 THEOREM (Distributional Laplacian of Green's functions)

In the sense of distributions,

$$-\Delta G_y = \delta_y, \tag{1}$$

where δ_y is Dirac's delta measure at y (often written as $\delta(x-y)$).

PROOF. To prove (1) we can take $y = 0$. We require

$$I := \int_{\mathbb{R}^n} (\Delta\phi) G_0 = -\phi(0)$$

when $\phi \in C_c^\infty(\mathbb{R}^n)$. Since $G_0 \in L^1_{\text{loc}}(\mathbb{R}^n)$, it suffices to show that

$$-\phi(0) = \lim_{r \to 0} I(r),$$

where

$$I(r) := \int_{|x|>r} \Delta\phi(x) G_0(x)\,dx.$$

We can also restrict the integration to $|x| < R$ for some R since ϕ has compact support. However, when $|x| > 0$, G_0 is infinitely differentiable and $\Delta G_0 = 0$. We can evaluate $I(r)$ by partial integration, and note that boundary integrals at $|x| = R$ vanish. Thus, denoting the set $\{x : r \leq |x| \leq R\}$ by A,

$$\begin{aligned} I(r) &= \int_A (\Delta\phi) G_0 = -\int_A \nabla\phi \cdot \nabla G_0 + \int_{|x|=r} G_0 \nabla\phi \cdot \nu \\ &= -\int_{|x|=r} \phi \nabla G_0 \cdot \nu + \int_{|x|=r} G_0 \nabla\phi \cdot \nu, \end{aligned} \tag{2}$$

where ν is the unit outward normal to A. On the sphere $|x| = r$, we have $\nabla G_0 \cdot \nu = |\mathbb{S}^{n-1}|^{-1} r^{-n+1}$, and therefore the penultimate integral in (2) is

$$-\int_{|x|=r} \phi \nabla G_0 \cdot \nu = -|\mathbb{S}^{n-1}|^{-1} \int_{\mathbb{S}^{n-1}} \phi(r\omega)\,d\omega,$$

which converges to $-\phi(0)$ as $r \to 0$, since ϕ is continuous. The last integral in (2) converges to zero as $r \to 0$ since $\nabla\phi \cdot \nu$ is bounded by some constant, while $||x|^{n-1} G_0(x)| < |x|^{1/2}$ for small $|x|$. Thus, (1) has been verified. ∎

6.21 THEOREM (Solution of Poisson's equation)

Let $f \in L^1_{\text{loc}}(\mathbb{R}^n)$, $n \geq 1$. Assume that for almost every x the function $y \mapsto G_y(x)f(y)$ is summable (here, G_y is Green's function given before 6.20) and define the function $u : \mathbb{R}^n \to \mathbb{C}$ by

$$u(x) = \int_{\mathbb{R}^n} G_y(x) f(y) \, dy. \tag{1}$$

Then u satisfies:

$$u \in L^1_{\text{loc}}(\mathbb{R}^n), \tag{2}$$

$$-\Delta u = f \quad \text{in } \mathcal{D}'(\mathbb{R}^n). \tag{3}$$

Moreover, the function u has a distributional derivative that is a function; it is given, for almost every x, by

$$\partial_i u(x) = \int_{\mathbb{R}^n} (\partial G_y / \partial x_i)(x) f(y) \, dy. \tag{4}$$

When $n = 3$, for example, the partial derivative is

$$(\partial G_y / \partial x_i)(x) = -\frac{1}{4\pi} |x - y|^{-3} (x_i - y_i). \tag{5}$$

REMARKS. (1) A trivial consequence of the theorem is that $\mathbb{R}^n$ can be replaced by any open set $\Omega \subset \mathbb{R}^n$. Suppose $f \in L^1_{\text{loc}}(\Omega)$ and $y \mapsto G_y(x) f(y)$ is summable over Ω for almost every $x \in \Omega$. Then (see Exercises)

$$u(x) := \int_{\Omega} G_y(x) f(y) \, dy \tag{6}$$

is in $L^1_{\text{loc}}(\Omega)$ and satisfies

$$-\Delta u = f \quad \text{in } \mathcal{D}'(\Omega). \tag{7}$$

(2) The summability condition in Theorem 6.21 is equivalent to the condition that the function $w_n(y) f(y)$ is summable. Here

$$w_n(y) = \begin{cases} (1 + |y|)^{2-n}, & n \geq 3, \\ \ln(1 + |y|), & n = 2, \\ |y|, & n = 1. \end{cases} \tag{8}$$

The easy proof of this equivalence is left to the reader as an exercise. (It proceeds by decomposing the integral in (1) into a ball containing x, and its complement in $\mathbb{R}^n$. The contribution from the ball is easily shown to be finite for almost every x in the ball, by Fubini's theorem.)

(3) It is also obvious that any solution to equation (7) has the form $u+h$, where u is defined by (6) and where $\Delta h = 0$. Hence h is a harmonic function on Ω (see Sect. 9.3). Since harmonic functions are infinitely differentiable (Theorem 9.4), it follows that *every* solution to (7) is in $C^k(\Omega)$ if and only if $u \in C^k(\Omega)$.

PROOF. To prove (2) it suffices to prove that $I_B := \int_B |u| < \infty$ for each ball $B \subset \mathbb{R}^n$. Since $|u(x)| \leq \int_{\mathbb{R}^n} |G_y(x) f(y)| \, dy$, we can use Fubini's theorem to conclude that

$$I \leq \int_{\mathbb{R}^n} H_B(y) |f(y)| \, dy \quad \text{with} \quad H_B(y) = \int_B |G_y(x)| \, dx.$$

It is easy to verify (by using Newton's Theorem 9.7, for example) that if B has center x_0 and radius R, then $H_B(y) = |B||G_y(x_0)|$ for $|y - x_0| \geq R$ for $n \neq 2$ and $H_B(y) = |B||G_y(x_0)|$ when $|y - x_0| \geq R + 1$ when $n = 2$ (in order to keep the logarithm positive). Moreover, $H_B(y)$ is bounded when $|y - x_0| < R$. From this observation it follows easily that $I_B < \infty$. (Note: Fubini's theorem allows us to conclude both that u is a measurable function and that this function is in $L^1_{\text{loc}}(\mathbb{R}^n)$.)

To verify (3) we have to show that

$$-\int u \Delta \phi = \int f \phi \tag{9}$$

for each $\phi \in C_c^\infty(\mathbb{R}^n)$. We can insert (1) into the left side of (9) and use Fubini's theorem to evaluate the double integral. But Theorem 6.20 states that $-\int_{\mathbb{R}^n} \Delta \phi(x) G_y(x) \, dx = \phi(y)$, and this proves (9).

To prove (4) we begin by verifying that the integral in (4) (call it $V_i(x)$) is well defined for almost every $x \in \mathbb{R}^n$. To see this note that $|(\partial G_y/\partial x_i)(x)|$ is bounded above by $c|x-y|^{1-n}$, which is in $L^1_{\text{loc}}(\mathbb{R}^n)$. The finiteness of $V_i(x)$ follows as in Remark (2) above. Next, we have to show that

$$\int_{\mathbb{R}^n} \partial_i \phi(x) u(x) \, dx = -\int_{\mathbb{R}^n} \phi(x) V_i(x) \, dx \tag{10}$$

for all $\phi \in C_c^\infty(\mathbb{R}^n)$. Since the function $(x,y) \to (\partial_i \phi)(x) G_y(x) f(y)$ is $\mathbb{R}^n \times \mathbb{R}^n$ summable, we can use Fubini's theorem to equate the left side of (10) to

$$\int_{\mathbb{R}^n} \left\{ \int_{\mathbb{R}^n} (\partial_i \phi)(x) G_y(x) \, dx \right\} f(y) \, dy. \tag{11}$$

A limiting argument, combined with integration by parts, as in 6.20(2), shows that the inner integral in (11) is

$$-\int_{\mathbb{R}^n} \phi(x) \partial G_y/\partial x_i)(x) \, dx$$

for every $y \in \mathbb{R}^n$. Applying Fubini's theorem again, we arrive at (4). ∎

● The next theorem may seem rather specialized, but it is useful in connection with the potential theory in Chapter 9. Its proof (which does not use Lebesgue measure) is an important exercise in measure theory. We shall leave a few small holes in our proof that we ask the reader to fill in as further exercises. Among other things, this theorem yields a construction of Lebesgue measure (Exercise 5).

6.22 THEOREM (Positive distributions are measures)

Let $\Omega \subset \mathbb{R}^n$ be open and let $T \in \mathcal{D}'(\Omega)$ be a positive distribution (meaning that $T(\phi) \geq 0$ for every $\phi \in \mathcal{D}(\Omega)$ such that $\phi(x) \geq 0$ for all x). We denote this fact by $T \geq 0$.

Our assertion is that there is then a unique, positive, regular Borel measure μ on Ω such that $\mu(K) < \infty$ for all compact $K \subset \Omega$ and such that for all $\phi \in \mathcal{D}(\Omega)$

$$T(\phi) = \int_\Omega \phi(x) \mu(\mathrm{d}x). \tag{1}$$

Conversely, any positive Borel measure with $\mu(K) < \infty$ for all compact $K \subset \Omega$ defines a positive distribution via (1).

REMARK. The representation (1) shows that a positive distribution can be extended from $C_c^\infty(\Omega)$-functions to a much larger class, namely the Borel measurable functions with compact support in Ω. This class is even larger than the continuous functions of compact support, $C_c(\Omega)$.

The theorem amounts to an extension, from $C_c(\Omega)$-functions to $C_c^\infty(\Omega)$-functions, of what is known as the **Riesz–Markov representation theorem**. See [Rudin, 1987].

PROOF. In the following, *all* sets are understood to be subsets of Ω. For a given open set $\mathcal{O}$ denote by $\mathcal{C}(\mathcal{O})$ the set of all functions $\phi \in C_c^\infty(\Omega)$ with $0 \leq \phi(x) \leq 1$ and $\operatorname{supp} \phi \subset \mathcal{O}$. Clearly, this set is not empty. (Why?) Next we define for any open set $\mathcal{O}$

$$\mu(\mathcal{O}) = \sup\{T(\phi) : \phi \in \mathcal{C}(\mathcal{O})\}. \tag{2}$$

For the empty set $\varnothing$ we set $\mu(\varnothing) = 0$. The nonnegative set function μ has the following properties:

(i) $\mu(\mathcal{O}_1) \leq \mu(\mathcal{O}_2)$ if $\mathcal{O}_1 \subset \mathcal{O}_2$,

(ii) $\mu(\mathcal{O}_1 \cup \mathcal{O}_2) \leq \mu(\mathcal{O}_1) + \mu(\mathcal{O}_2)$,

(iii) $\mu\left(\bigcup_{i=1}^\infty \mathcal{O}_i\right) \leq \sum_{i=1}^\infty \mu(\mathcal{O}_i)$ for every countable family of *open* sets $\mathcal{O}_i$.

Property (i) is evident. The second property follows from the following fact (F) whose proof we leave as an exercise for the reader:

(F) For any compact set K and open sets $\mathcal{O}_1, \mathcal{O}_2$ such that $K \subset \mathcal{O}_1 \cup \mathcal{O}_2$ there exist functions ϕ_1 and ϕ_2, both C^∞ in a neighborhood $\mathcal{O}$ of K, such that $\phi_1(x) + \phi_2(x) = 1$ for $x \in K$ and $\phi \cdot \phi_1 \in C_c^\infty(\mathcal{O}_1), \phi \cdot \phi_2 \in C_c^\infty(\mathcal{O}_2)$ for any function $\phi \in C_c^\infty(\mathcal{O})$.

Thus, any $\phi \in \mathcal{C}(\mathcal{O}_1 \cup \mathcal{O}_2)$ can be written as $\phi_1 + \phi_2$ with $\phi_1 \in \mathcal{C}(\mathcal{O}_1)$ and $\phi_2 \in \mathcal{C}(\mathcal{O}_2)$. Hence $T(\phi) = T(\phi_1) + T(\phi_2) \leq \mu(\mathcal{O}_1) + \mu(\mathcal{O}_2)$ and property (ii) follows. By induction we find that

$$\mu\left(\bigcup_{i=1}^m \mathcal{O}_i\right) \leq \sum_{i=1}^m \mu(\mathcal{O}_i).$$

To see property (iii) pick $\phi \in \mathcal{C}\left(\bigcup_{i=1}^\infty \mathcal{O}_i\right)$. Since ϕ has compact support, we have that $\phi \in \mathcal{C}\left(\bigcup_{i \in I} \mathcal{O}_i\right)$ where I is a finite subset of the natural numbers. Hence, by the above,

$$T(\phi) \leq \mu\left(\bigcup_{i \in I} \mathcal{O}_i\right) \leq \sum_{i \in I} \mu(\mathcal{O}_i) \leq \sum_{i=1}^\infty \mu(\mathcal{O}_i),$$

which yields property (iii).

For *every* set A define

$$\mu(A) = \inf\{\mu(\mathcal{O}) : \mathcal{O} \text{ open}, A \subset \mathcal{O}\}. \tag{3}$$

The reader should not be confused by this definition. We have defined a set function, μ, that measures *all* subsets of Ω, but only for a special subcollection will this function be a measure, i.e., be countably additive. This set function μ will now be shown to have the properties of an **outer measure**, as defined in Theorem 1.15 (constructing a measure from an outer measure), i.e.,

(a) $\mu(\varnothing) = 0$,

(b) $\mu(A) \leq \mu(B)$ if $A \subset B$,

(c) $\mu\left(\bigcup_{i=1}^\infty A_i\right) \leq \sum_{i=1}^\infty \mu(A_i)$ for *every* countable collection of sets $A_1, A_2, \ldots$.

The first two properties are evident. To prove (c) pick open sets $\mathcal{O}_1, \mathcal{O}_2, \ldots$ with $A_i \subset \mathcal{O}_i$ and $\mu(\mathcal{O}_i) \leq \mu(A_i) + 2^{-i}\varepsilon$ for $i = 1, 2, \ldots$. Now

$$\mu\left(\bigcup_{i=1}^\infty A_i\right) \leq \mu\left(\bigcup_{i=1}^\infty \mathcal{O}_i\right) \leq \sum_{i=1}^\infty \mu(\mathcal{O}_i)$$

by (b) and (iii), and hence

$$\mu\left(\bigcup_{i=1}^{\infty} A_i\right) \leq \sum_{i=1}^{\infty} \mu(A_i) + \varepsilon,$$

which yields (c) since ε is arbitrary. By Theorem 1.15 the sets A such that $\mu(E) = \mu(E \cap A) + \mu(E \cap A^c)$ for every set E form a sigma-algebra, Σ, on which μ is countably additive.

Next we have to show that all open sets are measurable, i.e., we have to show that for any set E and any open set $\mathcal{O}$

$$\mu(E) \geq \mu(E \cap \mathcal{O}) + \mu(E \cap \mathcal{O}^c). \tag{4}$$

The reverse inequality is obvious. First we prove (4) in the case where E is itself open; call it V.

Pick any function $\phi \in \mathcal{C}(V \cap \mathcal{O})$ such that $T(\phi) \geq \mu(V \cap \mathcal{O}) - \varepsilon/2$. Since $K := \operatorname{supp} \phi$ is compact, its complement, U, is open and contains $\mathcal{O}^c$. Pick $\psi \in \mathcal{C}(U \cap V)$ such that $T(\psi) \geq \mu(U \cap V) - \varepsilon/2$. Certainly

$$\mu(V) \geq T(\phi) + T(\psi) \geq \mu(V \cap \mathcal{O}) + \mu(V \cap U) - \varepsilon$$
$$\geq \mu(V \cap \mathcal{O}) + \mu(V \cap \mathcal{O}^c) - \varepsilon$$

and since ε is arbitrary this proves (4) in the case where E is an open set. If E is arbitrary we have for any open set V with $E \subset V$ that $E \cap \mathcal{O} \subset V \cap \mathcal{O}$, $E \cap \mathcal{O}^c \subset V \cap \mathcal{O}^c$, and hence $\mu(V) \geq \mu(E \cap \mathcal{O}) + \mu(E \cap \mathcal{O}^c)$. This proves (4). Thus we have shown that the sigma-algebra Σ contains all open sets and hence contains the Borel sigma-algebra. Hence the measure μ is a Borel measure.

By construction, this measure is outer regular (see (3) above). We show next that it is inner regular, i.e., for any measurable set A

$$\mu(A) = \sup\{\mu(K) : K \subset A, \ K \text{ compact}\}. \tag{5}$$

First we have to establish that compact sets have finite measure. We claim that for K compact

$$\mu(K) = \inf\{T(\psi) : \psi \in C_c^{\infty}(\Omega), \ \psi(x) = 1 \text{ for } x \in K\}. \tag{6}$$

The set on the right side is not empty. Indeed for K compact and $K \subset \mathcal{O}$ open there exists a C_c^{∞}-function ψ such that $\operatorname{supp} \psi \subset \mathcal{O}$ and $\psi := 1$ on K. (Such a ψ was constructed in Exercise 1.15 *without the aid of Lebesgue measure.*)

Now (6) follows from the following fact which we ask the reader to prove as an exercise: $\mu(K) \leq T(\psi)$ for any $\psi \in C_c^\infty(\Omega)$ with $\psi \equiv 1$ on K. Given this fact, choose $\varepsilon > 0$ and choose $\mathcal{O}$ open such $\mu(K) \geq \mu(\mathcal{O}) - \varepsilon$. Also pick $\psi \in C_c^\infty(\Omega)$ with $\operatorname{supp} \psi \subset \mathcal{O}$ and $\psi \equiv 1$ on K. Then $\mu(K) \leq T(\psi) \leq \mu(\mathcal{O}) \leq \mu(K) + \varepsilon$. This proves (6).

It is easy to see that for $\varepsilon > 0$ and every measurable set A with $\mu(A) < \infty$ there exists an open set $\mathcal{O}$ with $A \subset \mathcal{O}$ and $\mu(\mathcal{O} \sim A) < \varepsilon$. Using the fact that Ω is a countable union of closed balls, the above holds for any measurable set, i.e., even if A does not have finite measure. We ask the reader to prove this.

For $\varepsilon > 0$ and a measurable set A we can find $\mathcal{O}$ with $A^c \subset \mathcal{O}$ such that $\mu(\mathcal{O} \sim (A^c)) < \varepsilon$. But

$$\mathcal{O} \sim (A^c) = \mathcal{O} \cap A = A \sim (\mathcal{O}^c)$$

and $\mathcal{O}^c$ is closed. Thus for any measurable set A and $\varepsilon > 0$ one can find a closed set $\mathcal{C}$ such that $\mathcal{C} \subset A$ and $\mu(A \sim \mathcal{C}) < \varepsilon$. Since any closed set in $\mathbb{R}^n$ is a countable union of compact sets, the inner regularity is proven.

Next we prove the representation theorem. The integral $\int_\Omega \phi(x) \mu(\mathrm{d}x)$ defines a distribution R on $\mathcal{D}(\Omega)$. Our aim is to show that $T(\phi) = R(\phi)$ for all $\phi \in C_c^\infty(\Omega)$. Because $\phi = \phi_1 - \phi_2$ with $\phi_{1,2} \geq 0$ and $\phi_{1,2} \in C_c^\infty(\Omega)$ (as Exercise 1.15 shows), it suffices to prove this with the additional restriction that $\phi \geq 0$. As usual, if $\phi \geq 0$,

$$R(\phi) = \int_0^\infty m(a)\,\mathrm{d}a = \lim_{n\to\infty} \frac{1}{n} \sum_{j \geq 1} m(j/n) \tag{7}$$

where $m(a) = \mu(\{x : \phi(x) > a\})$. The integral in (7) is a Riemann integral; it always makes sense for nonnegative monotone functions (like m) and it always equals the rightmost expression in (7). For each n, the sum in (7) has only finitely many terms, since ϕ is bounded.

For n fixed we define compact sets K_j, $j = 0, 1, 2, \ldots$, by setting $K_0 = \operatorname{supp} \phi$ and $K_j = \{x : \phi(x) \geq j/n\}$ for $j \geq 1$. Similarly, denote by O^j the open sets $\{x : \phi(x) > j/n\}$ for $j = 1, 2, \ldots$. Let χ_j and χ^j denote the characteristic functions of K_j and O^j. Then, as is easily seen,

$$\frac{1}{n} \sum_{j \geq 1} \chi^j < \phi < \frac{1}{n} \sum_{j \geq 0} \chi_j.$$

Since ϕ has compact support, all the sets have finite measure by (6).

For $\varepsilon > 0$ and $j = 0, 1, \ldots$ pick U_j open such that $K_j \subset U_j$ and $\mu(U_j) \leq \mu(K_j) + \varepsilon$. Next pick $\psi_j \in C_c^\infty(\mathbb{R}^n)$ such that $\psi_j \equiv 1$ on K_j and $\operatorname{supp} \psi_j \subset$

U_j. We have shown above that such a function exists. Obviously $\phi \leq \frac{1}{n}\sum_{j\geq 0}\psi_j$ and hence

$$T(\phi) \leq \frac{1}{n}\sum_{j\geq 0} T(\psi_j) \leq \frac{1}{n}\sum_{j\geq 0} \mu(U_j) \leq \frac{1}{n}\sum_{j\geq 0} \mu(K_j) + \varepsilon.$$

By the inner regularity we can find, for every open set $\mathcal{O}^j$ of finite measure, a compact set $C^j \subset \mathcal{O}_j$ such that $\mu(C^j) \geq \mu(\mathcal{O}^j) - \varepsilon$ and, in the same fashion as above, conclude that $T(\phi) \geq \frac{1}{n}\sum_{j\geq 1}\mu(\mathcal{O}^j) - \varepsilon$. Since $\varepsilon > 0$ is arbitrary,

$$\frac{1}{n}\sum_{j\geq 1}\mu(\mathcal{O}^j) \leq T(\phi) \leq \frac{1}{n}\sum_{j\geq 0}\mu(K_j).$$

By noting that $K_j \subset \mathcal{O}^{j-1}$ for $j \geq 1$, we have

$$\frac{1}{n}\sum_{j\geq 1} m(j/n) \leq T(\phi) \leq \frac{1}{n}\sum_{j\geq 1} m(j/n) + \frac{2}{n}\mu(K_0),$$

which proves the representation theorem. The uniqueness part is left to the reader. ∎

Exercises for Chapter 6

1. Fill in the details in the last paragraph of the proof of Theorem 6.19, i.e.,
   (a) Construct the sequence χ^j that converges everywhere to $\chi_{\text{(interval)}}$;
   (b) Complete the dominated convergence argument.

2. Verify the summability condition in Remark (2), equation (8) of Theorem 6.21.

3. Prove fact (F) in Theorem 6.22.

4. Prove that for K compact, $\mu(K)$ (defined in 6.22(3)) satisifies $\mu(K) \leq T(\psi)$ for $\psi \in C_c^\infty(\Omega)$ and $\psi \equiv 1$ on K.

5. Notice that the proof of Theorem 6.22 (and its antecedents) used only the Riemann integral and not the Lebesgue integral. Use the conclusion of Theorem 6.22 to prove the existence of Lebesgue measure. See Sect. 1.2.

6. Prove that the distributional derivative of a monotone nondecreasing function on $\mathbb{R}$ is a Borel measure.

7. Let $\mathcal{N}_T$ be the null-space of a distribution, T. Show that there is a function $\phi_0 \in \mathcal{D}$ so that every element $\phi \in \mathcal{D}$ can be written as $\phi = \lambda \phi_0 + \psi$ with $\psi \in \mathcal{N}_T$ and $\lambda \in \mathbb{C}$. One says that the null-space $\mathcal{N}_T$ has 'codimension one'.

8. Show that a function f is in $W^{1,\infty}(\Omega)$ if and only if $f = g$ a.e. where g is a function that is **Lipschitz continuous** on Ω, i.e., there exists a constant C such that

$$|g(x) - g(y)| \leq C|x - y| \quad \text{for all } x, y \in \Omega.$$

9. Verify Remark (1) in Theorem 6.21 that in this theorem $\mathbb{R}^n$ can be replaced by any open subset of $\mathbb{R}^n$.

Chapter 7

The Sobolev Spaces H^1 and $H^{1/2}$

7.1 INTRODUCTION

Among the spaces $W^{1,p}$, particular importance attaches to $W^{1,2}$ because it is a Hilbert-space, i.e., its norm comes from an inner product. It is also important for the study of many differential equations; indeed, it is of central importance for quantum mechanics, which is the study of Schrödinger's partial differential equation. A similar Hilbert-space that is less often used is $H^{1/2}$ and it is discussed here as well. This is done for two reasons: it provides a good exercise in fractional differentiation, which means going beyond operators that, like the derivative, are purely local. Another reason is that the space can be used to describe a version of Schrödinger's equation that incorporates some features of Einstein's special theory of relativity.

We begin by recalling, for completeness, the basic definition of $W^{1,2}$ which we now call H^1 (but see the remarks below about the Meyers–Serrin Theorem 7.5).

7.2 DEFINITION OF $H^1(\Omega)$

Let Ω be an open set in $\mathbb{R}^n$. A function $f : \Omega \to \mathbb{C}$ is said to be in $H^1(\Omega)$ if $f \in L^2(\Omega)$ and if its distributional gradient, ∇f, is a function that is in $L^2(\Omega)$.

Recall from Chapter 6 that $\nabla f \in L^2(\Omega)$ means that there exist n functions $b_1, \ldots, b_n$ in $L^2(\Omega)$, collectively denoted by ∇f, such that for all ϕ in

$\mathcal{D}(\Omega)$

$$\int_\Omega f(x) \frac{\partial \phi}{\partial x_i}(x)\,dx = -\int_\Omega b_i(x)\phi(x)\,dx, \qquad i=1,\dots,n. \tag{1}$$

$H^1(\Omega)$ is a linear space since, with f_1, f_2 in $H^1(\Omega)$, the sum $f_1 + f_2$ is in $L^2(\Omega)$ and further, since in $\mathcal{D}'(\Omega)$

$$\nabla(f_1 + f_2) = \nabla f_1 + \nabla f_2,$$

the distributional gradient of $f_1 + f_2$ is an $L^2(\Omega)$-function. It is clear that for λ in $\mathbb{C}$ and f in $H^1(\Omega)$ the function λf is in $H^1(\Omega)$ too. $H^1(\Omega)$ can be endowed with the norm

$$\|f\|_{H^1(\Omega)} = \left(\int_\Omega |f(x)|^2\,dx + \int_\Omega |\nabla f(x)|^2\,dx \right)^{1/2}. \tag{2}$$

Obviously it is true that f is in $H^1(\Omega)$ if and only if $\|f\|_{H^1(\Omega)} < \infty$.

The last integral in (2), i.e., $\int_\Omega |\nabla f|^2$, is called the **kinetic energy** of f.

The next theorem and remark show that $H^1(\Omega)$ is, in fact, a Hilbert-space.

7.3 THEOREM (Completeness of $H^1(\Omega)$)

Let f^m be any Cauchy sequence in $H^1(\Omega)$, i.e.,

$$\|f^m - f^n\|_{H^1(\Omega)} \to 0 \quad as\ m, n \to \infty.$$

Then there exists a function $f \in H^1(\Omega)$ such that $\lim_{m\to\infty} f^m = f$ in $H^1(\Omega)$, i.e.,

$$\lim_{m\to\infty} \|f^m - f\|_{H^1(\Omega)} = 0.$$

PROOF. Since f^m is a Cauchy sequence in $H^1(\Omega)$, it is also a Cauchy sequence in $L^2(\Omega)$, which, by Theorem 2.7, is complete. Hence there exists a function $f \in L^2(\Omega)$ such that $\lim_{m\to\infty} \|f^m - f\|_{L^2(\Omega)} = 0$. In the same fashion we find functions $\mathbf{b} = (b_1, \dots, b_n) \in L^2(\Omega)$ such that $\lim_{m\to\infty} \|\nabla f^m - \mathbf{b}\|_{L^2(\Omega)} = 0$. We have to show that $\mathbf{b} = \nabla f$ in $\mathcal{D}'(\Omega)$. For any $\phi \in \mathcal{D}(\Omega)$

$$\int_\Omega \nabla\phi(x)f(x)\,\mathrm{d}x = \lim_{m\to\infty} \int_\Omega \nabla\phi(x)f^m(x)\,\mathrm{d}x,$$

which can be seen using the Schwarz inequality

$$\left|\int_\Omega \nabla\phi(x)(f(x)-f^m(x))\,\mathrm{d}x\right| \leq \|\nabla\phi\|_{L^2(\Omega)}\|f-f^m\|_{L^2(\Omega)},$$

where the right side tends to zero as $m \to \infty$. $\|\nabla\phi\|_{L^2(\Omega)}$ is finite since ϕ is in $\mathcal{D}(\Omega)$. In the same fashion it is established that

$$\int_\Omega \phi(x)\mathbf{b}(x)\,\mathrm{d}x = \lim_{m\to\infty} \int_\Omega \phi(x)\mathbf{b}^m(x)\,\mathrm{d}x.$$

Hence

$$\int_\Omega \nabla\phi(x)f(x)\,\mathrm{d}x = \lim_{m\to\infty} \int_\Omega \nabla\phi(x)f^m(x)\,\mathrm{d}x$$
$$:= -\lim_{m\to\infty}\int_\Omega \phi(x)\mathbf{b}^m(x)\,\mathrm{d}x = -\int_\Omega \phi(x)\mathbf{b}(x)\,\mathrm{d}x$$

where the middle equality holds because $f^m \in H^1(\Omega)$ for all m. ∎

REMARKS. (1) $H^1(\Omega)$ can be equipped with **an inner (or scalar) product**

$$(f,g)_{H^1(\mathbb{R}^n)} = \left(\int \overline{f(x)}g(x)\,\mathrm{d}x + \sum_i \int \overline{\frac{\partial f(x)}{\partial x_i}}\frac{\partial g(x)}{\partial x_i}\,\mathrm{d}x\right)$$

and thus becomes a Hilbert-space (thanks to Theorem 7.3).

(2) In Theorem 7.9 (Fourier characterization of $H^1(\mathbb{R}^n)$) we shall see that $H^1(\mathbb{R}^n)$ is really just an L^2-space on $\mathbb{R}^n$, but with a measure that differs from Lebesgue's. This fact, together with Theorem 2.7, yields an alternative proof of the completeness of $H^1(\mathbb{R}^n)$.

7.4 LEMMA (Multiplication by functions in $C^\infty(\Omega)$)

Let f be in $H^1(\Omega)$ and let ψ be a bounded function in $C^\infty(\Omega)$ with bounded derivatives. Then the pointwise product of ψ and f,

$$(\psi \cdot f)(x) = \psi(x)f(x),$$

is in $H^1(\Omega)$ and

$$\frac{\partial}{\partial x_i}(\psi \cdot f) = \frac{\partial \psi}{\partial x_i} \cdot f + \psi \cdot \frac{\partial f}{\partial x_i} \tag{1}$$

is in $\mathcal{D}'(\Omega)$.

PROOF. Recall that by the product rule 6.12(2), (1) above holds since ψ has bounded derivatives and the right side of (1) is in $L^2(\Omega)$. Therefore $\psi \cdot f$ is in $H^1(\Omega)$. ∎

7.5 REMARK ABOUT $H^1(\Omega)$ AND $W^{1,2}(\Omega)$

Our definition above of $H^1(\Omega)$ was called $W^{1,2}(\Omega)$ in Sect. 6.7 and in the literature (see [Adams], [Brézis], [Gilbarg–Trudinger], [Ziemer]). $H^1(\Omega)$ is normally defined differently as the completion of $C^\infty(\Omega)$ in the norm given by 7.2(2). That these two definitions are equivalent (and hence $H^1(\Omega) = W^{1,2}(\Omega)$) is the content of the following theorem.

7.6 THEOREM (Density of $C^\infty(\Omega)$ in $H^1(\Omega)$)

If f is in $H^1(\Omega)$, then there exists a sequence of functions f^m in $C^\infty(\Omega) \cap H^1(\Omega)$ such that

$$\|f - f^m\|_{H^1(\Omega)} \longrightarrow 0 \quad as\ m \to \infty. \tag{1}$$

Moreover, if $\Omega = \mathbb{R}^n$, then the functions f^m can be taken to be in $C_c^\infty(\mathbb{R}^n)$.

REMARKS. (1) This theorem is due to Meyers and Serrin [Meyers–Serrin] and a proof can also be found, e.g., in [Adams]. The analogous theorem holds for $W^{1,p}(\Omega)$, not just $W^{1,2}(\Omega)$. The proof for general open sets Ω is tricky because of difficulties caused by the boundary of Ω, which accounts for the fact that it took some time to identify the completion of $C^\infty(\Omega)$ in $W^{1,2}(\Omega)$ with $H^1(\Omega)$. Here we content ourselves with a proof for the case $\Omega = \mathbb{R}^n$.

(2) The density of $C_c^\infty(\mathbb{R}^n)$ in $H^1(\mathbb{R}^n)$ is useful because the test functions themselves can now be used to approximate functions in $H^1(\mathbb{R}^n)$.

(3) If $\Omega \neq \mathbb{R}^n$, then $C_c^\infty(\Omega) = \mathcal{D}(\Omega)$ is *not necessarily* dense in $H^1(\Omega)$. The completion of $C_c^\infty(\Omega)$ is a **subspace of** $H^1(\Omega)$ **called** $H_0^1(\Omega)$ and is the subspace one uses to discuss differential equations with 'zero boundary conditions' on $\partial\Omega$, the **boundary** of Ω.

PROOF OF THEOREM 7.6 FOR THE CASE $\Omega = \mathbb{R}^n$. Let $j : \mathbb{R}^n \to \mathbb{R}^+$ be in $C_c^\infty(\mathbb{R}^n)$ with $\int_{\mathbb{R}^n} j = 1$ and let $j_\varepsilon(x) := \varepsilon^{-n} j(x/\varepsilon)$ for $\varepsilon > 0$ as in Theorem 2.16. Then, since f and ∇f are $L^2(\mathbb{R}^n)$-functions, $f_\varepsilon := j_\varepsilon * f \to f$ and $g_\varepsilon := j_\varepsilon * \nabla f \to \nabla f$ strongly in $L^2(\mathbb{R}^n)$ as $\varepsilon \to 0$. Thus, we have that $f_\varepsilon \to f$ strongly in $H^1(\mathbb{R}^n)$ *provided* $g_\varepsilon = \nabla f_\varepsilon$. But this is true by 2.16(3), and Lemma 6.8(1).

The functions f_ε are in $C^\infty(\mathbb{R}^n)$ and our first goal, namely (1), is achieved by setting $\varepsilon = 1/m$. However, the f_ε do not necessarily have compact support and to achieve this we first take some function $k : \mathbb{R}^n \to [0, 1]$ in $C_c^\infty(\mathbb{R}^n)$ with $k(x) = 1$ for $|x| \le 1$. Then define $g^m(x) = k(x/m) f(x)$. By Lemma 7.4, g^m is in $H^1(\mathbb{R}^n)$. Furthermore g^m has compact support and

$$\|f - g^m\|_2 \le \int_{|x| \ge m} |f(x)|^2 \, \mathrm{d}x \to 0 \quad \text{as } m \to \infty$$

and

$$\|\nabla f - \nabla g^m\|_2^2 \le 2 \int_{|x| \ge m} |\nabla f|^2 \, \mathrm{d}x + \frac{C}{m^2} \int_{\mathbb{R}^n} |f(x)|^2 \, \mathrm{d}x \to 0 \quad \text{as } m \to \infty.$$

Thus, $g^m \to f$ strongly in $H^1(\mathbb{R}^n)$. Finally, we take

$$F^m(x) := k(x/m) f_{1/m}(x)$$

which is in $C_c^\infty(\mathbb{R}^n)$, and it is an easy exercise to prove that $F^m \to f$ strongly in $H^1(\mathbb{R}^n)$. ∎

7.7 THEOREM (Partial integration for functions in $H^1(\mathbb{R}^n)$)

Let u and v be in $H^1(\mathbb{R}^n)$. Then

$$\int_{\mathbb{R}^n} u \frac{\partial v}{\partial x_i} \, \mathrm{d}x = -\int_{\mathbb{R}^n} \frac{\partial u}{\partial x_i} v \, \mathrm{d}x \tag{1}$$

for $i = 1, \ldots, n$.

Suppose, in addition, that Δv is a function (which, by definition, is necessarily in $L^1_{\mathrm{loc}}(\mathbb{R}^n)$) and that v is real. If we assume that $u \Delta v \in L^1(\mathbb{R}^n)$, then

$$-\int_{\mathbb{R}^n} u \Delta v = \int_{\mathbb{R}^n} \nabla v \cdot \nabla u. \tag{2}$$

Alternatively, if we assume that Δv can be written as $\Delta v = f + g$ with $f \ge 0$ in $L^1_{\mathrm{loc}}(\mathbb{R}^n)$ and with g in $L^2(\mathbb{R}^n)$, then $u \Delta v \in L^1(\mathbb{R}^n)$ for all u in $H^1(\mathbb{R}^n)$, and hence (2) holds.

REMARKS. (1) The reader should note the distinction, in principle, between Δv as a function and Δv as a distribution. Here the distinction may appear to be pedantic, but at the end of Sect. 7.15, where $\sqrt{-\Delta}\, v$ is considered, this kind of distinction will be important.

(2) In general, $u\Delta v$ need not be in $L^1(\mathbb{R}^n)$. Here is an example in $\mathbb{R}^1$ due to K. Yajima: $v(x) = (1+x^2)^{-1}\cos(x^2)$ and $u(x) = (1+x^2)^{-1/2}$. Even if we assume $\Delta v \in L^1(\mathbb{R}^n)$, $u\Delta v$ need not be in $L^1(\mathbb{R}^n)$ for $n > 2$. Take $u = |x|^{-b}\exp[-x^2]$ and $v = \exp[i\,|x|^{-a} - x^2]$, with $a, b < (n-2)/2$ and with $2a + b \geq (n-2)$.

(3) Statement (2) is important in the study of the Schrödinger equation. There, we have a function $\psi \in H^1(\mathbb{R}^n)$ that solves Schrödinger's (time independent) equation

$$-\Delta\psi + V\psi = E\psi \quad \text{in } \mathcal{D}'(\mathbb{R}^n). \tag{3}$$

We shall want to multiply this equation by some $\phi \in H^1(\mathbb{R}^n)$ to obtain

$$\int_{\mathbb{R}^n} \nabla\phi \cdot \nabla\psi + \int_{\mathbb{R}^n} V\phi\psi = E \int_{\mathbb{R}^n} \phi\psi. \tag{4}$$

Equation (4) is correct with suitable assumptions on V, as will be seen in Sect. 11.9, and (2) is its justification.

PROOF. Notice that (1) makes sense since $u, v, \partial u/\partial x_i$ and $\partial v/\partial x_i$ are all in $L^2(\mathbb{R}^n)$. According to Theorem 7.6 there exists a sequence u^m in $C_c^\infty(\mathbb{R}^n)$ such that

$$\|u^m - u\|_{H^1(\mathbb{R}^n)} \to 0 \quad \text{as } m \to \infty. \tag{5}$$

Therefore, by the Schwarz inequality, we have

$$\left| \int_{\mathbb{R}^n} (u - u^m) \frac{\partial v}{\partial x_i} \, \mathrm{d}x \right| \leq \|u - u^m\|_2 \left\| \frac{\partial v}{\partial x_i} \right\|_2 \tag{6}$$

and

$$\left| \int_{\mathbb{R}^n} \left(\frac{\partial u}{\partial x_i} - \frac{\partial u^m}{\partial x_i} \right) v \, \mathrm{d}x \right| \leq \left\| \frac{\partial u}{\partial x_i} - \frac{\partial u^m}{\partial x_i} \right\|_2 \|v\|_2. \tag{7}$$

The right sides of both (6) and (7) tend to zero as $m \to \infty$ by (5). Hence

$$\int_{\mathbb{R}^n} u \frac{\partial v}{\partial x_i} \, \mathrm{d}x = \lim_{m\to\infty} \int_{\mathbb{R}^n} u^m \frac{\partial v}{\partial x_i} \, \mathrm{d}x$$

$$:= -\lim_{m\to\infty} \int_{\mathbb{R}^n} \frac{\partial u^m}{\partial x_i} v \, \mathrm{d}x = -\int_{\mathbb{R}^n} \frac{\partial u}{\partial x_i} v \, \mathrm{d}x,$$

using the fact that $u^m \in C_c^\infty(\mathbb{R}^n)$ for all m and the definition of the distributional derivative.

To prove (2), note first that the assumption $u\Delta v \in L^1(\mathbb{R}^n)$ implies that $(\operatorname{Re} u)_+(\Delta v)_+ \in L^1(\mathbb{R}^n)$, $(\operatorname{Re} u)_-(\Delta v)_+ \in L^1(\mathbb{R}^n)$, etc. By Corollary 6.18,

$(\operatorname{Re} u)_\pm$ are functions in $H^1(\mathbb{R}^n)$. Thus, it suffices to prove the theorem in the case in which u is real and nonnegative. Again, by Corollary 6.18, $f^j(x) := \min(u(x), j)$ is in $H^1(\mathbb{R}^n)$ and $f^j \to u$ in $H^1(\mathbb{R}^n)$. Pick $\phi \in C_c^\infty(\mathbb{R}^n)$ with ϕ radial and nonnegative and with $\phi(x) = 1$ for $|x| \leq 1$. By Lemma 7.4, the truncated functions $u^j(x) := \phi(x/j) f^j(x)$ are in $H^1(\mathbb{R}^n)$ and, as in Sect. 7.6, it follows that $u^j \to u$ monotonically in $H^1(\mathbb{R}^n)$. Clearly, $u^j(\Delta v)_\pm \leq u(\Delta v)_\pm \in L^1(\mathbb{R}^n)$ and hence

$$\lim_{j \to \infty} \int u^j (\Delta v)_\pm = \int u(\Delta v)_\pm$$

by dominated convergence. Thus, it suffices to prove (2) in the case in which u is bounded and has compact support. As in the proof of Theorem 7.6, we replace u by $u_\varepsilon := j_\varepsilon * u$ and note that $u_\varepsilon \in C_c^\infty(\mathbb{R}^n)$ is bounded, uniformly in ε, and its support lies in a fixed ball, independent of ε. Again, as in Sect. 7.6, $u_\varepsilon \to u$ in $H^1(\mathbb{R}^n)$ and, by Theorems 2.7 and 2.16, there exists a subsequence, which we again denote by u^k, such that $u^k \to u$ pointwise almost everywhere. Hence,

$$\int u \Delta v = \lim_{k \to \infty} \int u^k \Delta v = -\lim_{k \to \infty} \int \nabla u^k \cdot \nabla v = -\int \nabla u \cdot \nabla v.$$

The nonnegative truncated functions u^j can be used to prove the last assertion of the theorem. By the above, we know that $\int u^j \Delta v = \int \nabla u^j \cdot \nabla v$, since u^j is bounded and has bounded support. Clearly,

$$\int \nabla u^j \cdot \nabla v \to \int \nabla u \cdot \nabla v$$

and, by *monotone* convergence, $\int u^j f \to \int uf$. Likewise, $\int u^j g \to \int ug$. Since $\int ug < \infty$ (because $g \in L^2(\mathbb{R}^n)$), we must have that $\int uf < \infty$. Consequently, $u \Delta v \in L^1(\mathbb{R}^n)$, and thus (2) is proved. ∎

7.8 THEOREM (Convexity inequality for gradients)

Let f, g be real-valued functions in $H^1(\mathbb{R}^n)$. Then

$$\int_{\mathbb{R}^n} \left| \nabla \sqrt{f^2 + g^2} \right|^2 (x) \, dx \leq \int_{\mathbb{R}^n} \left(|\nabla f|^2(x) + |\nabla g|^2(x) \right) dx. \qquad (1)$$

If, moreover, $g(x) > 0$, then equality holds if and only if there exists a constant c such that $f(x) = c\, g(x)$ almost everywhere.

REMARKS. (1) $g > 0$ means, by definition, that for any compact $K \subset \mathbb{R}^n$ there is an $\varepsilon > 0$ such that the set $\{x \in K : g(x) < \varepsilon\}$ has measure zero.

(2) Inequality (1) is equivalent to
$$\int_{\mathbb{R}^n} |\nabla |F|(x)|^2 \leq \int_{\mathbb{R}^n} |\nabla F(x)|^2$$
for complex-valued functions.

PROOF. By Theorem 6.17 the function $\sqrt{f(x)^2 + g(x)^2}$ is in $H^1(\mathbb{R}^n)$ and
$$\left(\nabla \sqrt{f^2 + g^2}\right)(x) = \begin{cases} \dfrac{f(x)\nabla f(x) + g(x)\nabla g(x)}{\sqrt{f(x)^2 + g(x)^2}} & \text{if } f(x)^2 + g(x)^2 \neq 0, \\ 0 & \text{otherwise.} \end{cases}$$

Now, the following identity is obvious:
$$\int_{\mathbb{R}^n} \left|\nabla \sqrt{f^2 + g^2}\right|^2 dx + \int_{f^2 + g^2 > 0} \frac{|g\nabla f - f\nabla g|^2}{f^2 + g^2} dx \qquad (2)$$
$$= \int_{\mathbb{R}^n} (|\nabla f|^2 + |\nabla g|^2) \, dx,$$

from which (1) is immediate. Let us assume that $g > 0$ and that equality holds in (1). This means that
$$g(x)\nabla f(x) = f(x)\nabla g(x) \qquad (3)$$
a.e. on $\mathbb{R}^n$ by (2).

For ϕ in $C_c^\infty(\mathbb{R}^n)$ consider the function $h = \phi/g$. It is easy to see that h is in $H^1(\mathbb{R}^n)$ and that the following holds in $\mathcal{D}'(\mathbb{R}^n)$:
$$\nabla h(x) = -\frac{\nabla g(x)}{g(x)^2}\phi(x) + \frac{\nabla \phi(x)}{g(x)}. \qquad (4)$$

To prove this, one approximates $h(x)$ by
$$h_\delta(x) = \frac{\phi(x)}{\sqrt{g(x)^2 + \delta^2}}$$
and applies Theorem 6.16 and a simple limiting argument using the fact that $g > 0$. Thus $h_\delta \to h$ in $H^1(\mathbb{R}^n)$ as $\delta \to 0$.

Now
$$\int_{\mathbb{R}^n} f(x)\nabla h(x) \, dx = -\int_{\mathbb{R}^n} f(x)\frac{\nabla g(x)}{g(x)^2}\phi(x) \, dx + \int_{\mathbb{R}^n} \frac{f(x)}{g(x)}\nabla \phi(x) \, dx$$
$$= -\int_{\mathbb{R}^n} \nabla f(x) h(x) \, dx + \int_{\mathbb{R}^n} \frac{f(x)}{g(x)}\nabla \phi(x) \, dx,$$

since $f(x)\nabla g(x)/g(x)^2 = \nabla f(x)/g(x)$ almost everywhere, by (3).

By Theorem 7.7
$$\int_{\mathbb{R}^n} \nabla f(x) h(x)\,\mathrm{d}x = -\int_{\mathbb{R}^n} \nabla h(x) f(x)\,\mathrm{d}x,$$
from which we conclude that
$$\int_{\mathbb{R}^n} \frac{f(x)}{g(x)} \nabla \phi(x)\,\mathrm{d}x = 0.$$
Since $g > 0$, we conclude that $f(x)/g(x)$ is in $L^1_{\mathrm{loc}}(\mathbb{R}^n)$ and is therefore a distribution. Since ϕ is an arbitrary test function,
$$\nabla(f/g) = 0$$
in $\mathcal{D}'(\mathbb{R}^n)$ and, by Theorem 6.11, $f(x)/g(x)$ is constant almost everywhere. ∎

7.9 THEOREM (Fourier characterization of $H^1(\mathbb{R}^n)$)

Let f be in $L^2(\mathbb{R}^n)$ with Fourier transform $\widehat{f}$. Then f is in $H^1(\mathbb{R}^n)$ (i.e., the distributional gradient ∇f is an $L^2(\mathbb{R}^n)$ vector-valued function) if and only if the function $k \mapsto |k|\widehat{f}(k)$ is in $L^2(\mathbb{R}^n)$. If it is in $L^2(\mathbb{R}^n)$, then

$$\widehat{\nabla f}(k) = 2\pi i k \widehat{f}(k), \tag{1}$$

and therefore

$$\|f\|^2_{H^1(\mathbb{R}^n)} = \int |\widehat{f}(k)|^2 (1 + 4\pi^2 |k|^2)\,\mathrm{d}k. \tag{2}$$

PROOF. Suppose $f \in H^1(\mathbb{R}^n)$. By Theorem 7.6 there is a sequence f^m in $C_c^\infty(\mathbb{R}^n)$ for $m = 1, 2, \ldots$ that converges to f in $H^1(\mathbb{R}^n)$. Since $f^m \in C_c^\infty(\mathbb{R}^n)$, a simple integration by parts shows that $\widehat{\nabla f^m}(k) = 2\pi i k \widehat{f^m}(k)$. By Plancherel's Theorem 5.3, $\widehat{\nabla f^m}$ converges to $\widehat{\nabla f}$ and $\widehat{f^m}$ converges to $\widehat{f}$ in $L^2(\mathbb{R}^n)$. For a subsequence, we can also require that *both* of these convergences be pointwise. Therefore $k\widehat{f^m}(k) \to k\widehat{f}(k)$, pointwise a.e. Also, $2\pi i k \widehat{f^m}(k) \to \widehat{\nabla f}(k)$ pointwise a.e. Therefore, $\widehat{\nabla f}(k) = 2\pi i k \widehat{f}(k)$.

Now suppose $\widehat{h}(k) = 2\pi i k \widehat{f}(k)$ is in $L^2(\mathbb{R}^n)$. Let $h := (\widehat{h})^\vee$ and $\phi \in C_c^\infty(\mathbb{R}^n)$. Then

$$\int \nabla \phi \overline{f} = \int \widehat{\nabla \phi}\, \overline{\widehat{f}} = 2\pi i \int k \widehat{\phi}(k) \overline{\widehat{f}(k)}\,\mathrm{d}k = -\int \widehat{\phi}\overline{\widehat{h}} = -\int \phi \overline{h}. \tag{3}$$

The first and fourth equality is Parseval's formula 5.3(2). The second equality is the integration by parts formula for $\widehat{\nabla \phi}$ mentioned above. The distributional gradient of $\overline{f}$ is thus $\overline{h}$ (see 7.2(1)). ∎

Heat kernel.

Theorem 7.9 yields the following useful characterization of $\|\nabla f\|_2$ in 7.10(2). Define the heat kernel on $\mathbb{R}^n \times \mathbb{R}^n$ to be

$$e^{t\Delta}(x,y) = (4\pi t)^{-n/2} \exp\left\{-\frac{|x-y|^2}{4t}\right\}. \tag{4}$$

The action of the heat kernel on functions is, by definition,

$$(e^{t\Delta} f)(x) = \int_{\mathbb{R}^n} e^{t\Delta}(x,y) f(y)\, \mathrm{d}y. \tag{5}$$

If $f \in L^p(\mathbb{R}^n)$ with $1 \leq p \leq 2$, then, by Theorem 5.8,

$$\widehat{e^{t\Delta} f}(k) = \exp\{-4\pi^2 |k|^2 t\} \widehat{f}(k). \tag{6}$$

Equation (6) explains why the heat kernel is denoted by $e^{t\Delta}$. The action of Δ on Fourier transforms is multiplication by $-|2\pi k|^2$ (see Theorem 7.9), while the heat kernel is multiplication by $\exp[-t|2\pi k|^2]$.

From (4) and (5) it is evident that when $f \in L^p(\mathbb{R}^n)$ with $1 \leq p \leq \infty$, then the function g_t, defined by the right side of (5), is a differentiable function of t, in the sense that the limit

$$\frac{\mathrm{d}}{\mathrm{d}t} g_t := \lim_{\varepsilon \to 0} \varepsilon^{-1}[g_{t+\varepsilon} - g_t]$$

exists as a strong limit in $L^p(\mathbb{R}^n)$. This function satisfies the **heat equation**

$$\Delta g_t = \frac{\mathrm{d}}{\mathrm{d}t} g_t \tag{7}$$

in $\mathcal{D}'(\mathbb{R}^n)$ for $t > 0$, and with the 'initial condition'

$$\lim_{t \downarrow 0} g_t = f. \tag{8}$$

The heat equation is a model for heat conduction and g_t is the temperature distribution (as a function of $x \in \mathbb{R}^n$) at time t. The kernel, given by (4), satisfies (7) for each fixed $y \in \mathbb{R}^n$ (as can be verified by explicit calculation) and satisfies the initial condition

$$\lim_{t \downarrow 0} e^{t\Delta}(\,\cdot\,, y) = \delta_y \quad \text{in } \mathcal{D}'(\mathbb{R}^n). \tag{9}$$

7.10 THEOREM ($-\Delta$ is the infinitesimal generator of the heat kernel)

A function f is in $H^1(\mathbb{R}^n)$ if and only if it is in $L^2(\mathbb{R}^n)$ and

$$I^t(f) := \frac{1}{t}[(f,f) - (f, e^{t\Delta}f)] \tag{1}$$

is uniformly bounded in t. (Here $(\cdot,\cdot)$ is the L^2, not the H^1, inner product.) In that case

$$\sup_{t>0} I^t(f) = \lim_{t\to 0} I^t(f) = \|\nabla f\|_2^2. \tag{2}$$

PROOF. By Theorem 7.9 it is sufficient to show that $f \in L^2(\mathbb{R}^n)$ and $I^t(f)$ is uniformly bounded in t if and only if

$$\int_{\mathbb{R}^n} (1 + |2\pi k|^2)|\widehat{f}(k)|^2 \, dk < \infty. \tag{3}$$

Note that by Plancherel's Theorem 5.3

$$I^t(f) = \frac{1}{t} \int_{\mathbb{R}^n} [1 - \exp\{-4\pi^2|k|^2 t\}]|\widehat{f}(k)|^2 \, dk. \tag{4}$$

It is easy to check that $y^{-1}(1 - e^{-y})$ is a decreasing function of $y > 0$, and hence $1/t$ times the factor [ ] in (4) converges monotonically to $|2\pi k|^2$ as $t \to 0$. Thus if $f \in H^1(\mathbb{R}^n)$, $I^t(f)$ is uniformly bounded. Conversely if $I^t(f)$ is uniformly bounded, Theorem 1.6 (monotone convergence) implies that $\sup_{t>0} I^t(f) = \lim_{t\to 0} I^t(f) = \int_{\mathbb{R}^n} |2\pi k|^2 |\widehat{f}(k)|^2 \, dk < \infty$. By Theorem 7.9, $f \in H^1(\mathbb{R}^n)$. ∎

Theorem 7.10 motivates the following.

7.11 DEFINITION OF $H^{1/2}(\mathbb{R}^n)$

An $L^2(\mathbb{R}^n)$-function f is said to be in $H^{1/2}(\mathbb{R}^n)$ if and only if

$$\|f\|^2_{H^{1/2}(\mathbb{R}^n)} := \int_{\mathbb{R}^n} (1 + 2\pi|k|)|\widehat{f}(k)|^2 \, dk < \infty. \tag{1}$$

By combining (1) with 7.9(2) we have that

$$\frac{3}{2}\|f\|^2_{H^1(\mathbb{R}^n)} \geq \|f\|^2_{H^{1/2}(\mathbb{R}^n)} \tag{2}$$

since $2\pi|k| \leq \frac{1}{2}[(2\pi|k|)^2 + 1]$. This, in turn, leads to the basic fact of inclusion:
$$H^{1/2}(\mathbb{R}^n) \supset H^1(\mathbb{R}^n). \tag{3}$$

The space $H^{1/2}(\mathbb{R}^n)$ endowed with the inner product
$$(f,g)_{H^{1/2}(\mathbb{R}^n)} = \int_{\mathbb{R}^n} (1 + 2\pi|k|)\overline{\widehat{f}(k)}\widehat{g}(k)\,\mathrm{d}k \tag{4}$$

is easily seen to be a Hilbert-space. (The completeness proof is the same as for the usual $L^2(\mathbb{R}^n)$ space except that the measure is now $(1 + 2\pi|k|)\,\mathrm{d}k$ instead of $\mathrm{d}k$.) $H^{1/2}(\mathbb{R}^n)$ is important for relativistic systems, in which one considers three-dimensional 'kinetic energy' operators of the form
$$\sqrt{p^2 + m^2} \tag{5}$$

where p^2 is the physicist's notation for $-\Delta$, and m is the mass of the particle under consideration. The operator (5) is *defined* in Fourier space as multiplication by $\sqrt{|2\pi k|^2 + m^2}$, i.e.,
$$\left(\sqrt{p^2 + m^2}\,f\right)^{\widehat{}}(k) := \sqrt{|2\pi k|^2 + m^2}\,\widehat{f}(k). \tag{6}$$

The right side of (6) is the Fourier transform of an $L^2(\mathbb{R}^3)$-function (and thus $\sqrt{p^2 + m^2}$ makes sense as an operator on functions) *provided* $f \in H^1(\mathbb{R}^3)$ (*not* $H^{1/2}(\mathbb{R}^3)$). However, as in the case of $p^2 = -\Delta$, we are more interested in the energy, which is the quadratic form
$$\left(g, \sqrt{p^2 + m^2}\,f\right) := \int_{\mathbb{R}^3} \overline{\widehat{g}}(k)\widehat{f}(k)\sqrt{|2\pi k|^2 + m^2}\,\mathrm{d}k, \tag{7}$$

and this makes sense if f and g are in $H^{1/2}(\mathbb{R}^3)$.

The quadratic form $(g, |p|f)$ is defined by setting $m = 0$ in (7).

Note the inequalities
$$\sqrt{\sum_{i=1}^{N} A_i} \leq \sum_{i=1}^{N} \sqrt{A_i} \leq \sqrt{N}\sqrt{\sum_{i=1}^{N} A_i}$$

which hold for any positive numbers A_i. Consequently, a function f is in $H^{1/2}(\mathbb{R}^{3N})$ if and only if
$$\int_{\mathbb{R}^{3N}} |\widehat{f}(k_1, \ldots, k_N)|^2 \left(1 + \sum_{i=1}^{N} 2\pi|k_i|\right) \mathrm{d}k_1 \cdots \mathrm{d}k_N$$

is finite. This fact is always used when dealing with the relativistic many-body problem, i.e., the obvious requirement that the above integral be finite is no different from requiring that $f \in H^{1/2}(\mathbb{R}^{3N})$.

We wish now to derive analogues of Theorems 7.9 and 7.10 for $|p|$ in place of $|p|^2$. First, the analogue of the kernel $e^{t\Delta} = e^{-tp^2}$ is needed. This is the **Poisson kernel** [Stein–Weiss]

$$e^{-t|p|}(x,y) := (e^{-t|p|})^{\vee}(x-y) = \int_{\mathbb{R}^n} \exp[-2\pi|k|t + 2\pi i k \cdot (x-y)]\, dk. \quad (8)$$

This integral can be computed easily in three dimensions because the angular integration gives $4\pi|k|^{-1}|x-y|^{-1}\sin(|k||x-y|)$ and then the $|k|^2\, d|k|$ integration is just the integral of $|k|$ times an exponential function. The three-dimensional result is

$$e^{-t|p|}(x,y) = \frac{1}{\pi^2}\frac{t}{[t^2 + |x-y|^2]^2}, \qquad n = 3. \quad (9)$$

Remarkably, (8) can also be evaluated in n dimensions. The result is

$$e^{-t|p|}(x,y) = \Gamma\left(\frac{n+1}{2}\right) \pi^{-(n+1)/2} \frac{t}{[t^2 + |x-y|^2]^{(n+1)/2}}. \quad (10)$$

It can be found in [Stein–Weiss, Theorem 1.14], for example.

Another remarkable fact is that the kernel of $\exp\left\{-t\sqrt{p^2 + m^2}\right\}$ can also be computed explicitly in three dimensions. The result is [Erdelyi–Magnus–Oberhettinger–Tricomi]

$$e^{-t\sqrt{p^2+m^2}}(x,y) = \frac{m^2}{2\pi^2}\frac{t}{t^2 + |x-y|^2} K_2(m[|x-y|^2 + t^2]^{1/2}), \qquad n = 3, \quad (11)$$

where K_2 is the modified Bessel function of the third kind.

The kernels (9), (10) and (11) are positive, $L^1(\mathbb{R}^n)$-functions of $(x-y)$ and so, by Theorem 4.2 (Young's inequality), they map $L^p(\mathbb{R}^n)$ into $L^p(\mathbb{R}^n)$ for all $p \geq 1$ by integration, as in 7.9(5). In fact, for all n,

$$\int e^{-t\sqrt{p^2+m^2}}(x,y)\, dy = e^{-tm}, \quad (12)$$

since the left side of (12) is just the inverse Fourier transform of (11) evaluated at $k = 0$.

The analogues of Theorems 7.8 and 7.9 can now be stated.

7.12 THEOREM (Integral formulas for $(f, |p|f)$ and $(f, \sqrt{p^2 + m^2}\, f)$)

(i) *A function f is in $H^{1/2}(\mathbb{R}^n)$ if and only if it is in $L^2(\mathbb{R}^n)$ and*

$$I^t_{1/2}(f) = \lim_{t \to 0} \frac{1}{t}[(f,f) - (f, e^{-t|p|}f)] \tag{1}$$

is uniformly bounded, in which case

$$\sup_{t>0} I^t_{1/2}(f) = \lim_{t \to 0} I^t_{1/2}(f) = (f, |p|f). \tag{2}$$

(ii) *The formula (in which $(\cdot, \cdot)$ is the L^2 inner product)*

$$\frac{1}{t}[(f,f) - (f, e^{-t|p|}f)] = \frac{\Gamma\left(\frac{n+1}{2}\right)}{2\pi^{(n+1)/2}} \int_{\mathbb{R}^n} \int_{\mathbb{R}^n} \frac{|f(x) - f(y)|^2}{(t^2 + (x-y)^2)^{(n+1)/2}}\, dx\, dy \tag{3}$$

holds, which leads to

$$(f, |p|f) = \frac{\Gamma\left(\frac{n+1}{2}\right)}{2\pi^{(n+1)/2}} \int_{\mathbb{R}^n} \int_{\mathbb{R}^n} \frac{|f(x) - f(y)|^2}{|x-y|^{n+1}}\, dx\, dy. \tag{4}$$

(iii) *Assertion (i) holds with $|p|$ replaced by $\sqrt{p^2 + m^2}$ in (1) and (2), for any $m \geq 0$.*

(iv) *If $n = 3$, the analogue of (4) is*

$$(f, [\sqrt{p^2 + m^2} - m]f) = \frac{m^2}{4\pi^2} \int_{\mathbb{R}^3} \int_{\mathbb{R}^3} \frac{|f(x) - f(y)|^2}{|x-y|^2} K_2(m|x-y|)\, dx\, dy. \tag{5}$$

REMARK. Since $|a - b| \geq ||a| - |b||$ for all complex numbers a and b, (4) tells us that

$$f \in H^{1/2}(\mathbb{R}^n) \quad \text{implies} \quad |f| \in H^{1/2}(\mathbb{R}^n). \tag{6}$$

PROOF. The proofs of (i) and (iii) are virtually the same as for Theorem 7.10. Equation (3) is just a restatement of (1) obtained by using 7.11(8) and 7.11(10) with $m = 0$. Equation (4) is obtained from (3) by using (2) and monotone convergence. Equation (5) is derived similarly since K_2 is a monotone function. Equation 7.11(12) is used in (5). ∎

7.13 THEOREM (Convexity inequality for the relativistic kinetic energy)

Let f and g be real-valued functions in $H^{1/2}(\mathbb{R}^3)$ with $f \not\equiv 0$. Then, with $T(p) = \sqrt{p^2 + m^2} - m$, and $m \geq 0$, we have that

$$\left(\sqrt{f^2 + g^2},\, T(p)\sqrt{f^2 + g^2}\right) \leq (f, T(p)f) + (g, T(p)g). \tag{1}$$

Equality holds if and only if f has a definite sign and $g = cf$ a.e. for some constant c.

PROOF. Using formula 7.12(5) and the fact that K_2 is strictly positive, the Schwarz inequality

$$f(x)f(y) + g(x)g(y) \leq \sqrt{f(x)^2 + g(x)^2}\sqrt{f(y)^2 + g(y)^2} \tag{2}$$

yields (1). To discuss the cases of equality we square both sides of (2) to see that equality amounts to

$$f(x)g(y) = f(y)g(x) \tag{3}$$

for almost every point (x, y) in $\mathbb{R}^6$. By Fubini's theorem, for almost every $y \in \mathbb{R}^3$ equation (3) must hold for almost every x in $\mathbb{R}^3$. Picking y_0 such that $f(y_0) \neq 0$ equation (3) shows that $g(x) = \lambda f(x)$ for the constant $\lambda = g(y_0)/f(y_0)$. Inserting this back into (2) (with equality sign) we see that f must have a definite sign. ∎

• We continue this chapter by stating the analogue of Theorem 7.6 for $H^{1/2}(\mathbb{R}^n)$.

7.14 THEOREM (Density of $C_c^\infty(\mathbb{R}^n)$ in $H^{1/2}(\mathbb{R}^n)$)

If f is in $H^{1/2}(\mathbb{R}^n)$, then there exists a sequence of functions f^m in $C_c^\infty(\mathbb{R}^n)$ such that

$$\|f - f^m\|_{H^{1/2}(\mathbb{R}^n)} \to 0 \quad as\ m \to \infty.$$

PROOF. On account of Theorem 7.6 it suffices to show that $H^1(\mathbb{R}^n) \subset H^{1/2}(\mathbb{R}^n)$ densely and that the embedding is continuous (i.e., $\|f^m - f\|_{H^1(\mathbb{R}^n)} \to 0$ implies $\|f_m - f\|_{H^{1/2}(\mathbb{R}^n)} \to 0$). By definition, $f \in H^{1/2}(\mathbb{R}^n)$ iff $\widehat{f}$ satisfies

$$\int_{\mathbb{R}^n} (1 + 2\pi|k|)|\widehat{f}(k)|^2\, dk < \infty.$$

Pick $\widehat{f^m}(k) = e^{-k^2/m}\widehat{f}(k)$ and note that, by Theorem 7.9, $f^m \in H^1(\mathbb{R}^n)$. But, by dominated convergence,

$$\|f - f^m\|^2_{H^{1/2}(\mathbb{R}^n)}$$
$$= \int_{\mathbb{R}^n} (1 + 2\pi|k|)(1 - e^{-k^2/m})|\widehat{f}(k)|^2 \, dk \to 0 \quad \text{as } m \to \infty. \quad \blacksquare$$

7.15 ACTION OF $\sqrt{-\Delta}$ AND $\sqrt{-\Delta + m^2} - m$ ON DISTRIBUTIONS

If T is a distribution, then it has derivatives, and thus it makes sense to talk about ΔT. It is a bit more difficult to make sense of $\sqrt{-\Delta}\, T$ since, by definition, $\sqrt{-\Delta}\, T$ would be given by

$$\sqrt{-\Delta}\, T(\phi) := T(\sqrt{-\Delta}\, \phi) \qquad (1)$$

for ϕ in $\mathcal{D}(\mathbb{R}^n)$. This is not possible, however, since the function

$$(\sqrt{-\Delta}\, \phi)(x) = \int_{\mathbb{R}^n} e^{+2\pi i k \cdot x} |2\pi k| \widehat{\phi}(k) \, dk$$

is *not generally in* $C_c^\infty(\mathbb{R}^n)$. Therefore (1) does *not* define a distribution. As an amusing aside, note that the convolution

$$(\sqrt{-\Delta}\, \phi) * (\sqrt{-\Delta}\, \phi) = -\Delta(\phi * \phi)$$

is always in $C_c^\infty(\mathbb{R}^n)$ when ϕ is.

If T is a *suitable function*, however, then (1) does make sense. More precisely, *if $f \in H^{1/2}(\mathbb{R}^n)$, then $\sqrt{-\Delta}\, f$ (and $(\sqrt{-\Delta + m^2} - m)f$) are both distributions*, i.e., the mapping

$$\phi \mapsto \sqrt{-\Delta}\, f(\phi) := \int_{\mathbb{R}^n} |2\pi k|\widehat{f}(k)\widehat{\phi}(-k) \, dk$$

makes sense (since $\sqrt{|k|}\,\widehat{f} \in L^2(\mathbb{R}^n)$ by definition), and we assert that it *is continuous in* $\mathcal{D}(\mathbb{R}^n)$. To see this, consider a sequence $\phi^j \to \phi$ in $\mathcal{D}(\mathbb{R}^n)$. By the Schwarz inequality and Theorem 5.3

$$|\sqrt{-\Delta}\, f(\phi - \phi^j)| \le \|\widehat{f}\|_2 \left(\int_{\mathbb{R}^n} |2\pi k|^2 |\widehat{\phi}(k) - \widehat{\phi^j}(k)|^2 \, dk \right)^{1/2}$$
$$= \|f\|_2 \|\nabla(\phi - \phi^j)\|_2.$$

But $\|\nabla(\phi - \phi^j)\|_2 \to 0$ as $j \to \infty$; hence $\sqrt{-\Delta}\, f(\phi - \phi^j) \to 0$ as $j \to \infty$ and thus $\sqrt{-\Delta}\, f$ is in $\mathcal{D}'(\mathbb{R}^n)$. A similar definition and proof applies to $(\sqrt{-\Delta + m^2} - m)f$.

An important consequence of this discussion is the analogue of 7.7(3): *the modified 'Schrödinger' equation*

$$\sqrt{-\Delta}\,\psi + V\psi = E\psi \quad \text{in } \mathcal{D}'(\mathbb{R}^n) \tag{2}$$

makes sense when ψ is in $H^{1/2}(\mathbb{R}^n)$. (Alternatively, we can have $n = 3N$ and replace the first term by $\sum_{i=1}^{N} \sqrt{-\Delta_i}\,\psi$.)

Another important fact is that the density of $C_c^\infty(\mathbb{R}^n)$ in $H^{1/2}(\mathbb{R}^n)$ (Theorem 7.14) allows us to imitate the proof of 7.7(2) and conclude that

$$\int_{\mathbb{R}^n} u\sqrt{-\Delta}\, v = \int_{\mathbb{R}^n} |2\pi k|\widehat{v}(k)\widehat{u}(-k)\,\mathrm{d}k \tag{3}$$

when $\sqrt{-\Delta}\,v \in L^1_{\text{loc}}(\mathbb{R}^n)$ and $u\sqrt{-\Delta}\,v \in L^1(\mathbb{R}^n)$. The latter condition is guaranteed by the condition $\sqrt{-\Delta}\,v = f + g$ with $f \geq 0$ in $L^1_{\text{loc}}(\mathbb{R}^n)$ and with g in $L^2(\mathbb{R}^n)$. See Exercise 3.

7.16 THEOREM (Multiplication of $H^{1/2}$-functions by C^∞-functions)

Let ψ be a bounded function in $C^\infty(\mathbb{R}^n)$ with bounded derivatives, and let f be in $H^{1/2}(\mathbb{R}^n)$. Then the pointwise product of ψ and f,

$$(\psi \cdot f)(x) = \psi(x)f(x),$$

is also a function in $H^{1/2}(\mathbb{R}^n)$.

PROOF. It is obvious that $\psi \cdot f$ is in $L^2(\mathbb{R}^n)$. Using 7.12(4), it remains to show that

$$I := \int_{\mathbb{R}^n}\int_{\mathbb{R}^n} |(\psi \cdot f)(x) - (\psi \cdot f)(y)|^2 |x - y|^{-n-1}\, \mathrm{d}x\, \mathrm{d}y \tag{1}$$

is finite. Using

$$|ab - cd|^2 = \frac{1}{4}|(a-c)(b+d) + (a+c)(b-d)|^2$$
$$\leq |a-c|^2(|b|^2 + |d|^2) + (|a|^2 + |c|^2)|b-d|^2,$$

we have that

$$\frac{1}{2}I \leq \int_{\mathbb{R}^n}\int_{\mathbb{R}^n} |\psi(x)-\psi(y)|^2 |f(x)|^2 |x-y|^{-n-1}\,dx\,dy$$
$$+ \int_{\mathbb{R}^n}\int_{\mathbb{R}^n} |f(x)-f(y)|^2 |\psi(y)|^2 |x-y|^{-n-1}\,dx\,dy. \qquad (2)$$

Since ψ is uniformly bounded, the second term in (2) is bounded by a constant times $(f,|p|f)$. The first term is easily estimated by considering the regions in $\mathbb{R}^n \times \mathbb{R}^n$ where $|x-y| \leq 1$ and where $|x-y| > 1$ separately. In the former we use the estimate $|\psi(x)-\psi(y)|^2 \leq C|x-y|^2$ for some constant C (since ψ is differentiable with uniformly bounded derivative) to find the bound

$$C\int_{|x-y|\leq 1} |x-y|^{-n+1}|f(x)|^2\,dx\,dy = C\int_{|x|\leq 1} |x|^{-n+1}\,dx\,\|f\|_2^2.$$

In the other region we use the fact that ψ is uniformly bounded to get the estimate

$$C\int_{|x-y|\geq 1} |x-y|^{-n-1}|f(x)|^2\,dx\,dy = C\int_{|x|\geq 1} |x|^{-n-1}\,dx\,\|f\|_2^2,$$

which proves the theorem. ∎

• Next, we give one of the most important applications of the concept of symmetric-decreasing rearrangement expounded in Chapter 3.

7.17 LEMMA (Symmetric decreasing rearrangement decreases kinetic energy)

Let $f : \mathbb{R}^n \to \mathbb{R}$ be a nonnegative measurable function that vanishes at infinity (cf. 3.2) and let f^* denote its symmetric-decreasing rearrangement (cf. 3.3). Assume that ∇f, in the sense of distributions, is a function that satisfies $\|\nabla f\|_2 < \infty$. Then ∇f^* has the same property and

$$\|\nabla f\|_2 \geq \|\nabla f^*\|_2. \qquad (1)$$

Likewise if $(f,|p|f) < \infty$, then

$$(f,|p|f) \geq (f^*,|p|f^*). \qquad (2)$$

Note that it is not assumed that $f \in L^2(\mathbb{R}^n)$. The inequality in (2) is strict unless f is the translate of a symmetric-decreasing function.

REMARKS. (1) To define $(f, |p|f)$ for functions that are not in $L^2(\mathbb{R}^n)$, we use the right side of 7.12(4), which is always well defined (even if it is infinite).

(2) Equality can occur in (1) without $f = f^*$. However, the level sets, $\{x : f(x) > a\}$, must be balls [Brothers–Ziemer].

(3) Inequality (2) and its proof easily extend to $\sqrt{p^2 + m^2}$.

(4) It is also true, but much more difficult to prove, that (1) extends to gradients that are in $L^p(\mathbb{R}^n)$ instead of $L^2(\mathbb{R}^n)$, namely $\|\nabla f\|_p \geq \|\nabla f^*\|_p$, for $1 \leq p < \infty$ ([Hilden], [Sperner], [Talenti]), and to other integrals of the form $\int \Psi(|\nabla f|)$ for suitable convex functions Ψ (cf. [Almgren–Lieb], p. 698). Part of the assertion is that when $\Psi(|\nabla f|)$ is integrable, then ∇f^* is also a function and $\Psi(|\nabla f^*|)$ is integrable.

PROOF $\boxed{\text{PART 1, REDUCTION TO } L^2.}$ First we show that it suffices to prove (1) and (2) for functions in $L^2(\mathbb{R}^n)$. For any f satisfying the assumptions of our theorem we define

$$f_c(x) = \min[\max(f(x) - c, 0), 1/c]$$

for $c > 0$. It follows from the definition of the rearrangement that $(f_c)^* = (f^*)_c$. Since f vanishes at infinity, f_c is in $L^2(\mathbb{R}^n)$. By Theorem 6.19, $\nabla f_c(x) = 0$ except for those $x \in \mathbb{R}^n$ with $c < f(x) < 1/c + c$, and $\nabla f_c(x)$ equals $\nabla f(x)$ for such x. Thus, by the monotone convergence theorem $\lim_{c \to 0} \|\nabla f_c\|_2 = \|\nabla f\|_2$. Likewise,

$$\lim_{c \to 0} \|\nabla (f_c)^*\|_2 = \lim_{c \to 0} \|\nabla (f^*)_c\|_2 = \|\nabla f^*\|_2.$$

To verify the analogous statement for $(f, |p|f)$ we use that, by definition,

$$(f, |p|f) = \text{const.} \int \int |f(x) - f(y)|^2 / |x - y|^{n+1} \, dx \, dy$$

together with the fact (which follows easily from the definition of f_c) that $|f_c(x) - f_c(y)| \leq |f(x) - f(y)|$ for all $x, y \in \mathbb{R}^n$. Again by monotone convergence we have that $\lim_{c \to 0} (f_c, |p|f_c) = (f, |p|f)$ and the same holds for f^*, as above.

Thus, we have shown that it suffices to prove the theorem for f_c, which is a function in $H^1(\mathbb{R}^n)$, respectively $H^{1/2}(\mathbb{R}^n)$.

PART 2, PROOF FOR L^2. Inequality (1) is now a consequence of formula 7.10(1). Indeed, for $f \in H^1(\mathbb{R}^n)$ we have $\|\nabla f\|_2 = \lim_{t\to 0} I^t(f)$, where

$$I^t(f) = t^{-1}[(f,f) - (f, e^{\Delta t}f)].$$

The $L^2(\mathbb{R}^n)$ norm of f does not change under rearrangements and the second term increases by Theorem 3.7 (Riesz's rearrangement inequality). Thus $I^t(f^*) \leq I^t(f)$ and by Theorem 7.10 $I^t(f^*)$ converges to $\|\nabla f^*\|_2$.

Inequality (2) is a consequence of Theorem 7.12(4). We write the kernel $K(x-y) = |x-y|^{-n-1}$ as

$$K(x-y) = K_+(x-y) + K_-(x-y)$$

with

$$K_-(x) := (1+|x|^2)^{-(n+1)/2}.$$

It is easy to check that both K_+ and K_- are symmetric decreasing and K_- is strictly decreasing. Let $I_-(f)$ denote the integral in 7.12(4) with K replaced by K_-, and similarly for K_+. Since K_- is in $L^1(\mathbb{R}^n)$, $I_-(f)$ is the difference of two finite integrals. In the first $|f(x) - f(y)|^2$ is replaced by $2|f(x)|^2$ and in the second by $2f(x)f(y)$. The first does not change if f is replaced by f^* while the second strictly increases unless f is a translate of f^* by Theorem 3.9 (strict rearrangement inequality).

This proves the theorem if we can show that $I_+(f) \geq I_+(f^*)$. To do this we cut off K_+ at a large height c, i.e., $K_+^c(x) = \min(K_+(x), c)$. Since $K_+^c \in L^1(\mathbb{R}^n)$, the previous argument for K_- gives the desired result for K_+^c. The rest follows by monotone convergence as $c \to \infty$. ∎

7.18 WEAK LIMITS

As a final general remark about $H^1(\mathbb{R}^n)$ and $H^{1/2}(R^n)$ we mention the generalizations of the Banach–Alaoglu Theorem 2.18 (bounded sequences have weak limits), Theorem 2.11 (lower semicontinuity of norms) and Theorem 2.12 (uniform boundedness principle). To do so we first require knowledge of the dual spaces—which is easy to do given the Fourier characterization of the norms, 7.9(2) and 7.11(1). These formulas show that $H^1(\mathbb{R}^n)$ and $H^{1/2}(R^n)$ are just $L^2(\mathbb{R}^n, \mathrm{d}\mu)$ with

$$\mu(\mathrm{d}x) = (1 + 4\pi^2|x|^2)\,\mathrm{d}x \quad \text{for } H^1(\mathbb{R}^n)$$

and

$$\mu(\mathrm{d}x) = (1 + 2\pi|x|)\,\mathrm{d}x \quad \text{for } H^{1/2}(\mathbb{R}^n).$$

Thus, *a sequence f^j converges weakly to f in $H^1(\mathbb{R}^n)$ means that as $j \to \infty$*

$$\int_{\mathbb{R}^n} \left[\widehat{f^j}(k) - \widehat{f}(k)\right] \widehat{g}(k)(1 + 4\pi^2 |k|^2) \, dk \to 0 \tag{1}$$

for every $g \in H^1(\mathbb{R}^n)$. Similarly, for $H^{1/2}(\mathbb{R}^n)$,

$$\int_{\mathbb{R}^n} [\widehat{f^j}(k) - \widehat{f}(k)] \widehat{g}(k)(1 + 2\pi|k|) \, dk \to 0$$

for every $g \in H^{1/2}(\mathbb{R}^n)$.

The validity of Theorems 2.11, 2.12 and 2.18 for $H^1(\mathbb{R}^n)$ and $H^{1/2}(\mathbb{R}^n)$ are then immediate consequences of those three theorems applied to the case $p = 2$.

• The following topics, 7.19 onwards, can certainly be omitted on a first reading. They are here for two reasons: (a) As an exercise in manipulating some of the techniques developed in the previous parts of this chapter; (b) Because they are technically useful in quantum mechanics.

7.19 MAGNETIC FIELDS: THE H_A^1-SPACES

In differential geometry it is often necessary to consider **connections**, which are more complicated derivatives than ∇. The simplest example is a connection on a 'U(1) bundle' over $\mathbb{R}^n$, which merely means acting on complex-valued functions f by $(\nabla + iA(x))$, with $A(x) : \mathbb{R}^n \to \mathbb{R}^n$ being some preassigned, real vector field. The same operator occurs in the quantum mechanics of particles in external magnetic fields (with $n = 3$). The introduction of a magnetic field $B : \mathbb{R}^3 \to \mathbb{R}^3$ in quantum mechanics involves replacing ∇ by $\nabla + iA(x)$ (in appropriate units). Here A is called a **vector potential** and satisfies

$$\operatorname{curl} A = B.$$

In general A is not a bounded vector field, e.g., if B is the constant magnetic field (0,0,1), then a suitable vector potential A is given by $A(x) = (-x_2, 0, 0)$. Unlike in the differential geometric setting, A need not be smooth either, because we could add an arbitrary gradient to A, $A \to A + \nabla \chi$, and still get the same magnetic field B. This is called **gauge invariance**. The problem is that χ (and hence A) could be a wild function—even if B is well behaved.

For these reasons we want to find a large class of A's for which we can make (distributional) sense of $(\nabla + iA(x))$ and $(\nabla + iA(x))^2$ when acting on a suitable class of $L^2(\mathbb{R}^3)$-functions. It used to be customary to restrict attention to A's with components in $C^1(\mathbb{R}^3)$ but that is unnecessarily restrictive, as shown in [Simon] (see also [Leinfelder–Simader]).

For general dimension n, the appropriate condition on A, which we assume henceforth, is

$$A_j \in L^2_{\text{loc}}(\mathbb{R}^n) \quad \text{for } j = 1, \ldots, n. \tag{1}$$

Because of this condition the functions $A_j f$ are in $L^1_{\text{loc}}(\mathbb{R}^n)$ for every $f \in L^2_{\text{loc}}(\mathbb{R}^n)$. Therefore the expression

$$(\nabla + iA)f,$$

called **the covariant derivative** (with respect to A) of f, is a distribution for every $f \in L^2_{\text{loc}}(\mathbb{R}^n)$.

7.20 DEFINITION OF $H^1_A(\mathbb{R}^n)$

For each $A : \mathbb{R}^n \to \mathbb{R}^n$ satisfying 7.19(1), the space $H^1_A(\mathbb{R}^n)$ consists of all functions $f : \mathbb{R}^n \to \mathbb{C}$ such that

$$f \in L^2(\mathbb{R}^n) \quad \text{and} \quad (\partial_j + iA_j)f \in L^2(\mathbb{R}^n) \quad \text{for } j = 1, \ldots, n. \tag{1}$$

We do *not assume* that ∇f or Af are separately in $L^2(\mathbb{R}^n)$ (but (1) does imply that $\partial_j f$ is an $L^1_{\text{loc}}(\mathbb{R}^n)$-function).

The inner product in this space is

$$(f_1, f_2)_A = (f_1, f_2) + \sum_{j=1}^n ((\partial_j + iA_j)f_1, (\partial_j + iA_j)f_2), \tag{2}$$

where $(\cdot, \cdot)$ is the usual $L^2(\mathbb{R}^n)$ inner product. The second term on the right side of (2), in the case that $f_1 = f_2 = f$, is called the **kinetic energy** of f. It is to be compared to the usual kinetic energy $\|\nabla f\|_2^2$.

As in the case of $H^1(\mathbb{R}^n)$ (see 7.3), $H^1_A(\mathbb{R}^n)$ is complete, and thus is a Hilbert-space. If f^m is a Cauchy-sequence, then, by completeness of $L^2(\mathbb{R}^n)$, there exist functions f and b_j in $L^2(\mathbb{R}^n)$ such that

$$f^m \to f \quad \text{and} \quad (\partial_j + iA_j)f^m \to b_j$$

as $m \to \infty$. We have to show that

$$b_j = (\partial_j + iA_j)f \quad \text{in } \mathcal{D}'(\mathbb{R}^n).$$

The proof of this fact is the same as that of Theorem 7.3, and we leave the details to the reader. (Note that for any $\phi \in C_c^\infty(\mathbb{R}^n)$, $A_j \phi \in L^2(\mathbb{R}^n)$.)

Important Remark: If $\psi \in H^1_A(\mathbb{R}^n)$, then $(\nabla + iA)\psi$ is an $\mathbb{R}^n$-*valued* $L^2(\mathbb{R}^n)$-*function. Hence* $(\nabla + iA)^2 \psi$ *makes sense as a distribution.*

• If $f \in H^1_A(\mathbb{R}^n)$, it is not necessarily true that $f \in H^1(\mathbb{R}^n)$ (as we remarked just after the definition 7.20(1)). However, $|f|$ is *always* in $H^1(\mathbb{R}^n)$ as the following shows. Theorem 7.21 is called the diamagnetic inequality because it says that removing the magnetic field ($A = 0$) allows us to decrease the kinetic energy by replacing $f(x)$ by $|f|(x)$ (and at the same time leaving $|f(x)|^2$ unaltered). (Cf. [Kato].)

7.21 THEOREM (Diamagnetic inequality)

Let $A : \mathbb{R}^n \to \mathbb{R}^n$ *be in* $L^2_{\text{loc}}(\mathbb{R}^n)$ *and let* f *be in* $H^1_A(\mathbb{R}^n)$. *Then* $|f|$, *the absolute value of* f, *is in* $H^1(\mathbb{R}^n)$ *and the diamagnetic inequality,*

$$\bigl|\nabla |f|(x)\bigr| \leq \bigl|(\nabla + iA)f(x)\bigr|, \tag{1}$$

holds pointwise for almost every $x \in \mathbb{R}^n$.

PROOF. Since $f \in L^2(\mathbb{R}^n)$ and each component of A is in $L^2_{\text{loc}}(\mathbb{R}^n)$, the distributional gradient of f is in $L^1_{\text{loc}}(\mathbb{R}^n)$. Writing $f = R + iI$ we have, by, Theorem 6.17 (derivative of the absolute value), that the distributional derivatives are functions in $L^1_{\text{loc}}(\mathbb{R}^n)$, and furthermore

$$(\partial_j |f|)(x) = \begin{cases} \operatorname{Re}\left(\frac{\overline{f}}{|f|} \partial_j f\right)(x) & \text{if } f(x) \neq 0, \\ 0 & \text{if } f(x) = 0. \end{cases} \tag{2}$$

Here $\overline{f} = R - iI$ is the complex conjugate function of f. Since

$$\operatorname{Re}\left(\frac{\overline{f}}{|f|} iA_j f\right) = \operatorname{Re}(iA_j |f|) = 0,$$

we see that (2) can be replaced by

$$(\partial_j |f|)(x) = \begin{cases} \operatorname{Re}\left(\frac{\overline{f}}{|f|}(\partial_j + iA_j)f\right)(x) & \text{if } f(x) \neq 0, \\ 0 & \text{if } f(x) = 0. \end{cases} \tag{3}$$

Then (1) follows from the fact that $|z| \geq |\operatorname{Re} z|$. Since the right side of (1) is in $L^2(\mathbb{R}^n)$, so is the left side. ∎

7.22 THEOREM ($C_c^\infty(\mathbb{R}^n)$ is dense in $H_A^1(\mathbb{R}^n)$)

If $f \in H_A^1(\mathbb{R}^n)$, then there exists a sequence $f^m \in C_c^\infty(\mathbb{R}^n)$ such that

$$\|f - f^m\|_{L^2(\mathbb{R}^n)} \to 0 \quad \text{and} \quad \|(\nabla + iA)(f - f^m)\|_{L^2(\mathbb{R}^n)} \to 0$$

as $m \to \infty$. Moreover, $\|f^m\|_p \leq \|f\|_p$ for every $1 \leq p \leq \infty$ such that $f \in L^p(\mathbb{R}^n)$.

PROOF. *Step* 1. Assume first that f is bounded and has compact support. Then $\|f\|_A < \infty$ implies that f is in $H^1(\mathbb{R}^n)$. This follows simply from the fact that $A_i f \in L^2(\mathbb{R}^n)$. Now take $f^m = j_\varepsilon * f$ as in 2.16 with $\varepsilon = 1/m$ and with $j \geq 0$ and j having compact support. By passing to a subsequence (again denoted by m) we can assume that

$$f^m \to f, \quad \partial_i f^m \to \partial_i f \quad \text{in } L^2(\mathbb{R}^n)$$

and $f^m \to f$ pointwise a.e. Since $f^m(x)$ is again uniformly bounded in x, the conclusion follows by dominated convergence.

Step 2. Next we show that functions in $H_A^1(\mathbb{R}^n)$ with compact support are dense in $H_A^1(\mathbb{R}^n)$. Pick $\chi \in C_c^\infty(\mathbb{R}^n)$ with $0 \leq \chi \leq 1$ and $\chi \equiv 1$ in the unit ball $\{x \in \mathbb{R}^n : |x| \leq 1\}$, and consider $\chi_m(x) = \chi(x/m)$. Then, for any $f \in H_A^1(\mathbb{R}^n)$, $\chi_m f \to f$ in $L^2(\mathbb{R}^n)$. Further, by 6.12,

$$(\nabla + iA)\chi_m f = \chi_m (\nabla + iA)f - i(\nabla \chi_m)f,$$

and hence

$$\|(\nabla + iA)(f - \chi_m f)\|_2 \leq \|(1 - \chi_m)(\nabla + iA)f\|_2 + \frac{1}{m} \sup_x |\nabla \chi(x)| \|f\|_2.$$

Clearly both terms on the right tend to zero as $m \to \infty$.

Step 3. Given $f \in H_A^1(\mathbb{R}^n)$, we know by the previous step that it suffices to assume that f has compact support. We shall now show that this f can be approximated by a sequence, f^k, of bounded functions in $H_A^1(\mathbb{R}^n)$ such that $|f^k(x)| \leq |f(x)|$ for all x. This, with Step 1, will conclude the proof.

Pick $g \in C_c^\infty(\mathbb{R})$ with $g(t) \equiv 1$ for $|t| \leq 1$, $g(t) \equiv 0$ for $|t| \geq 2$ and define $g_k(t) := g(t/k)$ for $k = 1, 2, \ldots$. Consider the sequence $f^k(x) := f(x)g_k(|f|(x))$. The function f^k is bounded by $2k$. Assuming the formula

$$\partial_i f^k = g_k(|f|)\partial_i f + f g_k'(|f|)\partial_i |f| \quad \text{in } \mathcal{D}'(\mathbb{R}^n) \tag{1}$$

for the moment, we can finish the proof. First note that in $L^2(\mathbb{R}^n)$

$$g_k(|f|)(\partial_i + iA_i)f \to (\partial_i + iA_i)f$$

by dominated convergence. Furthermore,

$$|fg'_k(|f|)| = |f||g'_k(|f|)| \leq \chi^k \sup_t |g'(t)|,$$

where $\chi^k = 1$ if $|f| \geq k$ and zero otherwise. By Theorem 7.21 (diamagnetic inequality) $\partial_i|f| \in L^2(\mathbb{R}^n)$ and hence $\|f(g_k)'(|f|)\partial_i f\|_2 \to 0$ as $k \to \infty$.

The proof of (1) is a consequence of the chain rule (Theorem 6.16). If we write $f = R + iI$, then $f^k = (R + iI)g_k(\sqrt{R^2 + I^2}\,)$ which is a differentiable function of both R and I with bounded derivatives. By assumption, the functions R and I have distributional derivatives in $L^1_{\text{loc}}(\mathbb{R}^n)$. Therefore the chain rule can be applied and the result is (1). ■

Exercises for Chapter 7

1. Show that the characteristic function of a set in $\mathbb{R}^n$ having positive and finite measure is never in $H^1(\mathbb{R}^n)$, or even in $H^{1/2}(\mathbb{R}^n)$.

2. Suppose that f^1, f^2, f^3, ... is a sequence of functions in $H^1(\mathbb{R}^n)$ such that $f^j \rightharpoonup f$ and $(\nabla f^j)_i \rightharpoonup g_i$ for $i = 1, 2, \ldots, n$ weakly in $L^2(\mathbb{R}^n)$. Prove that f is in $H^1(\mathbb{R}^n)$ and that $g_i = (\nabla f)_i$.

3. Prove 7.15(3), noting especially the meaning of the two sides of this equation and the distinction between $\sqrt{-\Delta}\,v$ as a function and as a distribution. Cf. Theorem 7.7.

4. Suppose that $f \in H^1(\mathbb{R}^n)$. Show that for each $1 \leq i \leq n$

$$\int_{\mathbb{R}^n} |\partial_i f|^2 = \lim_{t \to 0} \frac{1}{t} \int_{\mathbb{R}^n} |f(x + t\mathbf{e}_i) - f(x)|^2 \, dx,$$

where $\mathbf{e}_i$ is the unit vector in the direction i.

5. Verify equations 7.9(7) and 7.9(8) about the solution of the heat equation.

Chapter 8

Sobolev Inequalities

8.1 INTRODUCTION

In general terms, a 'Sobolev inequality' has come to mean an estimation of lower order derivatives of a function in terms of its higher order derivatives. Such estimates, valid for all functions in certain classes, have become a standard tool in existence and regularity theories for solutions of partial differential equations, in the calculus of variations, in geometric measure theory and in many other branches of analysis. The ideas go back to [Bliss], in one dimension, but achieved their true prominence through the work of [Sobolev], [Morrey] and others. Over the years many variations on the original theme have been produced, but here we shall mention only the most basic ones and, among these, will prove only the simplest.

The foremost example of a Sobolev inequality is the one relating the $L^2(\mathbb{R}^n)$ norm of the gradient of a function, f, defined on $\mathbb{R}^n$, $n \geq 3$, with an $L^q(\mathbb{R}^n)$ norm of f for some suitable q, i.e.,

$$\|\nabla f\|_2^2 \geq S_n \|f\|_q^2, \qquad q = \frac{2n}{n-2}, \tag{1}$$

where S_n is a universal constant depending only on n.

A similar inequality holds for the 'relativistic kinetic energy' '$|p|$' for $n \geq 2$,

$$(f, |p|f) \geq S'_n \|f\|_q^2, \qquad q = \frac{2n}{n-1}, \tag{2}$$

where, again, S'_n is a universal constant.

As an example of their usefulness, we shall exploit these two inequalities in Chapter 11 to prove the existence of a ground state for the one-particle Schrödinger equation.

A first and important step in understanding (1) and (2) is to note that the exponents q are the only exponents for which such inequalities can hold. Under dilations of $\mathbb{R}^n$,

$$x \mapsto \lambda x \quad \text{and} \quad f(x) \mapsto f(x/\lambda),$$

the multiplication operators p and $|p|$ in Fourier space multiply by λ^{-1}, while the n-dimensional integrals multiply by λ^n. Thus, the left sides of (1) and (2) are proportional to λ^{n-2} and λ^{n-1} respectively. The right sides multiply by $\lambda^{2n/q}$. Plainly, the two sides can only be compared when they scale similarly, which leads to $q = 2n/(n-2)$ or $q = 2n/(n-1)$, respectively.

Another thing to note is that (1) is only valid for $n \geq 3$ and (2) for $n \geq 2$. Hence, the question arises what inequalities should replace (1) in dimensions one and two and (2) in dimension one? There are many different answers, the usual ones being

$$\|\nabla f\|_2^2 + \|f\|_2^2 \geq S_{2,q} \|f\|_q^2 \quad \text{for all } 2 \leq q < \infty \quad \text{for } n = 2 \qquad (3)$$

(but not $q = \infty$) and

$$\left\|\frac{df}{dx}\right\|_2^2 + \|f\|_2^2 \geq S_1 \|f\|_\infty^2 \qquad \text{for } n = 1. \qquad (4)$$

For the relativistic case we shall consider an inequality of the form

$$(f, |p|f) + \|f\|_2^2 \geq S'_{1,q} \|f\|_q^2 \quad \text{for all } 2 \leq q < \infty \text{ and } n = 1. \qquad (5)$$

In this chapter we shall prove inequalities (1)–(5).

As mentioned before, inequalities (1) and (2) stand apart from inequalities (3)–(5). The main point is that (1) and to some lesser extent (2) have geometrical meaning which is manifest through their invariance under conformal transformations. See, e.g., Theorem 4.5 (conformal invariance of the HLS inequality) for a related statement.

The main point of these inequalities, however, is that they all serve as **uncertainty principles**, i.e., they effectively bound an average gradient of a function from below in terms of the 'spread' of the function. These principles can be extended to higher derivatives than the first as will be briefly mentioned later.

A related subject, which is of great importance in applications, is the Rellich–Kondrashov theorem, 8.6 and 8.9. Suppose $\mathcal{B}$ is a ball in $\mathbb{R}^n$ and suppose $f^1, f^2, \ldots$ is a sequence of functions in $L^2(\mathcal{B})$ with uniformly bounded $L^2(\mathcal{B})$ norms. As we know from the Banach–Alaoglu Theorem 2.18 there exists a weakly convergent subsequence. A strongly convergent subsequence *need not exist*. If, however, our sequence is uniformly bounded in $H^1(\mathcal{B})$ (i.e., $\int_\mathcal{B} |\nabla f^j|^2 \, dx < C$), then any weakly convergent subsequence is *also* strongly convergent in $L^2(\mathcal{B})$. This is the Rellich–Kondrashov theorem. By Theorem 2.7 (completeness of L^p-spaces) we can now pass to a further subsequence and thereby achieve pointwise convergence. This fact is very useful because when combined with the dominated convergence theorem one can infer the convergence of certain integrals involving the f^j's. It is remarkable that some crude bound on the average behavior of the gradient permits us to reach all these conclusions.

In Chapter 11 we shall illustrate these concepts with an application to the calculus of variations.

● Let us begin with a useful, technical remark about function spaces. In Sect. 7.2 we defined $H^1(\mathbb{R}^n)$ to consist of functions that, together with their distributional first derivatives, are in $L^2(\mathbb{R}^n)$. Most treatments of Sobolev inequalities use the fact that the functions are in $L^2(\mathbb{R}^n)$ but this, it turns out, is not the natural choice. The only relevant points are the facts that ∇f is in $L^2(\mathbb{R}^n)$ and that $f(x)$ goes to zero, in some sense, as $|x| \to \infty$. Therefore, we begin with a definition. A very similar definition applies to $W^{1,p}(\mathbb{R}^n)$.

8.2 DEFINITION OF $D^1(\mathbb{R}^n)$ AND $D^{1/2}(\mathbb{R}^n)$

A function $f : \mathbb{R}^n \to \mathbb{C}$ is in $D^1(\mathbb{R}^n)$ if it is in $L^1_{\text{loc}}(\mathbb{R}^n)$, if its distributional derivative, ∇f, is a function in $L^2(\mathbb{R}^n)$ and if f vanishes at infinity as in 3.2, i.e., $\{x : f(x) > a\}$ has finite measure for all $a > 0$. Similarly, $f \in D^{1/2}(\mathbb{R}^n)$ if f is in $L^1_{\text{loc}}(\mathbb{R}^n)$, f vanishes at infinity and if the integral 7.12(4) is finite.

REMARKS. (1) Obviously, this definition can be extended to $D^{1,p}$ or $D^{1/2,p}$ by replacing the exponent 2 for the derivatives by the exponent p. The integrand in 7.12(4) is then replaced by $[f(x) - f(y)]^p |x - y|^{-n-p/2}$. We shall not prove this, however.

(2) Note that this definition describes precisely the conditions under which the rearrangement inequalities for kinetic energies (Lemma 7.17) can be proved. In other words, Lemma 7.17 holds for functions in $D^1(\mathbb{R}^n)$ and $D^{1/2}(R^n)$.

(3) The notion of weak convergence in $D^1(\mathbb{R}^n)$ is obvious. The sequence f^j converges weakly to $f \in D^1(\mathbb{R}^n)$ if $\partial_i f^j \rightharpoonup \partial_i f$ weakly in $L^2(\mathbb{R}^n)$ for $i = 1, \ldots, n$. In $D^{1/2}(\mathbb{R}^n)$ the corresponding notion is the following: f^j

converges weakly in $D^{1/2}(\mathbb{R}^n)$ to f in $D^{1/2}(\mathbb{R}^n)$ if

$$\lim_{j\to\infty} \int_{\mathbb{R}^n}\int_{\mathbb{R}^n} \left(f^j(x) - f^j(y)\right)\left(\overline{g(x)} - \overline{g(y)}\right)|x-y|^{-n-1}\,\mathrm{d}x\,\mathrm{d}y$$
$$= \int_{\mathbb{R}^n}\int_{\mathbb{R}^n} \left(f(x) - f(y)\right)\left(\overline{g(x)} - \overline{g(y)}\right)|x-y|^{-n-1}\,\mathrm{d}x\,\mathrm{d}y$$

for every $g \in D^{1/2}(\mathbb{R}^n)$. By Schwarz's inequality all integrals are well defined.

In both cases the principle of uniform boundedness and the Banach–Alaoglu theorem are immediate consequences of their L^p counterparts, Theorem 2.12 and Theorem 2.18. The same holds for the weak lower semicontinuity of the norms (see Theorem 2.11). The easy proofs are left to the reader.

8.3 THEOREM (Sobolev's inequality for gradients)

For $n \geq 3$ let $f \in D^1(\mathbb{R}^n)$. Then $f \in L^q(\mathbb{R}^n)$ with $q = 2n/(n-2)$ and the following inequality holds:

$$\|\nabla f\|_2^2 \geq S_n \|f\|_q^2 \tag{1}$$

where

$$S_n = \frac{n(n-2)}{4}|\mathbb{S}^n|^{2/n} = \frac{n(n-2)}{4} 2^{2/n} \pi^{1+1/n} \Gamma\left(\frac{n+1}{2}\right)^{-2/n}. \tag{2}$$

There is equality in equation (1) if and only if f is a multiple of the function $(\mu^2 + (x-a)^2)^{-(n-2)/2}$ with $\mu > 0$ and with $a \in \mathbb{R}^n$ arbitrary.

REMARK. A similar inequality holds for L^p norms of ∇f for all $1 < p < n$, namely

$$\|\nabla f\|_p \geq C_{p,n}\|f\|_q \quad \text{with} \quad q = \frac{np}{n-p}. \tag{3}$$

The sharp constants $C_{p,n}$ and the cases of equality were derived by [Talenti].

PROOF. There are several ways to prove this theorem. One way is by competing symmetries as we did for Theorem 4.3 (HLS inequality). Another way is to minimize the quotient $\|\nabla f\|_2/\|f\|_q$ solely with the aid of rearrangement inequalities. Technically this is difficult because it is first necessary to prove the existence of an f that minimizes this ratio; this is done in [Lieb, 1983].

The route we shall follow here is to show that this theorem is the dual of the HLS inequality, 4.3, with the dual index p, where $1/q + 1/p = 1$.

Recall that $G_y(x) = [(n-2)|\mathbb{S}^{n-1}|]^{-1}|x-y|^{2-n}$ is the Green's function for the Laplacian, i.e., $-\Delta G_y(x) = \delta_y$ (see Sect. 6.20). We shall use the notation
$$(G * g)(x) = \int_{\mathbb{R}^n} G_y(x) g(y) \, dy$$
and (f, g) denotes $\int_{\mathbb{R}^n} \overline{f(x)} g(x) \, dx$. Our aim is the inequality, for pairs of functions f and g,
$$|(f,g)|^2 \leq \|\nabla f\|_2^2 \, (g, G * g), \tag{4}$$
which expresses the duality between the Sobolev inequality and the HLS inequality. Assuming (4) we have, by Theorem 2.3 (Hölder's inequality), $\|f\|_q = \sup\{|(f,g)| : \|g\|_p \leq 1\}$, and hence $\|f\|_q^2 \leq \|\nabla f\|_2^2 \sup\{(g, G*g) : \|g\|_p \leq 1\}$, which is finite by Theorem 4.3 (HLS inequality), and which leads immediately to (1).

We prove inequality (4) first for $g \in L^p(\mathbb{R}^n) \cap L^2(\mathbb{R}^n)$ and $f \in H^1(\mathbb{R}^n) \cap L^q(\mathbb{R}^n)$. Since f and g are in $L^2(\mathbb{R}^n)$, Parseval's formula yields
$$(f, g) = (\widehat{f}, \widehat{g}) = \int_{\mathbb{R}^n} \{|k|\overline{\widehat{f}(k)}\}\{|k|^{-1}\widehat{g}(k)\} \, dk. \tag{5}$$

By Corollary 5.10(1) of 5.9 (Fourier transform of $|x|^{\alpha-n}$), we have
$$h(k) := c_{n-1}(|x|^{1-n} * g)^{\vee}(k) = c_1 |k|^{-1} \widehat{g}(k).$$

By Plancherel's theorem and by the HLS inequality, h is square integrable, and thus we can apply the Schwarz inequality to the two functions $\{\ \}, \{\ \}$ in (5) to obtain the upper bound
$$\left(\int_{\mathbb{R}^n} |k|^2 |\widehat{f}(k)|^2 \, dk \right)^{1/2} \left(\int_{\mathbb{R}^n} |k|^{-2} |\widehat{g}(k)|^2 \, dk \right)^{1/2}.$$

The first factor equals $(2\pi)^{-1}\|\nabla f\|_2$ by Theorem 7.9 (Fourier characterization of $H^1(\mathbb{R}^n)$), and the second factor equals $2\pi(g, G*g)^{1/2}$ by Corollary 5.10. Thus we have (4) for all $f \in H^1(\mathbb{R}^n) \cap L^q(\mathbb{R}^n)$ and $g \in L^p(\mathbb{R}^n) \cap L^2(\mathbb{R}^n)$.

A simple approximation argument using the HLS inequality then shows that (4) holds for all $g \in L^p(\mathbb{R}^n)$. Now setting $g = f^{q-1} \in L^p(\mathbb{R}^n)$, one obtains from (4) and 4.3(1), (2) that
$$\|f\|_q^{2q} \leq \|\nabla f\|_2^2 \, (f^{q-1}, G * f^{q-1}) \leq d_n \|\nabla f\|_2^2 \|f\|_q^{2(q-1)}, \tag{6}$$
where
$$d_n := S_n^{-1} = [(n-2)|\mathbb{S}^{n-1}|]^{-1} \pi^{n/2-1} [\Gamma(n/2+1)]^{-1} \{\Gamma(n/2)/\Gamma(n)\}^{-2/n}.$$

Using the fact that $|\mathbb{S}^{n-1}| = 2\pi^{n/2}[\Gamma(n/2)]^{-1}$ together with the duplication formula for the Γ-function, i.e., $\Gamma(2z) = (2\pi)^{-1/2}2^{2z-1/2}\Gamma(z)\Gamma(z+1/2)$, we obtain (1) and (2) for $f \in H^1(\mathbb{R}^n) \cap L^q(\mathbb{R}^n)$.

To show that (1) holds for $f \in D^1(\mathbb{R}^n)$ we first note that by Theorem 7.8 (convexity inequality for gradients) f can be assumed to be a nonnegative function. Replace f by

$$f_c(x) = \min[\max(f(x) - c, 0), 1/c],$$

where $c > 0$ is a constant. Since f_c is bounded and the set where it does not vanish has finite measure, it follows that $f_c \in L^q(\mathbb{R}^n)$. Further by Corollary 6.18, $\nabla f_c(x) = \nabla f(x)$ for all x such that $c < f(x) < c+1/c$, and $\nabla f_c(x) = 0$ otherwise. By Theorem 1.6 (monotone convergence) it follows that

$$\|\nabla f\|_2^2 = \lim_{c \to 0} \|\nabla f_c\|_2^2 \geq S_n \lim_{c \to 0} \|f_c\|_q^2 = S_n \|f\|_q^2,$$

which shows that $f \in L^q(\mathbb{R}^n)$ and satisfies (1). The same argument shows that (6) holds for all nonnegative functions in $D^1(\mathbb{R}^n)$.

The validity of (6) for $D^1(\mathbb{R}^n)$ can then be used to establish all the cases of equality in (1). To have equality in (1) and hence in (6), it is necessary that f^{q-1} yields equality in the HLS inequality part of (6), i.e., f must be a multiple of $(\mu^2 + (x-a)^2)^{(n-2)/2}$ (see Sect. 4.3). A direct computation shows that functions of this type indeed yield equality in (1). ∎

8.4 THEOREM (Sobolev's inequality for $|p|$)

For $n \geq 2$ let $f \in D^{1/2}(\mathbb{R}^n)$. Then $f \in L^q(\mathbb{R}^n)$ with $q = 2n/(n-1)$ and the following inequality holds:

$$(f, |p|f) \geq S'_n \|f\|_q^2, \tag{1}$$

where

$$S'_n = \frac{n-1}{2}|\mathbb{S}^n|^{1/n} = \frac{n-1}{2} 2^{1/n} \pi^{(n+1)/2n} \Gamma\left(\frac{n+1}{2}\right)^{-1/n}. \tag{2}$$

There is equality in (1) if and only if f is a multiple of the function $(\mu^2 + (x-a)^2)^{-(n-1)/2}$ with $\mu > 0$ and with $a \in \mathbb{R}^n$ arbitrary.

PROOF. Analogously to the proof of the previous theorem, the inequality

$$|(f,g)|^2 \leq \tfrac{1}{2}\pi^{-(n+1)/2}\Gamma\left(\frac{n-1}{2}\right)(f,|p|f)(g,|x|^{1-n}*g) \qquad (3)$$

is seen to hold for all functions $f \in H^{1/2}(\mathbb{R}^n) \cap L^q(\mathbb{R}^n)$ and $g \in L^p(\mathbb{R}^n)$ ($1/p + 1/q = 1$). Setting $g = f^{q-1}$ and using Theorem 4.3 (HLS inequality) we obtain

$$\|f\|_q^{2q} \leq [\sqrt{\pi}\,(n-1)]^{-1}\left\{\frac{\Gamma(n)}{\Gamma(n-1)}\right\}^{1/n}(f,|p|f)\|f\|_q^{2q-2}, \qquad (4)$$

which yields (1) and (2) for $f \in H^{1/2}(\mathbb{R}^n) \cap L^q(\mathbb{R}^n)$. Note that there can only be equality in (4) if f^{q-1} saturates the HLS inequality, i.e., if f is of the form given in the statement of the theorem. A tedious calculation shows that for such functions there is indeed equality in (1). Finally we have to show that (1) holds under the weaker assumption that $f \in D^{1/2}(\mathbb{R}^n)$. As in the proof of Theorem 8.3 it suffices to show this for f nonnegative. This follows from Theorem 7.13. Next replace f by $f_c(x) = \min(\max(f(x) - c, 0), 1/c)$ for $c > 0$ some constant. It is a simple exercise to see that $|f_c(x) - f_c(y)| \leq |f(x) - f(y)|$ and hence by the definition of $(f,|p|f)$, 7.12(4), we see that $(f_c,|p|f_c) \leq (f,|p|f)$. Now $f_c \in L^q(\mathbb{R}^n)$ and hence, by Theorem 1.6 (monotone convergence), $f \in L^q(\mathbb{R}^n)$ because

$$(S'_n)^{1/2}\|f\|_q = (S'_n)^{1/2}\lim_{c\to 0}\|f_c\|_q \leq \lim_{c\to 0}(f_c,|p|f_c) = (f,|p|f). \qquad \blacksquare$$

8.5 THEOREM (Sobolev inequalities in 1 and 2 dimensions)

(i) Any $f \in H^1(\mathbb{R})$ is bounded and satisfies the estimate

$$\left\|\frac{df}{dx}\right\|_2^2 + \|f\|_2^2 \geq 2\|f\|_\infty^2 \qquad (1)$$

with equality if and only if f is a multiple of $\exp[-|x-a|]$ for some $a \in \mathbb{R}$. Moreover, f is equivalent to a continuous function that satisfies the estimate

$$|f(x) - f(y)| \leq \left\|\frac{df}{dx}\right\|_2 |x-y|^{1/2} \qquad (2)$$

for all $x, y \in \mathbb{R}$.

(ii) *For $f \in H^1(\mathbb{R}^2)$ the inequality*

$$\|\nabla f\|_2^2 + \|f\|_2^2 \geq S_{2,q}\|f\|_q^2 \tag{3}$$

holds for all $2 \leq q < \infty$ with a constant that satisfies

$$S_{2,q} > q^{1-2/q}(q-1)^{-1+1/q}((q-2)/8\pi)^{1/2-1/q}.$$

(iii) *For $f \in H^{1/2}(\mathbb{R})$ the inequality*

$$(f, |p|f) + \|f\|_2^2 \geq S'_{1,q}\|f\|_q^2 \tag{4}$$

holds for all $2 \leq q < \infty$ with a constant that satisfies

$$S'_{1,q} > (q-1)^{-1/2+1/2q}(q(q-2)/2\pi)^{1/2-1/q}.$$

PROOF. For $f \in H^1(\mathbb{R})$, by Theorem 7.6 (density of C_c^∞ in $H^1(\Omega)$) there exists a sequence $f^j \in C_c^\infty(\mathbb{R})$ that converges to f in $H^1(\mathbb{R})$. Now

$$(f^j(x))^2 = \int_{-\infty}^x f^j(y)(\mathrm{d}f^j/\mathrm{d}x)(y)\,\mathrm{d}y - \int_x^\infty f^j(y)(\mathrm{d}f^j/\mathrm{d}x)(y)\,\mathrm{d}y$$

by the fundamental theorem of calculus. Since $f^j \to f$ and $\mathrm{d}f^j/\mathrm{d}x \to \mathrm{d}f/\mathrm{d}x$ in $L^2(\mathbb{R})$, we see that the right side converges to $\int_{-\infty}^x f(y)f'(y)\,\mathrm{d}y - \int_x^\infty f(y)f'(y)\,\mathrm{d}y$. Using Theorem 2.7 (completeness of L^p-spaces) we can assume, by passing to a subsequence, that $f^j(x) \to f(x)$ pointwise for almost every x. Thus we have that for a.e. $x \in \mathbb{R}$

$$f(x)^2 = \int_{-\infty}^x f(y)f'(y)\,\mathrm{d}y - \int_x^\infty f(y)f'(y)\,\mathrm{d}y \tag{5}$$

for functions in $H^1(\mathbb{R})$. Now

$$|f(x)|^2 \leq \int_{-\infty}^x |f||f'| + \int_x^\infty |f||f'| = \int_{-\infty}^\infty |f||f'|$$

which, by Schwarz's inequality, yields

$$\|f\|_\infty^2 \leq \|f'\|_2 \|f\|_2. \tag{6}$$

Inequality (1) is now an immediate consequence of (6), by using the arithmetic-geometric mean inequality $2ab < a^2 + b^2$.

Inequality (2) is proved similarly. (2) shows that f is equivalent to a continuous, indeed Hölder continuous, function that we also denote by f (see the second remark in Sect. 10.1).

By Theorem 7.8 (convexity inequality for gradients) there can only be equality in (1) if f is real. Since f is continuous and vanishes at infinity, there exists $a \in \mathbb{R}$ such that $(f(a))^2 = \|f\|_\infty^2$. Hence

$$\|f\|_\infty^2 = 2 \int_{-\infty}^a ff' \leq 2 \left(\int_{-\infty}^a f^2 \right)^{1/2} \left(\int_{-\infty}^a (f')^2 \right)^{1/2}$$

and similarly

$$\|f\|_\infty^2 \leq 2 \left(\int_a^\infty f^2 \right)^{1/2} \left(\int_x^\infty (f')^2 \right)^{1/2}.$$

Equality in (1) implies equality in the above two expressions and in particular equality in the application of Schwarz's inequality (see Theorem 2.3). Hence $f'(x) = cf(x)$ for some constant $c > 0$ if $x \leq a$ and, therefore, $f(x) = \|f\|_\infty \exp[c(x - a)]$ for $x < a$, with $c > 0$. The reader might object that the equation $f' = cf$ holds only in the sense of distributions. This equation is, however, equivalent to the equation $(e^{cx} f)' = 0$ and the result follows by Theorem 6.11. Similarly $f(x) = \|f\|_\infty \exp[d(x - a)]$ for $x > a$, with $d < 0$. Equality in (1) implies that $c = -d = 1$, and thus we have proved that equality in (1) implies $f(x) = \exp[-|x - a|]$ for some a.

The proofs of (3) and (4) follow a different line, for they use the Fourier transform. By Theorem 7.9 (Fourier characterization of $H^1(\mathbb{R}^n)$) the left side of (3) equals

$$\int_{\mathbb{R}^2} (1 + 4\pi^2 |k|^2) |\widehat{f}(k)|^2 \, dk.$$

Let $p < 2$ be the dual index of $q > 2$, i.e., $1/p + 1/q = 1$. Now by Theorem 2.3 (Hölder's inequality)

$$\|\widehat{f}\|_p = \left(\int_{\mathbb{R}^2} |\widehat{f}(k)(1 + 4\pi^2 |k|^2)^{1/2}|^p (1 + 4\pi^2 |k|^2)^{-p/2} \, dk \right)^{1/p}$$
$$\leq K(\|f\|_2^2 + \|\nabla f\|_2^2),$$

where $K = \left(\int_{\mathbb{R}^2} (1 + 4\pi^2 |k|^2)^{-q/(q-2)} \, dk \right)^{(q-2)/2q}$, which is finite for $2 < q < \infty$. In fact $K = [(q-2)/8\pi]^{(q-2)/2q}$. Finally using Theorem 5.7 (sharp Hausdorff-Young inequality) $\|f\|_q \leq C_p \|\widehat{f}\|_p$ with $C_p = (p^{1/p} q^{-1/q})$, which yields (3).

The proof of (4), starting from

$$\|f\|_2^2 + (f, |p|f) = \int_\mathbb{R} (1 + 2\pi |k|) |\widehat{f}(k)|^2 \, dk,$$

is a word for word translation of the previous proof. ∎

8.6 THEOREM (Weak convergence implies strong convergence on small sets)

Let $f^1, f^2 \ldots$, be a sequence of functions in $D^1(\mathbb{R}^n)$ such that ∇f^j converges weakly in $L^2(\mathbb{R}^n)$ to some vector-valued function, v. If $n = 1, 2$ we also assume that f^j converges weakly in $L^2(\mathbb{R}^n)$. Then $v = \nabla f$ for some unique function $f \in D^1(\mathbb{R}^n)$.

Now let $A \subset \mathbb{R}^n$ be any set of finite measure and let χ_A be its characteristic function. Then

$$\chi_A f^j \to \chi_A f \text{ strongly in } L^p(\mathbb{R}^n) \tag{1}$$

for every $p < 2n/(n-2)$ when $n \geq 3$, every $p < \infty$ when $n = 2$ and every $p \leq \infty$ when $n = 1$. In fact, for $n = 1$ the convergence is pointwise and uniform.

An analogous theorem for functions in $D^{1/2}(\mathbb{R}^n)$ also holds, i.e., assume that f^j converges weakly to $f \in D^{1/2}(\mathbb{R}^n)$ in the sense of Remark (3) in Sect. 8.2. Then (1) holds for every $p < 2n/(n-1)$ when $n \geq 2$. In one dimension the same conclusion holds for all $p < \infty$ if we assume, in addition, that f^j converges weakly to f in $L^2(\mathbb{R}^n)$.

PROOF. For $n \geq 3$ we first note that the sequence f^j is uniformly bounded in $L^q(\mathbb{R}^n)$, $q = 2n/(n-2)$. This follows from Theorem 2.12 (uniform boundedness principle), which implies that the sequence $\|\nabla f^j\|_2$ is uniformly bounded, and from Theorem 8.3 (Sobolev's inequality for gradients). For $n = 1$ or 2 the sequence f^j is bounded in $L^2(\mathbb{R}^n)$. By Theorem 2.18 (bounded sequences have weak limits) there exists a subsequence $f^{j(k)}, k = 1, 2, \ldots$, such that $f^{j(k)}$ converges weakly in $L^q(\mathbb{R}^n)$ to some function $f \in L^q(\mathbb{R}^n)$. We wish to prove that the entire sequence converges weakly to f so, supposing the contrary, let $f^{i(k)}$ be some other subsequence that converges to, say, g weakly in $L^q(\mathbb{R}^n)$. Since for any vector-valued function $\phi \in C_c^\infty(\mathbb{R}^n)$

$$\begin{aligned} -\int_{\mathbb{R}^n} f \partial_i \phi \, \mathrm{d}x &= -\lim_{k \to \infty} \int_{\mathbb{R}^n} f^{j(k)} \partial_i \phi \, \mathrm{d}x \\ &= \lim_{k \to \infty} \int_{\mathbb{R}^n} \partial_i f^{j(k)} \phi \, \mathrm{d}x = \int_{\mathbb{R}^n} v_i \phi \, \mathrm{d}x, \end{aligned} \tag{2}$$

and similarly for g we conclude that $\int_{\mathbb{R}^n} (f-g) \partial_i \phi \, \mathrm{d}x = 0$, i.e., $\partial_i(f-g) = 0$ in $\mathcal{D}'(\mathbb{R}^n)$ for all i. By Theorem 6.11, $f_1 - f_2$ is constant and, since both f and g are in $L^q(\mathbb{R}^n)$, this constant is zero. Since every subsequence of f^j that has a weak limit has the same weak limit, $f \in L^q(\mathbb{R}^n)$, this implies that $f^j \rightharpoonup f$ in $L^q(\mathbb{R}^n)$. (This is a simple exercise using the Banach–Alaoglu theorem.) By (2), $\nabla f = v$ in $\mathcal{D}'(\mathbb{R}^n)$. The argument for $n = 1, 2$ is precisely the same. For the sequence f^j in $D^{1/2}(\mathbb{R}^n)$ we note that by the Banach–Alaoglu theorem (see Remark (3) in Sect. 8.2) the sequence $(f^j, |p|f^j)$ is uniformly bounded.

Now let $A \subset \mathbb{R}^n$ be any set of finite measure and let χ_A denote its characteristic function. We claim that for any $f \in D^1(\mathbb{R}^n)$

$$\|\chi_A(f - e^{\Delta t}f)\|_2 \leq \|\nabla f\|_2 \sqrt{t} \tag{3}$$

where, as in 7.9(5),

$$(e^{\Delta t}f)(x) = (2\pi t)^{-n/2} \int \exp[-(x-y)^2/2t] f(y) \, dy. \tag{4}$$

For $f \in H^1(\mathbb{R}^n)$, (3) follows from Theorem 5.3 (Plancherel's theorem),

$$\|f - e^{\Delta t}f\|_2^2 = \int_{\mathbb{R}^n} |\widehat{f}(k)|^2 (1 - \exp[-4\pi^2 |k|^2 t])^2 \, dk,$$

the fact that $1 - \exp[-4\pi^2 |k|^2 t] \leq \min(1, 4\pi^2 |k|^2 t)$, and by using Theorem 7.9 (Fourier characterization of $H^1(\mathbb{R}^n)$). By considering the real and imaginary parts of f, and among those the positive and negative parts separately, it suffices to show (3) for $f \in D^1(\mathbb{R}^n)$ nonnegative. Replacing f by $f_c(x) = \min(\max(f(x)-c, 0), 1/c)$, as in the proof of Theorem 8.3, we see that $\|\nabla f_c\|_2$ converges to $\|\nabla f\|_2$ as $c \to 0$ and, since $f_c \in L^2(\mathbb{R}^n)$, $\liminf_{c \to 0} \|f_c - e^{\Delta t} f_c\|_2 \geq \|f - e^{\Delta t} f\|_2$ by Theorem 1.7 (Fatou's lemma). Thus, we have proved (3). Note, as an aside, that we have proved that

$$f - e^{\Delta t} f \in L^2(\mathbb{R}^n) \quad \text{for any } f \in D^1(\mathbb{R}^n).$$

In precisely the same fashion one proves that for $f \in D^{1/2}(\mathbb{R}^n)$

$$\|\chi_A(f - e^{t|p|}f)\|_2 \leq (f, |p|f)^{1/2} \sqrt{t} \tag{5}$$

where, according to 7.11(10),

$$(e^{-t|p|}f)(x) = \Gamma\left(\frac{n+1}{2}\right) \pi^{-(n+1)/2} t \int (t^2 + (x-y)^2)^{-(n+1)/2} f(y) \, dy. \tag{6}$$

Consider now the sequence f^j and note that, since $\|\nabla f^j\|_2 < C$ independent of j, we have that $\|\chi_A(f^j - e^{\Delta t}f^j)\|_2 \leq C\sqrt{t}$. Assuming for the moment that for every $t > 0$, $g^j := \chi_A e^{\Delta t} f^j$ converges strongly in $L^2(\mathbb{R}^n)$ to $g := \chi_A e^{\Delta t} f$, we now show that $\chi_A f^j$ also converges strongly to $\chi_A f$. Simply note that

$$\|\chi_A(f^j - f)\|_2 \leq \|\chi_A(f^j - g^j)\|_2 + \|\chi_A(g^j - g)\|_2 + \|\chi_A(g - f)\|_2.$$

The first and the last term are bounded by $C\sqrt{t}$, since $\liminf_{j\to\infty} \|\nabla f^j\|_2 \geq \|\nabla f\|_2$ by Theorem 2.11 (lower semicontinuity of norms). Thus

$$\|\chi_A(f^j - f)\|_2 \leq 2C\sqrt{t} + \|\chi_A(g^j - g)\|_2.$$

For $\varepsilon > 0$ given, first choose $t > 0$ (depending on ε) such that $2C\sqrt{t} < \varepsilon/2$ and then j (also depending on ε) large enough such that $\|\chi_A(g^j - g)\|_2 < \varepsilon/2$, and hence $\|\chi_A(f^j - f)\|_2 < \varepsilon$ for $j > j(\varepsilon)$.

It remains to prove that $\chi_A g^j \to \chi_A g$ strongly in $L^2(\mathbb{R}^n)$. To see this note that by (4) and Hölder's inequality

$$|g^j(x)| \leq (4\pi t)^{-n/2} \left(\int_{\mathbb{R}^n} \exp[-x^2 p/2t] \, dx \right)^{1/p} \|f^j\|_q \chi_A(x)$$

with $1/p = 1 - 1/q$. Using Theorem 8.3 (Sobolev's inequality for gradients), $\|f^j\|_q \leq S_n^{-1/2} \|\nabla f^j\|_2 \leq S_n^{-1/2} C$. Hence $\chi_A g^j$ is dominated by a constant multiple of the square integrable function $\chi_A(x)$. On the other hand, $g^j(x)$ converges pointwise for every $x \in \mathbb{R}^n$ since, for every fixed x, $\exp[-(x-y)^2/2t]$ is in the dual of $L^q(\mathbb{R}^n)$ and $f^j \rightharpoonup f$ weakly in $L^q(\mathbb{R}^n)$. The result follows from Theorem 1.8 (dominated convergence).

The proof of the corresponding result in dimensions 1 and 2 is the same, in fact it is simpler since the sequence is uniformly bounded in $L^2(\mathbb{R}^n)$ by assumption.

The proof for $D^{1/2}(\mathbb{R}^n)$ is the same with minor modifications which are left to the reader. Thus the strong convergence of $\chi_A f^j$ is proved for $p = 2$.

The inequality

$$\|\chi_A(f - f^j)\|_p \leq \|\chi_A\|_r \|\chi_A(f - f^j)\|_2$$

for $1/p = 1/r + 1/p$ proves the theorem for $1 \leq p \leq 2$. Again by Hölder's inequality

$$\|\chi_A(f - f^j)\|_p \leq \|\chi_A(f - f^j)\|_2^\alpha \|\chi_A(f - f^j)\|_q^{1-\alpha}$$

with $\alpha = (1/p - 1/q)/(1/2 - 1/q)$, which is strictly positive if $p < q$. If $f^j \in D^1(\mathbb{R}^n)$ and $n \geq 3$, then

$$\|\chi_A(f - f^j)\|_q \leq \|f - f^j\|_q$$
$$\leq S_n^{-1/2}(\|\nabla f\|_2 + \|\nabla f^j\|_2) \leq C' = \text{some constant,}$$

by Theorem 8.3 (Sobolev's inequality for gradients). Thus $\|\chi_A(f - f^j)\|_p \leq C^{1-\alpha} \|\chi_A(f - f^j)\|_2^\alpha \to 0$ as $j \to \infty$.

The proof for $f^j \in D^{1/2}(\mathbb{R}^n)$ is precisely the same. The reader can easily prove the theorem in the remaining cases $n = 1, 2$ using the corresponding Sobolev inequalities (Theorem 8.5). The case that needs special attention is the statement that $f^j \to f$ pointwise uniformly on bounded sets if $\frac{d}{dx} f^j \mapsto \frac{d}{dx} f$ in $L^2(\mathbb{R})$.

To see that $f^j(x)$ converges to $f(x)$ pointwise we first note that by Theorem 6.9 (fundamental theorem of calculus for distributions)

$$f^j(x) - f^j(0) = \int_0^x (df^j/dx)(s) \, ds$$

converges pointwise to $f(x) - f(0)$. Next by Theorem 8.5 (Sobolev inequalities in 1 and 2 dimensions) the sequence $f^j(x)$ is pointwise uniformly bounded and thus, for $g \in L^2(\mathbb{R}) \cap L^1(\mathbb{R})$, we have that

$$\lim_{j \to \infty} \int_{\mathbb{R}} (f^j(x) - f^j(0)) g(x) \, dx = \int_{\mathbb{R}} (f(x) - f(0)) g(x) \, dx$$

by Theorem 1.8 (dominated convergence). However,

$$\lim_{j \to \infty} \int_{\mathbb{R}} f^j(x) g(x) \, dx = \int_{\mathbb{R}} f(x) g(x) \, dx,$$

by assumption, and thus $f^j(0)$, and hence $f^j(x)$, converges pointwise.

Next we show that the convergence is uniform on any closed, bounded interval. First note that by the fundamental theorem of calculus

$$|f(x) - f(y)| = \left| \int_y^x (df/dx)(s) \, ds \right| \le \|f'\|_2 |x - y|^{1/2}.$$

Thus we can assume that the functions f^j and f are continuous. Moreover since $\|f^j\|_2$ is uniformly bounded, the previous estimate is uniform, i.e., $|f^j(x) - f^j(y)| \le C|x - y|^{1/2}$ with C independent of j. Suppose I is a closed, bounded interval for which the convergence is not uniform. Then there exists an $\varepsilon > 0$ and a sequence of points x_j such that $|f^j(x_j) - f(x_j)| > \varepsilon$. By passing to a subsequence we can assume that x_j converges to $x \in I$. Now

$$|f^j(x_j) - f(x_j)| \le |f^j(x_j) - f^j(x)| + |f^j(x) - f(x)| + |f(x) - f(x_j)|.$$

The first term is bounded by $C|x - x_j|^{1/2}$ with C independent of j and hence vanishes as $j \to \infty$. The second tends to zero since $f^j \to f$ pointwise. The last also tends to zero since f is continuous. Thus we have obtained a contradiction. ∎

REMARK. It is worth noting that statement (1) with $p = 2$ was derived without using Theorems 8.4 and 8.5 (Sobolev inequalities). The only thing that was used were equations (3) and (5). The theorem and its proof can be extended to any $r < p$ for which we know a-priori that $\|f^j\|_p < C$. The only role of the Sobolev inequality in Theorem 8.6 was to establish such a bound for $p = 2n/(n-2)$, etc.

8.7 COROLLARY (Weak convergence implies a.e. convergence)

Let $f^1, f^2, \ldots$ be any sequence satisfying the assumptions of Theorem 8.6. Then there exists a subsequence $n(j)$, i.e., $f^{n(1)}(x), f^{n(2)}(x), \ldots$, that converges to $f(x)$ for almost every $x \in \mathbb{R}^n$.

REMARK. The point, of course, is the convergence on *all* of $\mathbb{R}^n$, not merely on a set of finite measure.

PROOF. Consider the sequence B_k of balls centered at the origin with radius $k = 1, 2, \ldots$. By the previous theorem and Theorem 2.7 we can find a subsequence $f^{n_2(j)}$ that converges to f almost everywhere in B_1. From that sequence we choose another subsequence $f^{n_2(j)}$ that converges a.e. in B_2 to f, and so forth. The subsequence $f^{n_j(j)}$ obviously converges to f for a.e. $x \in \mathbb{R}^n$ since, for every $x \in \mathbb{R}^n$, there is a k such that $x \in B_k$. ∎

• The material, presented so far, can be generalized in several ways. First, one replaces the first derivatives by higher derivatives and the L^2-norms by L^p-norms, i.e., we replace $H^1(\mathbb{R}^n)$ by $W^{m,p}(\mathbb{R}^n)$. One can expect, essentially by iteration, that theorems similar to 8.3–8.6 continue to hold. Another generalization is to replace $\mathbb{R}^n$ by more general domains (open sets) $\Omega \subset \mathbb{R}^n$, i.e., by considering $W^{m,p}(\Omega)$.

As explained in Sect. 7.6, $H_0^1(\Omega)$ is the space of functions in $H^1(\Omega)$ that can be approximated in the $H^1(\Omega)$ norm by functions in $C_c^\infty(\Omega)$. We define $W_0^{1,2}(\Omega) := H_0^1(\Omega)$. For the space $W_0^{1,2}(\Omega)$ it is obvious that Theorems 8.3, 8.5 and 8.6 continue to hold. For general $1 \leq p < \infty$, $W_0^{1,p}(\Omega)$ is defined similarly as the closure of $C_c^\infty(\Omega)$ in the $W^{1,p}(\Omega)$ norm. Corresponding theorems are valid for $W_0^{1,p}(\Omega)$, which we summarize in the remarks in Sect. 8.8.

The spaces $W^{m,p}(\Omega)$ (defined in Sect. 6.7) are more delicate. We remind the reader that an $f \in W^{m,p}(\Omega)$ is required to be in $L^p(\Omega)$. A Sobolev inequality for these functions will require some additional conditions on Ω.

To see this, consider a 'horn', i.e., a domain in $\mathbb{R}^3$ given by the following inequalities:

$$0 < x_1 < 1, \quad (x_2^2 + x_3^2)^{1/2} < x_1^\beta, \quad \text{with } \beta \geq 1.$$

Note that the function $|x|^{-\alpha}$ has a square integrable gradient for all $\alpha < \beta - 1/2$ but its L^6-norm is finite only if $\alpha < \beta/3 + 1/6$. The computations are elementary using cylindrical coordinates. Thus, if we consider the 'horn' Ω given by $\beta = 2$ the function $|x|^{-1}$ is in $H^1(\Omega)$ but not in $L^6(\Omega)$ and thus the Sobolev inequality cannot hold.

It is interesting to note that the above example is consistent with the Sobolev inequality if $\beta = 1$, i.e., if the 'horn' becomes a 'cone'. It is a fact that the Sobolev inequality does, indeed, hold in this cone case. Our immediate task is to define a suitable class of domains that generalizes a cone and for which the Sobolev inequality holds.

Consider the cone

$$\{x \in \mathbb{R}^n : x \neq 0, \ 0 < x_n < |x|\cos\theta\}.$$

This is a cone with vertex at the origin and with opening angle θ. If one intersects this cone with a ball of radius r centered at zero one obtains a **finite cone** $K_{\theta,r}$ with vertex at the origin. A domain $\Omega \subset \mathbb{R}^n$ is said to have the **cone property** if there exists a fixed finite cone $K_{\theta,r}$ such that for every $x \in \Omega$ there is a finite cone K_x, congruent to $K_{\theta,r}$, that is contained in Ω and whose vertex is x. This cone property is essential in the next theorem.

The Sobolev inequalities are summarized in the following list. The proofs are omitted but the interested reader may consult [Adams] for details. In the following, $W^{0,p}(\Omega) \equiv L^p(\Omega)$.

8.8 THEOREM (Sobolev inequalities for $W^{m,p}(\Omega)$)

Let Ω be a domain in $\mathbb{R}^n$ that has the cone property for some θ and r. Let $1 \leq p \leq q$, $m \geq 1$ and $k \leq m$. The following inequalities hold for $f \in W^{m,p}(\Omega)$ with a constant C depending on m, k, q, p, θ, r, but not otherwise on Ω or on f.

(i) *If $kp < n$, then*

$$\|f\|_{W^{m-k,q}(\Omega)} \leq C\|f\|_{W^{m,p}(\Omega)} \quad \text{for } p \leq q \leq \frac{np}{n-kp}. \tag{1}$$

(ii) *If $kp = n$, then*

$$\|f\|_{W^{m-k,q}(\Omega)} \leq C\|f\|_{W^{m,p}(\Omega)} \quad \text{for } p \leq q < \infty. \tag{2}$$

(iii) *If $kp > n$, then*

$$\max_{0\leq|\alpha|\leq m-k} \sup_{x\in\Omega} |D^\alpha f(x)| \leq C\|f\|_{W^{m,p}(\Omega)}. \qquad (3)$$

REMARKS. (1) Inequalities (iii) state that a function in a sufficiently 'high' Sobolev space is continuous—or even differentiable. These inequalities are due to [Morrey]. In three dimensions, for example, a function in $W^{1,2} = H^1$ is not necessarily continuous, but it is continuous if it has two derivatives in L^2, i.e., if $f \in W^{2,2} =: H^2$.

(2) A simple, but important remark concerns $W_0^{1,p}(\Omega)$. Since $\|\nabla f\|_{L^p(\mathbb{R}^n)} = \|\nabla f\|_{L^p(\Omega)}$ and $\|f\|_{L^q(\mathbb{R}^n)} = \|f\|_{L^q(\Omega)}$ for each p and q, two theorems are true about $W_0^{1,p}(\Omega)$. One is 8.8 with $m = 1$ and $k = 1$ and there are three cases depending on whether $p < n$, $p = n$ or $p > n$. In Theorem 8.8 q is constrained, but not fixed. The second theorem is 8.3(3) with the same $C_{p,n}$. Here q is fixed to be $np/(n-p)$ and $p < n$. The important difference is that only $\|\nabla f\|_p$ appears in 8.3(3), while $\|f\|_{W^{1,p}(\Omega)}$ appears in 8.8. The cone condition is *not* needed for either theorem since $\|f\|_{W^{1,p}(\Omega)} = \|f\|_{W^{1,p}(\mathbb{R}^n)}$, and since $\mathbb{R}^n$ has the cone property.

● The next question to address is whether Theorem 8.6 (weak convergence implies strong convergence on small sets) carries over to the spaces $W^{m,p}(\Omega)$ and $W_0^{1,p}(\Omega)$. The following theorem provides the extension of Theorem 8.6 and again we shall state it without proof. The interested reader can consult [Adams].

8.9 THEOREM (Rellich–Kondrashov theorem)

Suppose that Ω has the cone property for some θ and r, and let $f^1, f^2, \ldots$ be a sequence in $W^{m,p}(\Omega)$ that converges weakly in $W^{m,p}(\Omega)$ to a function $f \in W^{m,p}(\Omega)$. Here $1 \leq p < \infty$ and $m \geq 1$. Fix $q \geq 1$ and $1 \leq k \leq m$. Let $\omega \subset \Omega$ be any open bounded set. Then

(i) *If $kp < n$ and $q < \frac{np}{n-kp}$, then $\lim_{j\to\infty} \|f^j - f\|_{W^{m-k,q}(\omega)} = 0$.*

(ii) *If $kp = n$, then $\lim_{j\to\infty} \|f^j - f\|_{W^{m-k,q}(\omega)} = 0$ for all $q < \infty$.*

(iii) *If $kp > n$, then f^j converges to f in the norm*

$$\max_{0\leq|\alpha|\leq m-k} \sup_{x\in\Omega} |(D^\alpha f)(x)|.$$

Exercises for Chapter 8

1. Let Ω be an open subset of $\mathbb{R}^n$ that is not equal to $\mathbb{R}^n$. For functions in $H_0^1(\Omega)$ (see Sect. 7.6) show that a Sobolev inequality 8.3(1) holds and that the sharp constant is the same as that given in 8.3(2). Show also that in distinction to the $\mathbb{R}^n$ case, there is no function in $H_0^1(\Omega)$ for which equality holds.

2. One might expect that the space $H_0^1(\Omega)$ consists of all those functions in $H^1(\mathbb{R}^n)$ that vanish outside the set Ω. Show that this is not so.

   ▶ *Hint.* Consider $H_0^1(\Omega)$ where $\Omega = (-1, 1) \sim \{0\}$ and find a function $f \in H^1(\mathbb{R}^n)$ that is not in $H_0^1(\Omega)$ but vanishes outside Ω.

Chapter 9

Potential Theory and Coulomb Energies

9.1 INTRODUCTION

The subject of potential theory harks back to Newton's theory of gravitation and the mathematical problems associated with the potential function, Φ, of a source function, f, in three dimensions, given by

$$\Phi(x) = \int_{\mathbb{R}^3} |x-y|^{-1} f(y) \, dy. \tag{1}$$

The generalization from $\mathbb{R}^3$ to $\mathbb{R}^n$ replaces $|x-y|^{-1}$ by $|x-y|^{2-n}$ for $n \geq 3$ and by $\ln|x-y|$ for $n=2$ (cf. 6.20 (distributional Laplacian of Green's functions)). In the gravitational case $f(x)$ is interpreted as the *negative* of the mass density at x. If we move up a century, we can let $f(x)$ be the electric charge density at x, and $\Phi(x)$ is then the **Coulomb potential** of f (in Gaussian units).

Associated with Φ is a **Coulomb energy** which we define for $\mathbb{R}^n, n \geq 3$, and for complex-valued functions f and g in $L^1_{\text{loc}}(\mathbb{R}^n)$ by

$$D(f,g) := \frac{1}{2} \int_{\mathbb{R}^n} \int_{\mathbb{R}^n} \overline{f}(x) g(y) |x-y|^{2-n} \, dx \, dy. \tag{2}$$

We assume that either the above integral is absolutely convergent or that

$f > 0$ and $g > 0$, in which case $D(f,g)$ is well defined although it might be $+\infty$.

As far as physical interpretation of (2) for $n = 3$ goes, $D(f,f)$ is the true physical energy of a *real* charge density f. It is the energy needed to assemble f from 'infinitesimal' charges. In the gravitational case the physical energy is $-GD(f,f)$, with G being Newton's gravitational constant and f the mass density.

We defer the study of Φ and $D(f,g)$ to Sect. 9.6 and begin, instead, with the definition and properties of sub- and superharmonic functions. This is the natural class in which to view Φ; the study of such functions is called potential theory.

9.2 DEFINITION OF HARMONIC, SUBHARMONIC AND SUPERHARMONIC FUNCTIONS

Let Ω be an open subset of $\mathbb{R}^n, n \geq 1$, and let $f : \Omega \to \mathbb{R}$ be an $L^1_{\text{loc}}(\Omega)$-function. Here we are speaking of a definite, Borel measurable function, not an equivalence class. For each open ball $B_{x,R} \subset \Omega$ of radius R, center $x \in \mathbb{R}^n$ and volume $|B_{x,R}|$, let

$$\langle f \rangle_{x,R} := |B_{x,R}|^{-1} \int_{B_{x,R}} f(y) \, dy \tag{1}$$

denote the average of f in $B_{x,R}$. If, for almost every $x \in \Omega$,

$$f(x) \leq \langle f \rangle_{x,R} \tag{2}$$

for every R such that $B_{x,R} \subset \Omega$, we say that f is **subharmonic** (on Ω). If inequality (2) is reversed (i.e., $-f$ is subharmonic), f is said to be **superharmonic**. If (2) is an equality, i.e., $f(x) = \langle f \rangle_{x,R}$ for almost every x, then f is **harmonic**.

Since f is Borel measurable, f restricted to a sphere is ($n-1$-dimensional) measurable on the sphere. Let $S_{x,R} = \partial B_{x,R}$ denote the sphere of radius R centered at x. If f is summable over $S_{x,R} \subset \Omega$, we denote its mean by

$$[f]_{x,R} = |S_{x,R}|^{-1} \int_{S_{x,R}} f(y) \, dy = |\mathbb{S}^{n-1}|^{-1} \int_{\mathbb{S}^{n-1}} f(x + R\omega) \, d\omega. \tag{3}$$

Here $\mathbb{S}^{n-1}$ is the sphere of unit radius in $\mathbb{R}^n$ and $|\mathbb{S}^{n-1}|$ is its $n-1$-dimensional area; $|S_{x,R}|$ is the area of $S_{x,R}$.

By Fubini's theorem (and with the help of polar coordinates), we have that for every $x \in \Omega$ the function f is indeed summable on the sphere for almost every R. For each $x \in \Omega$ we define R_x to be $\sup\{R : B_{x,R} \subset \Omega\}$. The function $[f]_{x,r}$, defined for $0 < r < R_x$, is a summable function of r.

Recall the definition of upper- and lower-semicontinuous functions in Sect. 1.5 and Exercise 1.2. Recall, also, the meaning of $\Delta f \geq 0$ in $\mathcal{D}'(\Omega)$ from Sect. 6.22.

9.3 THEOREM (Properties of harmonic, subharmonic, and superharmonic functions)

Let $f \in L^1_{\text{loc}}(\Omega)$ with $\Omega \subset \mathbb{R}^n$ open. Then the distributional Laplacian satisfies

$$\Delta f \geq 0 \quad \text{if and only if } f \text{ is subharmonic.} \tag{1}$$

In case f is subharmonic, there exists a unique function $\widetilde{f} : \Omega \to \mathbb{R} \cup \{-\infty\}$ satisfying

- $\widetilde{f}(x) = f(x)$ for almost every $x \in \Omega$.
- $\widetilde{f}(x)$ is upper semicontinuous. (Note that even if f is bounded there need not exist a continuous function that agrees with f a.e.)
- $\widetilde{f}$ is subharmonic for all $x \in \Omega$, i.e., $\widetilde{f}$ satifies 9.2(2) for all x, R such that $B_{x,R} \subset \Omega$.

In addition,

(i) $\widetilde{f}$ is bounded above on compact sets although $\widetilde{f}(x)$ might be $-\infty$ for some x's.
(ii) $\widetilde{f}$ is summable on every sphere $S_{x,R}$ for which $B_{x,R} \subset \Omega$.
(iii) For each fixed $x \in \Omega$ the function $r \mapsto [\widetilde{f}]_{x,r}$, defined for $0 < r < R_x$, is a continuous, nondecreasing function of r satisfying

$$\widetilde{f}(x) = \lim_{r \to 0} [\widetilde{f}]_{x,r}. \tag{2}$$

REMARKS. (1) An obvious consequence of Theorem 9.3 is that $\widetilde{f}$ then has the property (called the **mean value inequality**) that

$$[\widetilde{f}]_{x,r} \geq \langle \widetilde{f} \rangle_{x,r} \geq \widetilde{f}(x). \tag{3}$$

(2) If f is superharmonic, the above results are reversed in the obvious way. If f is harmonic, both sets of conclusions apply; in particular, the inequalities in (2) become equalities and therefore $[\widetilde{f}]_{x,r} = \widetilde{f}(x)$ is independent of r. By definition, equation (1) implies that

$$\Delta f = 0 \quad \text{if and only if } f \text{ is harmonic}, \tag{4}$$
$$\Delta f \leq 0 \quad \text{if and only if } f \text{ is superharmonic}. \tag{5}$$

(3) One new feature appears in the harmonic case: $\tilde{f}$ is not only continuous, it is also infinitely differentiable. We leave the proof of this fact as an exercise.

(4) In $\mathbb{R}^1$, the condition $\Delta f \geq 0$ is the same as the condition that a $L^1_{\text{loc}}(\mathbb{R}^1)$-function be convex. In $\mathbb{R}^n$, however, subharmonicity is similar to, but weaker than, convexity. The relation is the following. We can define the symmetric $n \times n$ **Hessian matrix**

$$H_{ij}(x) = \partial^2 f(x)/\partial x^i \partial x^j$$

(in the distributional sense); convexity is the condition that $H(x)$ be positive semidefinite for all x while subharmonicity requires only Trace $H(x) \geq 0$. There is, however, some convexity inherent in subharmonicity. With $r(t)$ defined by

$$r(t) = \begin{cases} t, & 0 < t < R_x & \text{if } n = 1, \\ e^t, & -\infty < t < \ln R_x & \text{if } n = 2, \\ t^{-1/(n-2)}, & R_x^{2-n} < t < \infty & \text{if } n \geq 3, \end{cases} \quad (6)$$

the function

$$t \mapsto [\tilde{f}]_{x,r(t)} \quad (7)$$

is convex. The proof of this convexity is left as an exercise.

(5) Despite the fact that the original definition 9.2(2) defines subharmonic as a global property (i.e., 9.2(2) must hold for all balls), (1) above shows that it really is only a local property, i.e., it suffices to check $\Delta f \geq 0$, and for this purpose it suffices to check 9.2(2) on balls whose radius is less than any arbitrarily small number. There is some similarity here with complex analytic functions; indeed, if $\Omega \subset \mathbb{C}$ and $f : \Omega \to \mathbb{C}$ is analytic, then $|f| : \Omega \to \mathbb{R}^+$ is subharmonic.

PROOF. *Step* 1. At first we assume that $f \in C^\infty(\Omega)$, so that we can integrate by parts freely. Let

$$g_{x,r} := \int_{S_{x,r}} \nabla f \cdot \nu = r^{n-1} \int_{\mathbb{S}^{n-1}} \nabla f(x + r\omega) \cdot \omega \, d\omega, \quad (8)$$

where ν is the unit outward normal vector. If $\Delta f \geq 0$, then, by Gauss's theorem,

$$0 \leq \int_{B_{x,r}} \Delta f = g_{x,r}, \quad (9)$$

and hence $g_{x,r}$ is a nondecreasing, nonnegative, continuous function of r. Using the right hand formula in 9.2(3), we can differentiate under the integral to find that

$$\frac{\mathrm{d}}{\mathrm{d}r}[f]_{x,r} = |\mathbb{S}^{n-1}|^{-1} r^{1-n} g_{x,r} \, . \tag{10}$$

From (10) we see that $r \mapsto [f]_{x,r}$ is continuous and nondecreasing, and (3) is an elementary consequence of that fact. If we choose $\widetilde{f} \equiv f$, then all the assertions of our theorem about $\widetilde{f}$ are easily seen to hold with the exception of uniqueness which we shall prove at the end.

Next, we show that $\Delta f \geq 0$ when f is subharmonic. If not, then $h := \Delta f$ is in $C^\infty(\Omega)$, and h is negative in some open set $\Omega' \subset \Omega$. By the previous result, f is *superharmonic* in Ω', i.e., $f(x) > [f]_{x,R}$ when $B_{x,R} \subset \Omega'$. (The reason we can write $>$ instead of merely $\geq$ is that (9) and (10) show that $-[f]_{x,R}$ has a strictly positive derivative.) This relation implies $f(x) > \langle f \rangle_{x,R}$ in Ω', which contradicts the subharmonicity assumption. This proves (1) for $f \in C^\infty(\Omega)$.

Step 2. Now we remove the $C^\infty(\Omega)$ assumption. Choose some $h \in C_c^\infty(\mathbb{R}^n)$ such that $h \geq 0$, $\int h = 1$, $h(x) = 0$ for $|x| \geq 1$, and h is spherically symmetric. Let us also define $h_\varepsilon(x) = \varepsilon^{-n} h(x/\varepsilon)$ for $\varepsilon > 0$. Then the function

$$f_\varepsilon := h_\varepsilon * f \tag{11}$$

is well defined in the set $\Omega_\varepsilon = \{x : \mathrm{dist}(x, \partial\Omega) > \varepsilon\} \subset \Omega$ and $f_\varepsilon \in C^\infty(\Omega_\varepsilon)$. As usual $*$ denotes the convolution of two functions. Also, $\Delta f_\varepsilon \geq 0$ if $\Delta f \geq 0$, in fact

$$\Delta f_\varepsilon = h_\varepsilon * \Delta f.$$

For this see Theorem 2.16, where it was also shown that there exists a sequence $\varepsilon_1 > \varepsilon_2 > \cdots$ tending to zero such that as $i \to \infty$, $f_{\varepsilon_i}(x) \to f(x)$ for a.e. x and $f_{\varepsilon_i} \to f$ in $L^1(K)$ for any compact set $K \subset \Omega$. Henceforth, we denote this $i \to \infty$ limit simply by $\lim_{\varepsilon \to 0}$. In the following we shall frequently introduce integrals, such as in (11), with the implicit understanding that they are defined only if ε is small enough or x is not too close to $\partial\Omega$, etc.

If $\Delta f \geq 0$, then $\Delta f_\varepsilon \geq 0$, and then (by Step 1) f_ε is subharmonic in Ω_ε. By definition

$$f_\varepsilon(x) \leq |B_{x,R}|^{-1} \int_{B_{x,R}} f_\varepsilon$$

for small ε. As $\varepsilon \to 0$ the right side converges to $|B_{x,R}|^{-1} \int_{B_{x,R}} f$ while the left side converges to $f(x)$ for a.e. x. Thus, f is subharmonic as well. Conversely, suppose that f is subharmonic. Then f_ε is subharmonic in Ω_ε because f subharmonic $\Leftrightarrow |B|f \leq \chi_B * f$, where χ_B is the characteristic function of a ball, and hence

$$\chi_B * f_\varepsilon = \chi_B * (h_\varepsilon * f) = h_\varepsilon * (\chi_B * f) \geq |B| h_\varepsilon * f = |B| f_\varepsilon.$$

However, f_ε subharmonic $\Rightarrow \Delta f_\varepsilon \geq 0 \Rightarrow \int f_\varepsilon \Delta \phi \geq 0$ for any nonnegative ϕ in $C_c^\infty(\Omega)$ and for sufficiently small ε. As $\varepsilon \to 0$ this integral converges to $\int f \Delta \phi$, so $\Delta f \geq 0$. This proves (1) for $f \in L^1_{\text{loc}}(\Omega)$.

Step 3. It remains to prove the existence of a unique $\widetilde{f}$ with the stated properties, under the assumptions $f \in L^1_{\text{loc}}(\Omega)$ and f subharmonic.

To see uniqueness, let g be any function satisfying the same three properties as $\widetilde{f}$. Since $\langle f \rangle_{x,r} = \langle g \rangle_{x,r} \geq g(x)$ for all x we see that g is bounded *above* on compact sets, in particular there is a constant C independent of r such that $g \leq C$ on all of $B_{x,r}$ for r sufficiently small. The function $C - g$ is positive and lower semicontinuous. This, together with Fatou's lemma, implies that $\limsup_{r \to 0} \langle g \rangle_{x,r} \leq g(x)$. Since g is subharmonic everywhere, $\liminf_{r \to 0} \langle g \rangle_{x,r} \geq g(x)$ and therefore $\lim_{r \to 0} \langle f \rangle_{x,r} = \lim_{r \to 0} \langle g \rangle_{x,r} = g(x)$. Obviously the same is true for $\widetilde{f}$ which proves uniqueness.

An important fact, which we show next, is that $\varepsilon \mapsto f_\varepsilon(x)$ is a nondecreasing function of ε. If $f \in C^\infty(\Omega)$, a simple calculation shows that

$$f_\varepsilon(x) = \int_{|y| \leq 1} h(y) [f]_{x, |y|\varepsilon} \, dy, \tag{12}$$

and this is monotone increasing in ε, by virtue of (10), which holds for $f \in C^\infty(\Omega)$. If $f \notin C^\infty(\Omega)$, define $g_{\varepsilon,\mu} = h_\varepsilon * f_\mu$. By the foregoing, this function is monotone in ε for each fixed μ and, as $\mu \to 0$, $(h_\varepsilon * f_\mu)(x) \to h_\varepsilon * f(x) = f_\varepsilon(x)$ for all x because $f_\mu \to f$ in $L^1(K)$ for every compact $K \subset \Omega$. Therefore f_ε is monotone, even if $f \notin C^\infty(\Omega)$ because a pointwise limit of monotone functions is monotone.

Armed with this information, we define

$$\widetilde{f}(x) := \inf \{ f_\varepsilon(x) : \Omega_\varepsilon \subset \Omega \}. \tag{13}$$

This $\widetilde{f}$ is upper semicontinuous (because it is the infimum of continuous functions). For any compact $K \subset \Omega$, there is an $\varepsilon_K > 0$ such that $f_{\varepsilon_K}(x)$ is defined for all $x \in K$ (Why?); we then have $\widetilde{f}(x) \leq f_{\varepsilon_K}(x)$ and hence $\widetilde{f}$ is bounded above on K by a $C^\infty(\Omega)$-function. Moreover, $\widetilde{f}(x) = f(x)$ for almost every $x \in \Omega$ because the monotonicity implies

$$\widetilde{f}(x) = \lim_{\varepsilon \to 0} f_\varepsilon(x) \tag{14}$$

for all $x \in \Omega$, but this limit equals $f(x)$ almost everywhere as stated above. Now, with the usual definition of $f_\pm(x)$, we have $\widetilde{f}_\pm(x) = \lim_{\varepsilon \to 0} f_{\pm,\varepsilon}(x)$, by (14). If $S_{x,R} \subset K$, then

$$\lim_{\varepsilon \to 0} \int_{S_{x,R}} f_{+,\varepsilon} = \int_{S_{x,R}} \widetilde{f}_+ \tag{15}$$

by dominated convergence (since $0 \leq \widetilde{f}_+ \leq f_{+,\varepsilon_K}$), while

$$\lim_{\varepsilon \to 0} \int_{S_{x,R}} f_{-,\varepsilon} = \int_{S_{x,R}} \widetilde{f}_- \tag{16}$$

by monotone convergence (the monotonicity follows from (12)). While the limit in (15) is finite, the limit in (16) could conceivably be $+\infty$. This cannot happen, however, because if it were $+\infty$, then the integral would have to be $+\infty$ for all $r < R$ (since $[f_\varepsilon]_{x,r}$ is nondecreasing in r). This would contradict the fact that $\widetilde{f} \in L^1_{\text{loc}}(\Omega)$.

We have arrived at the conclusion that $[\widetilde{f}]_{x,r}$ is defined and finite for all $r < R$ such that $B_{x,R} \subset \Omega$, and it equals $\lim_{\varepsilon \to 0}[f_\varepsilon]_{x,r} \geq \lim_{\varepsilon \to 0} f_\varepsilon(x) = \widetilde{f}(x)$. Moreover, $[\widetilde{f}]_{x,r}$ is the pointwise limit of nondecreasing functions, and hence it is itself nondecreasing. Since $\langle \widetilde{f} \rangle_{x,r}$ is an integral over spherical averages, we have shown that $\widetilde{f}$ is subharmonic at every point $x \in \Omega$.

Next, we show that

$$J(x) := \lim_{r \to 0} [\widetilde{f}]_{x,r} = \widetilde{f}(x).$$

This limit, $J(x)$, exists for every x since $[\widetilde{f}]_{x,r}$ is nondecreasing in r (although it could be $-\infty$), and by the foregoing we know that $J(x) \geq \widetilde{f}(x)$. Suppose there is a point y such that $J(y) \geq \widetilde{f}(y) + C$ with $C > 0$. Then, for all small r's, there must be an $x(r) \in \Omega$ such that $\widetilde{f}(x(r)) \geq \widetilde{f}(y) + C$ (because the *average* of $\widetilde{f}$ on $S_{y,r}$ exceeds $\widetilde{f}(y) + C$). But $\widetilde{f}$ is upper semicontinuous and hence $\limsup_{r \to 0} \widetilde{f}(x(r)) \leq \widetilde{f}(y)$; this is a contradiction, and hence $J(y) = \widetilde{f}(y)$.

The continuity of the function $r \mapsto [\widetilde{f}]_{x,r}$ follows now from the convexity properties stated in (6) and the fact that a convex function defined on an open interval is continuous. ∎

9.4 THEOREM (The strong maximum principle)

Let $\Omega \subset \mathbb{R}^n$ be open and connected. Let $f : \Omega \to \mathbb{R}$ be subharmonic and assume $f = \widetilde{f}$, where $\widetilde{f}$ is the unique representative of f with the properties given in Theorem 9.3. Suppose that

$$F := \sup\{f(x) : x \in \Omega\} \tag{1}$$

is finite. Then there are two possibilities. Either

    (i) *$f(x) < F$ for all $x \in \Omega$*

or else

    (ii) *$f(x) = F$ for all $x \in \Omega$.*

If f is superharmonic, then the sup *in* (1) *is replaced by* inf *and the inequality in* (i) *is reversed. If f is harmonic, then f achieves neither its supremum nor infimum unless f is constant.*

REMARKS. (1) The 'weak' maximum principle would eliminate (ii) and replace (i) by $f(x) \leq F$.

(2) If f is subharmonic and continuous in Ω and has a continuous extension to $\overline{\Omega}$, the closure of Ω, then Theorem 9.4 states that *f has its maximum on $\partial\Omega$, the boundary of Ω* (which is defined to be $\overline{\Omega} \cap \overline{\Omega^c}$) or at infinity (if Ω is unbounded).

(3) The strong maximum principle is well known for the absolute value of analytic functions on $\mathbb{C}$.

(4) One obvious consequence of the strong maximum principle is known as **Earnshaw's theorem** in the physics literature (cf. [Earnshaw], [Thomson]). It states that there can be no stable equilibrium for static point charges. This implies that atoms must be dynamic objects, and it was one of the observations that eventually led to the quantum theory.

PROOF. We have to prove that $f(y) = F$ for some $y \in \Omega$ implies that $f(x) = F$ for all $x \in \Omega$. Let $B \subset \Omega$ be a ball with y as its center. Then, by 9.2(2), we have that

$$|B|F \leq \int_B f \leq \int_B F = |B|F,$$

and hence $f(x) = F$ for almost every x in B. Pick any point x in B. Since sets of full Lebesgue measure are dense, there exists a sequence x_j in B converging to x such that $f(x_j) = F$. By the upper semicontinuity of f it follows that

$$F = \lim_{j \to \infty} f(x_j) \leq f(x) \leq F,$$

and hence $f(x) = F$. Thus we conclude that $f(x) = F$ for *every* $x \in B$.

Now let x be an arbitrary point of Ω and let Γ be a continuous curve connecting y to x (which exists, since Ω is connected). This curve can be defined by a continuous function $\gamma : [0,1] \to \Omega$ such that $\gamma(0) = y$ and $\gamma(1) = x$. Let $T \in [0,1]$ be the largest t such that $f(\gamma(t)) = F$. (The reader should check, as above, that the existence of this T follows from the continuity of γ and the upper semicontinuity of f.) We claim that $T = 1$, and hence that $f(x) = F$, as asserted in the theorem. Indeed, if $0 \leq T < 1$, then there is some ball $B_T \subset \Omega$ centered at $\gamma(T) \in \Omega$ (since Ω is open); by the preceding paragraph, $f(z) = F$ for all $z \in B_T$. By the continuity of γ, B_T contains some points $\gamma(s)$ with $s > T$. But then $f(\gamma(s)) = F$, which contradicts the assumption that $f(\gamma(t)) < F$ for all $t > T$. ∎

● The following inequality is of great use, for it quantifies the maximum principle by setting bounds on the possible variation of a nonnegative harmonic function. This version of Harnack's inequality is very far from being the best of its genre but the proof is simple.

9.5 THEOREM (Harnack's inequality)

Suppose f is a nonnegative harmonic function on the open ball $B_{z,R} \subset \mathbb{R}^n$. Then, for every x and $y \in B_{z,R/3}$

$$3^{-n} f(x) \leq 2^{-n} f(z) \leq f(y). \tag{1}$$

A corollary of (1) is that when f is harmonic on $\mathbb{R}^n$ and, for some constant C, either $f(x) \leq C$ for all x, or $f(x) \geq C$ for all x, then f is a constant function. Therefore, the only semi-bounded harmonic functions on $\mathbb{R}^n$ are the constant functions.

PROOF. Without loss, assume that $R = 3$. If $y \in B_{z,1}$, then we have

$$f(y) = \langle f \rangle_{y,2} \geq 2^{-n} \langle f \rangle_{z,1} = 2^{-n} f(z)$$

since $B_{z,3} \supset B_{y,2} \supset B_{z,1}$. On the other hand,

$$f(z) = \langle f \rangle_{z,3} \geq 3^{-n} 2^n \langle f \rangle_{x,2} = 3^{-n} 2^n f(x) \ .$$

To prove the corollary, note that (1) holds for every pair $x, y \in \mathbb{R}^n$. Assuming $f \geq C$, let $F = \inf\{f(x) : x \in \mathbb{R}^n\}$, which is finite. Let $g(x) := f(x) - F \geq 0$. Given $\varepsilon > 0$ there is a $y \in \mathbb{R}^n$ such that $0 \leq g(y) \leq \varepsilon$. Then (1) implies $g(x) \leq 3^n \varepsilon$ for all x. This obviously implies $g(x) \equiv 0$, i.e., $f(x) \equiv F$. ∎

● Now we return to the Coulomb potentials and energies discussed in the introduction, 9.1.

9.6 THEOREM (Subharmonic functions are potentials)

Let $n \geq 3$ and let

$$G_y(x) := [(n-2)|\mathbb{S}^{n-1}|]^{-1}|x-y|^{2-n} \tag{1}$$

be the Green's function given before Sect. 6.20. Let $f : \mathbb{R}^n \to [-\infty, 0]$ be a nonpositive subharmonic function. By Theorem 9.3, $\mu := \Delta f \geq 0$ in $\mathcal{D}'(\mathbb{R}^n)$ and, by Theorem 6.22 (positive distributions are measures), μ is a positive measure on $\mathbb{R}^n$.

Our new assertion is that $(1+|x|)^{2-n}$ is μ-summable and that

$$f^\dagger(x) := -\int_{\mathbb{R}^n} G_y(x) \mu(\mathrm{d}y) \tag{2}$$

is finite for almost every x. In fact, there is a constant $C \geq 0$ such that

$$\widetilde{f} = f^\dagger - C$$

is the unique $\widetilde{f}$ representative of f given in Theorem 9.3.

Conversely, if μ is any positive Borel measure on $\mathbb{R}^n$ such that $(1+|x|)^{2-n}$ is μ-summable, then the integral in (2) defines a subharmonic function $f^\dagger : \mathbb{R}^n \to [-\infty, 0]$ with $\Delta f^\dagger = \mu$ in $\mathcal{D}'(\mathbb{R}^n)$.

REMARKS. (1) When $n = 1$ or 2 there are no nonpositive subharmonic functions (on *all* of $\mathbb{R}^n$) other than the constant functions. For $n = 1$ this follows from the fact that such a function must be convex. For $n = 2$ this follows from Theorem 9.3(6) which says that the circular average, $[f]_{0,\exp(t)}$, must be convex in t on the whole line $-\infty < t < \infty$.

(2) Obviously, the theorem holds for superharmonic functions by reversing the signs in obvious places.

(3) The condition $f(x) \leq 0$ may seem peculiar. What it really means, in general, is that when f is subharmonic (without the $f(x) \leq 0$ condition) then, with $\Delta f = \mu$, we can write

$$f = f^\dagger + H, \tag{3}$$

with $f^\dagger$ given by equation (2) and with H harmonic, *provided there exists some harmonic function $\widetilde{H}$ with the property that $\widetilde{H}(x) \geq f(x)$ for all $x \in \mathbb{R}^n$.* As a counterexample, let $f(x_1, x_2, x_3) := |x_1|$. This f is subharmonic but there is no $\widetilde{H}$ that dominates f. In this case the integral in equation (2) is infinite for all y since Δf is a 'delta-function' on the two-dimensional plane $x_1 = 0$.

PROOF. *Step* 1. Assume first that $\Delta f = m$ and m is a nonnegative $C_c^\infty(\mathbb{R}^n)$-function. Clearly, $(1+|x|)^{2-n}m(x)$ is summable and we have

$$f^\dagger(y) = -\int_{\mathbb{R}^n} G_y(x)m(x)\,dx = -(G_0 * m)(y) = -(m * G_0)(y),$$

recalling that $G_y(x) = G_0(y-x)$ and that convolution is commutative. By Theorem 2.16, $\Delta f^\dagger = -m * (\Delta G_0)$. But $\Delta G_0 = -\delta_0$ by Theorem 6.20, so $\Delta f^\dagger = m$. We conclude that $\phi(x) := f(x) - f^\dagger(x)$ is harmonic (since $\Delta \phi = 0$). Moreover, $|f^\dagger(x)|$ is obviously bounded (by Hölder's inequality, for example) and therefore $\phi(x)$ is bounded above (since $f(x) \leq 0$). By Theorem 9.5, $\phi(x) = -C$. Clearly $f^\dagger(x) \to 0$ as $x \to \infty$, so $C \geq 0$. Finally, $f^\dagger \in C^\infty(\mathbb{R}^n)$ by Theorem 2.16, so $f^\dagger - C$ is the unique $\widetilde{f}$ of Theorem 9.3.

Conversely, if $m \in C_c^\infty(\mathbb{R}^n)$ then, by 6.21 (solution of Poisson's equation), $f^\dagger$, defined by (2) with $\mu(dx) = m(x)\,dx$, satisfies $\Delta f^\dagger = m$.

Step 2. Now assume that $\Delta f = m$ and $m \in C^\infty(\mathbb{R}^n)$, but m does not have compact support. Choose some $\chi \in C_c^\infty(\mathbb{R}^n)$ that is spherically symmetric and radially decreasing and satisfies $\chi(x) = 1$ for $|x| \leq 1$. Define $\chi_R(x) = \chi(x/R)$ and set $m_R(x) := \chi_R(x)m(x)$. Clearly $m_R \in C_c^\infty(\mathbb{R}^n)$. Let $f_R^\dagger = -G_0 * m_R$ as in (2), and let $\widetilde{f}$ be as in Theorem 9.3. Then, as proved in Step 1, $\Delta f_R^\dagger = m_R$, and so $\phi_R := \widetilde{f} - f_R^\dagger$ is subharmonic because $\Delta \phi_R = m - m_R \geq 0$. Since $m_R(x)$ is an increasing function of R (because χ is radially decreasing), $f_R^\dagger(y)$ is a decreasing function of R for each y. Also, $f_R^\dagger \in C^\infty(\mathbb{R}^n)$, as proved in Step 1, and $f_R^\dagger(x) \to 0$ as $|x| \to \infty$.

Several conclusions can be drawn.

(i) $f_R^\dagger(x) \geq \widetilde{f}(x)$ a.e. Otherwise, $\phi_R(x)$ would be a subharmonic function that is positive on a set of positive measure but that satisfies $\lim_{|x|\to\infty}\phi_R(x) \leq 0$ uniformly. (Why?) This is impossible by Theorem 9.3.

(ii) Since, by monotone convergence,

$$\int_{\mathbb{R}^n}(1+|x|)^{2-n}m(x)\,dx = \lim_{R\to\infty}\int_{\mathbb{R}^n}(1+|x|)^{2-n}m_R(x)\,dx,$$

we can conclude from (i) and the definition of $f_R^\dagger$ that the above integral on the left is finite. In fact, for the same reason, $f^\dagger(y) = \lim_{R\to\infty}f_R^\dagger(y)$ and, since the limit is monotone, $f^\dagger$ is upper semicontinuous.

(iii) If we define $\phi = \widetilde{f} - f^\dagger$, then, since $m(x) - m_R(x) \to 0$ as $R \to \infty$ for each x, we have $\Delta\phi = 0$. (Note that $\Delta\phi$ is defined by $\int h\Delta\phi = \int \phi\Delta h$ for $h \in C_c^\infty(\mathbb{R}^n)$; but $\int \phi\Delta h = \lim_{R\to\infty}\int \phi_R \Delta h$ (dominated convergence) $= \lim_{R\to\infty}\int \Delta\phi_R h = \lim_{R\to\infty}\int h(m_R - m) = 0$.) Thus, ϕ is harmonic and $\phi \leq 0$ a.e. (since $f_R^\dagger \geq \widetilde{f}$ a.e.), so $\phi = -C$.

Finally, if $m \in C^\infty(\mathbb{R}^n)$ is given with $(1+|x|)^{2-n} m(x)$ summable, then $f^\dagger$ is subharmonic and $\Delta f^\dagger = m$. To prove this, introduce m_R and $f_R^\dagger$ as above and take the limit $R \to \infty$.

Step 3. The last step is the general case that $\Delta f = \mu$, a measure. With $h_\varepsilon \in C_c^\infty(\mathbb{R}^n)$ as in the proof of Theorem 9.3, we consider $f_\varepsilon := h_\varepsilon * f \in C^\infty(\mathbb{R}^n)$. f_ε satisfies the hypotheses of the theorem, and also $f_\varepsilon \geq f$ (by the subharmonicity of f as in 9.3(12)). Moreover, it is easy to check that $\Delta f_\varepsilon = m_\varepsilon \in C^\infty(\mathbb{R}^n)$ with $m_\varepsilon(y) = \int h_\varepsilon(y-x)\mu(\mathrm{d}x)$. If $f_\varepsilon^\dagger$ is given by (2) with $\mu(\mathrm{d}x) = m_\varepsilon(x)\,\mathrm{d}x$, then $f_\varepsilon = f_\varepsilon^\dagger - C_\varepsilon$, with $C_\varepsilon \geq 0$. As $\varepsilon \to 0$ (through an appropriate subsequence), $f_\varepsilon \to f$ a.e. and monotonically and also $f_\varepsilon^\dagger \to f^\dagger$ a.e. (by using $f_\varepsilon^\dagger = -G_0 * (h_\varepsilon * \mu) = -h_\varepsilon * (G_0 * \mu)$, which follows from Fubini's theorem). Again, $f^\dagger$ is a monotone limit of $f_\varepsilon^\dagger$, as in 9.3(12)–(14), and $f_\varepsilon^\dagger \geq f_\varepsilon \geq f$, so $f^\dagger$ is upper semicontinuous. It is also easy to check, as above, that $\Delta(f - f^\dagger) = 0$. Since $f - f^\dagger \leq 0$, we conclude that $f = f^\dagger - C$.

The converse is left to the reader. ∎

• The following theorem of [Newton] is fundamental. Today we consider it simple but it is one of the high points of seventeenth century mathematics. We prove it for measures, μ. Equation (3) says (gravitationally speaking) that away from Earth's surface, all of Earth's mass appears to be concentrated at its center.

9.7 THEOREM (Spherical charge distributions are 'equivalent' to point charges)

Let μ_+ and μ_- be (positive) Borel measures on $\mathbb{R}^n$ and set $\mu := \mu_+ - \mu_-$. Assume that $\nu := \mu_+ + \mu_-$ satisfies $\int_{\mathbb{R}^n} w_n(x)\mathrm{d}\nu(x) < \infty$, where $w_n(x)$ is defined in 6.21(8). Define

$$V(x) := \int_{\mathbb{R}^n} G_y(x)\mu(\mathrm{d}y). \tag{1}$$

Then, the integral in (1) is absolutely integrable (i.e., $G_y(x)$ is ν-summable) for almost every x in $\mathbb{R}^n$ (with respect to Lebesgue measure). Hence, $V(x)$ is well defined almost everywhere; in fact, $V \in L_{\mathrm{loc}}^1(\mathbb{R}^n)$.

Now assume μ is spherically symmetric (i.e., $\mu(A) = \mu(\mathcal{R}A)$ for any Borel set A and any rotation $\mathcal{R}$). Then

$$|V(x)| \leq |G_0(x)| \int_{\mathbb{R}^n} \mathrm{d}\nu. \tag{2}$$

If B_R denotes the closed ball of radius R centered at 0 and if $\mu(A) = 0$ whenever $A \cap B_R = \emptyset$, then, for all $|x| \geq R$, we have **Newton's theorem**:

$$V(x) = G_0(x) \int_{\mathbb{R}^n} \mathrm{d}\mu. \tag{3}$$

PROOF. The proof will be carried out for $n \geq 3$ but the statement holds in general. Let $P(x) := \int_{\mathbb{R}^n} |x-y|^{2-n} \nu(\mathrm{d}y)$. To show that $P \in L^1_{\mathrm{loc}}(\mathbb{R}^n)$ it suffices to show that $\int_B P(x)\,\mathrm{d}x < \infty$ for any ball centered at 0. By Fubini's theorem we can do the x integration before the y integration, for which purpose we need the formula

$$J(r,y) = |\mathbb{S}^{n-1}|^{-1} \int_{\mathbb{S}^{n-1}} |r\omega - y|^{2-n}\,\mathrm{d}\omega = \min(r^{2-n}, |y|^{2-n}). \tag{4}$$

In case $n = 3$ this formula follows by an elementary integration in polar coordinates. The general case is a bit more difficult and we prove (4) in a different fashion. We note that $J(r,y)$ is the average of the function $|x-y|^{2-n}$ in x over the sphere of radius r. The function $x \mapsto |x-y|^{2-n}$ is harmonic as a function of x in the ball $\{x : |x| < |y|\}$ and hence, by the mean value property (cf. 9.3(3) with equalities), $J(r,y) = J(0,y) = |y|^{2-n}$. J depends only on $|y|$ and r and is a *symmetric* function of these variables. Thus, (4) follows for $r \neq |y|$. It is left to the reader to show that $J(r,y)$ is continuous in r and y, and hence that (4) is true for $r = |y|$.

It is easy to check that $\int_0^R \min(r^{2-n}, |y|^{2-n}) r^{n-1}\,\mathrm{d}r \leq C(R)(1+|y|)^{2-n}$, where $C(R)$ depends on R but not on $|y|$. Thus, by using polar coordinates, we have that

$$\int_B |x-y|^{2-n}\,\mathrm{d}x \leq C(R)(1+|y|)^{2-n},$$

and our integrability hypothesis about μ guarantees that $\int_B P < \infty$. Since $P \in L^1_{\mathrm{loc}}(\mathbb{R}^n)$, P is finite a.e. and the same holds for V since $|V| \leq P$.

To prove (2) we observe that V is spherically symmetric (i.e., $V(x_1) = V(x_2)$ when $|x_1| = |x_2|$) so, for each fixed x, $V(x) = V(|x|\omega)$ for all $\omega \in \mathbb{S}^{n-1}$. We can then compute the average of $V(|x|\omega)$ over $\mathbb{S}^{n-1}$ and, using (4), we conclude (2). To prove (3) we do the same computation with μ instead of ν (which is allowed by the absolute integrability) and find that

$$V(x) = |x|^{2-n} \int_{|y| \leq |x|} \mu(\mathrm{d}y) + \int_{|y| > |x|} |y|^{2-n} \mu(\mathrm{d}y), \tag{5}$$

from which (3) follows if $\nu(\{y : |y| > |x|\}) = 0$. ∎

9.8 THEOREM (Positivity properties of the Coulomb energy)

If $f : \mathbb{R}^n \to \mathbb{C}$ satisfies $D(|f|, |f|) < \infty$, then

$$D(f, f) \geq 0. \tag{1}$$

There is equality if and only if $f \equiv 0$. Moreover, if $D(|g|, |g|) < \infty$, then

$$|D(f, g)|^2 \leq D(f, f) D(g, g), \tag{2}$$

with equality for $g \not\equiv 0$ if and only if $f = cg$ for some constant c. The map $f \mapsto D(f, f)$ is strictly convex, i.e., when $f \neq g$ and $0 < \lambda < 1$

$$D(\lambda f + (1-\lambda)g, \lambda f + (1-\lambda)g) < \lambda D(f, f) + (1-\lambda) D(g, g). \tag{3}$$

REMARK. Theorem 9.8 could have been stated in greater generality by omitting the restriction $n \geq 3$ and by replacing the exponent $2 - n$ in the definition 9.1(2) of $D(f, g)$ by any number $\gamma \in (-n, 0)$. See Theorem 4.3 (Hardy–Littlewood–Sobolev inequality). The reason for choosing $2 - n$ is, of course, that $|x - y|^{2-n}$ has a potential theoretic significance as the Green's function of the Laplacian (cf. Sects. 6.20 and 9.7).

PROOF. By a simple consideration of the real and imaginary parts of f, one sees that to prove (1) it suffices to assume that f is real-valued. Let $h \in C_c^\infty(\mathbb{R}^n)$ with $h(x) \geq 0$ for all x and with h spherically symmetric, i.e., $h(x) = h(y)$ when $|x| = |y|$. Let k be the convolution $k(x) := (h * h)(x) = K(|x|)$. By multiplying h by a suitable constant, we can assume henceforth that $\int_0^\infty t^{n-3} K(t) \, dt = \frac{1}{2}$. By the simple scaling $t \mapsto t|x|^{-1}$,

$$I(x) := \int_0^\infty t^{n-3} k(tx) \, dt = |x|^{2-n} \int_0^\infty t^{n-3} K(t) \, dt = \frac{1}{2} |x|^{2-n}. \tag{4}$$

However, $I(x - y)$ can also be written as

$$I(x - y) := \int_0^\infty t^{2n-3} \int_{\mathbb{R}^n} h(t(z - y)) h(t(z - x)) \, dz \, dt,$$

where $h(x) = h(-x)$ has been used. Using Fubini's theorem (the hypothesis $D(|f|, |f|) < \infty$ is needed here),

$$D(f, f) = \int_{\mathbb{R}^n} \int_{\mathbb{R}^n} \overline{f}(x) f(y) I(x - y) \, dx \, dy = \int_0^\infty t^{-3} \int_{\mathbb{R}^n} |g_t(z)|^2 \, dz \, dt, \tag{5}$$

with $g_t(z) = t^n \int_{\mathbb{R}^n} h(t(z-x))f(x)\,dx = h_t * f(z)$ and $h_t(y) := t^n h(ty)$. The inequality $D(f,f) \geq 0$ is evident from (5).

Now assume that $D(f,f) = 0$. We must show $f \equiv 0$. From (5) we see that $g_t \equiv 0$ for almost every $t \in (0, \infty)$. Suppose h has support in the ball B_R of radius R, so that the support of h_t is also in B_R for all $t \geq 1$. Then, if $\chi_{w,2R}$ is the characteristic function of the ball $B_{w,2R}$ of radius $2R$ centered at w, and if $f_w(x) = \chi_{w,2R}(x)f(x)$, we have that if $t \geq 1$ and $|x-w| \leq R$, then $(h_t * f_w)(x) = (h_t * f)(x) = g_t(x) = 0$. However, $f_w \in L^1(\mathbb{R}^n)$, and we can use Theorem 2.16 (approximation by C^∞-functions) (noting that $C := \int h_t$ is independent of t) to conclude that $h_t * f_w \to C f_w$ in $L^1(\mathbb{R}^n)$ as $t \to \infty$ through a sequence of t's such that $g_t \equiv 0$. Thus, as $t \to \infty$, $0 \equiv g_t \to f$ in $L^1(B_{w,R})$. Hence $f(x) = 0$ a.e. in $B_{w,R}$ and, since w was arbitrary, $f \equiv 0$.

The last two statements are trivial consequences of the first two. Inequality (2) is proved by considering $D(F,F)$ with $F = f - \lambda g$ and $\lambda = D(g,f)/D(g,g)$. To prove (3) note that the right side minus the left side is just $\lambda(1-\lambda)D(f-g, f-g)$. ∎

● We have seen that $\Delta f \geq 0$ implies the mean value inequalities 9.3(3). As an aid in finding effective lower bounds for positive solutions to Schrödinger's equation (see Sect. 9.10) it is useful to extend the foregoing Theorem 9.5 to functions that satisfy the weaker condition $\Delta f \geq \mu^2 f$, *without* requiring $f \geq 0$.

9.9 THEOREM (Mean value inequality for $\Delta - \mu^2$)

Let $\Omega \subset \mathbb{R}^n$ be open, let $\mu > 0$ and let $f \in L^1_{\text{loc}}(\Omega)$ satisfy
$$\Delta f - \mu^2 f \geq 0 \quad \text{in } \mathcal{D}'(\Omega). \tag{1}$$
Then there is a unique upper semicontinuous function $\widetilde{f}$ on Ω that agrees with f almost everywhere and satisfies
$$\widetilde{f}(x) \leq \frac{1}{J(R)}[\widetilde{f}]_{x,R}, \tag{2}$$
and, moreover, the right side of (2) is a monotone nondecreasing function of R. The spherical average $[\widetilde{f}]_{x,R}$ is defined in 9.2(3) and $J : [0, \infty) \to (0, \infty)$ is the solution to
$$(\Delta - \mu^2)J(|x|) = 0 \tag{3}$$
with $J(0) = 1$.

Inequality (2) can be integrated over r to yield
$$\widetilde{f}(x) \leq \left\langle \frac{\widetilde{f}}{J} \right\rangle_{x,R} \leq \frac{1}{J(R)}[\widetilde{f}]_{x,R}. \tag{4}$$

REMARKS. (1) Note that $J(R)$ depends only on the product μR. This function can be expressed in terms of Bessel functions and it is strictly positive. When $n = 3$
$$J(r) = \frac{\sinh \mu r}{\mu r}. \tag{5}$$

(2) If the inequality is reversed in equation (1), then clearly (2) and (4) are reversed and the corresponding $\tilde{f}$ is lower semicontinuous.

PROOF. We shall largely imitate the proof of Theorem 9.3.

Step 1. Assume that $f \in C^\infty(\Omega)$, in which case (1) holds as a pointwise inequality. We shall show that $[f]_{x,r}/J(r)$ is an increasing function of r. Let K denote the C^∞-function $x \mapsto J(|x|)$, and note that (1) implies
$$\operatorname{div}(K\nabla f - f\nabla K) \geq 0. \tag{6}$$

(Here, $\operatorname{div} V = \sum_1^n \partial V_i/\partial x_i$.) Integrate (6) over $B_{x,r}$ to obtain
$$J(r)\frac{\mathrm{d}}{\mathrm{d}r}[f]_{x,r} - [f]_{x,r}\frac{\mathrm{d}}{\mathrm{d}r}J(r) \geq 0$$
which, in turn, implies
$$\frac{\mathrm{d}}{\mathrm{d}r}\frac{[f]_{x,r}}{J(r)} \geq 0. \tag{7}$$
This immediately implies (2) for C^∞-functions.

Step 2. For the general case, let j, as usual, be a spherically symmetric, nonnegative $C_c^\infty(\mathbb{R}^n)$-function with support in the unit ball and let $j_m(x) = m^n j(mx)$ for $m = 1, 2, 3, \ldots$. Define $h_m(x) = j_m(x)/J(|x|)$, which is also in $C_c^\infty(\mathbb{R}^n)$, and let
$$f_m = h_m * f,$$
which is C^∞ on
$$\Omega_M := \{x \in \Omega : x + y \in \Omega \quad \text{when } |y| \leq 1/M\}$$
provided $m > M$. Then f_m satisfies (1) pointwise in Ω_M and we want to show that $f_m(x)$ is a nonincreasing function of m for each x. As before we consider $f_{l,m} := h_l * f_m = h_m * f_l$ with m and l large. For $x \in \Omega_M$ this is a decreasing function of m when m and l are large since, for each large l,
$$f_{lm}(x) = \int_{\mathbb{R}^n} \frac{j_m(y)}{J(|y|)} f_l(x-y)\,\mathrm{d}y = \int_{\mathbb{R}^n} \frac{j(y)}{J(|y|/m)}[f_l]_{x,|y|/m}\,\mathrm{d}y.$$

This is nonincreasing in m because f_l is C^∞ and $[f_l]_{x,r}/J(r)$ is nondecreasing in r for each x and, as $l \to \infty$, $f_{lm}(x) \to f_m(x)$ for all x. From this we conclude that $\widetilde{f}(x) = \lim_{m\to\infty} f_m(x)$ exists and is an upper semicontinuous function.

The rest of the proof is as in 9.3 with some slight modifications. One is that the assertion in 9.3(iii) that $[f]_{x,r}$ is increasing in r has to be replaced by $[f]_{x,r}/J(r)$ is increasing in r, according to (7). The other is that a minor modification of the proof of Theorem 2.16 (approximation by C^∞-functions) shows that $h_m * f \to f$ in $L^1_{\text{loc}}(\Omega_M)$ as $m \to \infty$. Both modifications are trivial and rely on the facts that J is C^∞ and $J(r) \to 1$ as $r \to 0$. ∎

• We shall use Theorem 9.9 to prove a generalization of Harnack's inequality to solutions of Schrödinger's equation. This is a big topic of which the following only scratches the surface. The subject has a long history.

9.10 THEOREM (Lower bounds on Schrödinger 'wave' functions)

Let $\Omega \subset \mathbb{R}^n$ be open and connected, let $\mu > 0$ and let $W : \Omega \to \mathbb{R}$ be a measurable function such that $W(x) \leq \mu^2$ for all $x \in \Omega$. No lower bound is imposed on W. Suppose that $f : \Omega \to [0, \infty)$ is a nonnegative $L^1_{\text{loc}}(\Omega)$-function such that $Wf \in L^1_{\text{loc}}(\Omega)$ and such that the inequality

$$-\Delta f + Wf \geq 0 \quad \text{in } \mathcal{D}'(\Omega) \tag{1}$$

is satisfied.

Our conclusion is that there is a unique lower semicontinuous $\widetilde{f}$ that satisfies (1) and agrees with f almost everywhere. $\widetilde{f}$ has the following property: For each compact set $K \subset \Omega$ there is a constant $C = C(K, \Omega, \mu)$ depending only on K, Ω and μ but not on $\widetilde{f}$, such that

$$\widetilde{f}(x) \geq C \int_K f(x)\,dx \tag{2}$$

for each $x \in K$.

REMARKS. (1) The f in (1) should be compared with $-f$ in 9.9. Thus, upper semicontinuous there becomes lower semicontinuous here, etc. The signs in 9.9 and 9.10 have been chosen to agree with convention.

(2) Our hypothesis on W and our conclusion are far from optimal. The situation was considerably improved in [Aizenman–Simon] and then in [Fabes–Stroock], [Chiarenza–Fabes–Garofalo], [Hinz–Kalf].

PROOF. The existence of an $\tilde{f}$ is guaranteed by Theorem 9.9. Our problem here is to prove (2). We set $f = \tilde{f}$.

Since K is compact, there is a number $3R > 0$ such that $B_{x,3R} \subset \Omega$ for all $x \in K$. Moreover K, being compact, can be covered by finitely many, say N, balls $B_i := B_{x_i,R}$ with $x_i \in K$. Set $F_i = \int_{B_i} f$. At least one of these numbers, say F_1, satisfies $F_i \geq N^{-1} \int_K f$.

As in the proof of Theorem 9.6 we have, using 9.9(4), that for every $w \in B_i$
$$f(w) \geq \delta F_i, \tag{3}$$
with $\delta = [J(2R)|B_{0,2R}|]^{-1}$. Now let $y \in K$ and let γ be a continuous curve connecting y to x_1. This curve is covered by balls B_i, say $B_2, B_3, \ldots, B_M$ with $B_i \cap B_{i+1}$ nonempty for $i = 1, 2, \ldots, M-1$. We then have that for $z \in B_{i+1}$
$$F_{i+1} \geq \int_{B_i \cap B_{i+1}} f \geq \delta |B_i \cap B_{i+1}| F_i \tag{4}$$
since each $w \in B_i \cap B_{i+1}$ satisfies (3). From (4), with $\alpha := \min\{|B_i \cap B_j| : B_i \cap B_j \text{ nonempty}\} > 0$, we conclude that
$$F_{i+1} \geq \delta \alpha F_i. \tag{5}$$

We also conclude, by iterating (5) and using (3), that
$$f(y) \geq \delta(\delta\alpha)^{M-1} F_1 \geq \delta(\delta\alpha)^{M-1} N^{-1} \int_K f.$$

Obviously $M \leq N$, and the theorem is proved with $C = \delta^N \alpha^{N-1}/N$. ∎

Exercises for Chapter 9

1. Referring to Remark 3 after Theorem 9.3, prove that harmonic functions are infinitely differentiable. Use only the harmonicity property $f(x) = \langle f \rangle_{x,R}$ for every x.

2. Prove **Weyl's lemma**: Let T be a *distribution* that satisfies $\Delta T = 0$ in $\mathcal{D}'(\Omega)$. Show that T is a harmonic *function*.

3. Prove the assertion made in Remark 4 after Theorem 9.3, namely the function $t \mapsto [\widetilde{f}]_{x,r(t)}$, defined by 9.3(7), is convex.

4. Let $f^1, f^2, \ldots$ be a sequence of subharmonic functions on the open set $\Omega \subset \mathbb{R}^n$ and consider $g(x) = \sup_{1 \le i < \infty} f^i(x)$ for every $x \in \Omega$. Show that g is also subharmonic. Consider the analogous statement for superharmonic functions.

Chapter 10

Regularity of Solutions of Poisson's Equation

10.1 INTRODUCTION

Theorem 6.21 states that Poisson's equation

$$-\Delta u = f \quad \text{in } \mathcal{D}'(\mathbb{R}^n) \tag{1}$$

has a solution for any $f \in L^1_{\text{loc}}(\mathbb{R}^n)$ satisfying some mild integrability condition at infinity, e.g., $y \mapsto w_n(y)f(y)$ is summable (see 6.21(8) for the definition of $w_n(y)$). A solution is then given for almost every $x \in \mathbb{R}^n$ by

$$K_f(x) = \int_{\mathbb{R}^n} G_y(x) f(y) \, dy, \tag{2}$$

and any other solution to (1) is given by

$$u = K_f + h, \tag{3}$$

where h is an arbitrary harmonic function. The same is true when $\mathbb{R}^n$ is replaced by an open set Ω; in that case we merely replace $\mathbb{R}^n$ by Ω in (2).

The function K_f is an $L^1_{\text{loc}}(\mathbb{R}^n)$-function. It is not necessarily classically differentiable—or even continuous—but it does have a distributional derivative that is a function. The questions to be addressed in this chapter are the following. What additional conditions on f will insure that K_f is twice continuously differentiable, or even once continuously differentiable, or—most modestly—even continuous? Note that the harmonic function h in (3) is always infinitely differentiable (Theorem 9.3 Remark 3), so the above questions about K_f apply to the general solution in (3). These questions will be answered, but, before doing so, some general remarks are in order.

(1) Our treatment here barely scratches the surface of a larger subject called **elliptic regularity theory**. There, the Laplacian Δ is replaced by more general second order differential operators

$$L = \sum_{i,j=1}^{n} a_{ij}(x)\partial^2/\partial x_i \partial x_j + \sum_{i=1}^{n} b_i(x)\partial/\partial x_i + c(x).$$

The word elliptic stems from the fact that the symmetric matrix $a_{ij}(x)$ is required to be positive definite for each x. Furthermore, one considers domains Ω other than $\mathbb{R}^n$ and inquires about regularity (i.e., differentiability, etc.) up to the boundary of Ω. Questions of this type are difficult and we ignore them here by taking $\mathbb{R}^n$ as our domain. An alternative way to state this is that we can consider arbitrary domains (see (2)) but we concern ourselves only with **interior regularity**. The book [Gilbarg–Trudinger] can be consulted for more information about elliptic regularity. In particular, the last part of our proof of Theorem 10.2 is based on [Gilbarg–Trudinger, Lemma 4.5]. For more information about singular integrals, see [Stein].

(2) In the present context, a more useful notion than mere continuity (or even something stronger like continuous derivative) is local Hölder continuity (or locally Hölder continuous derivative). A function g defined on a domain $\Omega \subset \mathbb{R}^n$ is said to be **locally Hölder continuous** of order α (with $0 < \alpha \leq 1$) if, for each compact set K in Ω, there is a constant $b(K)$ such that

$$|f(x) - f(y)| \leq b(K)|x - y|^\alpha$$

for all x and y in K. The special case $\alpha = 1$ is also called **Lipschitz continuity**. The set of functions on Ω that are k-fold differentiable and whose k-fold derivatives are locally Hölder continuous of order α are denoted by

$$C^{k,\alpha}_{\text{loc}}(\Omega).$$

Here are two examples that demonstrate the inadequacy of ordinary continuity when $n > 1$.

EXAMPLE 1. Let $B \subset \mathbb{R}^3$ be the ball of radius $1/2$ centered at the origin and let $u(x) = w(r) := \ln[-\ln r]$ with $r = |x|$. By computing Δu in the usual way, i.e., $f(x) = -\Delta u(x) = -w''(r) - 2w'(r)/r$, we find that f is in $L^{3/2}(B)$. (It is easy to check, as in Sect. 6.20, that the above formula correctly gives Δu in the sense of distributions.) Now the interesting point is this: f is in $L^{3/2}(B)$ but u is not continuous; it is not even bounded. But Theorem 10.2 states that if $f \in L^{3/2+\varepsilon}(B)$ for any $\varepsilon > 0$, then u is automatically *Hölder continuous* for every exponent less than $4\varepsilon/(3+2\varepsilon)$.

EXAMPLE 2. With B as above, let $u(x) = w(r)Y_2(x/r)$ with $w(r) = r^2\ln[-\ln r]$ and $Y_2(x/r)$ the second spherical harmonic x_1x_2/r^2. Again, as is easily checked,

$$f(x) = -\Delta u(x) = [-w''(r) - 2r^{-1}w'(r) + 6r^{-2}w(r)]Y_2(x/r),$$

and f is continuous. f behaves as $-5(\ln r)^{-1}Y_2(x/r)$ near the origin and hence vanishes there. However, u is not twice differentiable at the origin, and $\partial^2 u/\partial x_1 \partial x_2$ even goes to infinity as $r \to 0$. Thus, continuity of f does not imply that $u \in C^2(\Omega)$, as might have been expected, but Theorem 10.3 states that if f is locally Hölder continuous of some order $\alpha < 1$, then $u \in C^{2,\alpha}_{\text{loc}}(\Omega)$.

(3) Regularity questions are purely local and, as a consequence of this fact, we can always assume in our proofs that f has compact support. The reason is that if we wish to investigate u and f near some point $x_0 \in \Omega$, we can fix some function $j \in C_c^\infty(\Omega)$ such that $j(x) = 1$ for x in some ball $B_1 \subset \Omega$ centered at x_0 and $0 \leq j(x) \leq 1$ for all $x \in \Omega$. Then write

$$f = jf + (1-j)f := f_1 + f_2, \tag{4}$$

whence $K_f = K_{f_1} + K_{f_2}$. The function K_{f_1} will be the object of our study. On the other hand, K_{f_2} is a function that, according to Theorem 6.21, satisfies $-\Delta K_{f_2} = f_2 = 0$ in B_1. Since K_{f_2} is harmonic in B_1, it is infinitely differentiable there and hence K_{f_2} and K_{f_1} have the same continuity and differentiability properties. In conclusion, we learn that the regularity properties of K_f in any open set $\omega \subset \Omega$ are completely determined by f inside ω alone. The term **hypoelliptic** is used to denote those operators, L, that, like $-\Delta$, have the property that whenever f is infinitely differentiable in some $\omega \subset \Omega$ all solutions, u, to $Lu = f$ in $\mathcal{D}'(\Omega)$ are also infinitely differentiable in ω.

A typical application of the theorems below is the so-called '**bootstrap**' **process**. As an example, consider the equation

$$-\Delta u = Vu \quad \text{in} \quad \mathcal{D}'(\mathbb{R}^n), \tag{5}$$

where $V(x)$ is a $C^\infty(\mathbb{R}^n)$-function. Since $u \in L^1_{\text{loc}}(\mathbb{R}^n)$, by definition, $Vu \in L^1_{\text{loc}}(\mathbb{R}^n)$. (In any case, Vu must be in $L^1_{\text{loc}}(\mathbb{R}^n)$ in order for (5) to make sense in $\mathcal{D}'(\mathbb{R}^n)$.) By equation (3), the preceding Remark (3) and Theorem 10.2, we have that $u \in L^{q_0}_{\text{loc}}(\mathbb{R}^n)$ with $q_0 = n/(n-2) > 1$. Thus $Vu \in L^{q_0}_{\text{loc}}(\mathbb{R}^n)$ and, repeating the above step, $u \in L^{q_1}_{\text{loc}}(\mathbb{R}^n)$ with $q_1 = n/(n-4)$. Eventually, we have $Vu \in L^p(\mathbb{R}^n)$ with $p > n/2$. By Theorem 10.2, u is in $C^{0,\alpha}(\mathbb{R}^n)$ with $\alpha > 0$. Then, using Theorem 10.3, $u \in C^{2,\alpha}(\mathbb{R}^n)$. Iterating this, we reach the final conclusion that $u \in C^\infty(\mathbb{R}^n)$.

10.2 THEOREM (Continuity and first differentiability of solutions of Poisson's equation)

Let f be in $L^p(\mathbb{R}^n)$ for some $1 \leq p \leq \infty$ with compact support, and let K_f be given by 10.1(2).

(i) K_f is continuously differentiable for $n = 1$. For $n \geq 2$ and for $1 \leq p < n/2$

$$K_f \in L^q_{\text{loc}}(\mathbb{R}^2) \quad \text{for all } q < \infty, \quad \text{for } p=1, n=2,$$
$$K_f \in L^q_{\text{loc}}(\mathbb{R}^n) \quad \text{for all } q < \frac{n}{n-2} \quad \text{for } p=1, n \geq 3,$$
$$K_f \in L^q(\mathbb{R}^n), \quad q = \frac{pn}{n-2p} \quad \text{for } p > 1, n \geq 3.$$

(ii) If $n/2 < p \leq n$, then K_f is Hölder continuous of every order $\alpha < 2 - n/p$,

$$|K_f(x) - K_f(y)| \leq C_n(\alpha, p)|x-y|^\alpha \|f\|_p (\mathcal{L}^n(\text{supp}\{f\}))^{\frac{2-\alpha}{n} - \frac{1}{p}}. \quad (1)$$

(iii) If $n < p$, then K_f has a derivative, given by 6.21(4),

$$\partial_i K_f(x) = \int_{\mathbb{R}^n} (\partial G_y / \partial x_i)(x) f(y) \, dy,$$

which is Hölder continuous of every order $\alpha < 1 - n/p$, i.e.,

$$|\partial_i K_f(x) - \partial_i K_f(y)| \leq D_n(\alpha, p)|x-y|^\alpha \|f\|_p (\mathcal{L}^n(\text{supp}\{f\}))^{\frac{1-\alpha}{n} - \frac{1}{p}}. \quad (2)$$

Here $D_n(\alpha, p)$ and $C_n(\alpha, p)$ are universal constants depending only on α and p.

PROOF. We shall treat only $n \geq 2$ and leave the simple $n = 1$ case to the reader. First we prove part (i). For $n = 2$ we can use the fact that for every $\varepsilon > 0$ and for x and y in a fixed ball of radius R in $\mathbb{R}^2$, there are constants c and d such that $|\ln|x-y|| \leq c|x-y|^{-\varepsilon} + d := h(x-y)$. Now we can apply Young's inequality, 4.2(4), to the pair $f(y)$ and $H(x) = h(x)\chi_{2R}(x)$, where χ is the characteristic function of the ball of radius $2R$. Since $H \in L^r(\mathbb{R}^2)$ for all $r < 2/\varepsilon$, we have that $K_f \in L^q_{\text{loc}}$ with $1 + 1/q = 1/p + 1/r > 1/p + \varepsilon/2$.

For $n \geq 3$ and $p = 1$, we use the fact that $|x|^{2-n}\chi_{2R}(x) \in L^r(\mathbb{R}^n)$ for all $r < n/n - 2$, and proceed as above. If $1 < p < n/2$, we appeal to the Hardy–Littlewood–Sobolev inequality, Sect. 4.3.

For part (ii) we first note that if $b > 1$ and $0 < \alpha < 1$, we have (using Hölder's inequality) that for $m \geq 1$

$$\frac{1}{m}(1 - b^{-m}) = \int_1^b t^{-m-1}\,dt$$
$$\leq \left(\int_1^b dt\right)^\alpha \left(\int_1^\infty t^{-(m+1)/(1-\alpha)}\,dt\right)^{1-\alpha} \leq (b-1)^\alpha.$$

Likewise,

$$\ln(b) = \int_1^b t^{-1}\,dt \leq (b-1)^\alpha \left(\int_1^\infty t^{-1/(1-\alpha)}\,dt\right)^{1-\alpha} \leq \frac{1}{\alpha}(b-1)^\alpha.$$

Substituting b/a for b, we find (for $a > 0$) that

$$|b^{-m} - a^{-m}| \leq m|b-a|^\alpha \max(a^{-m-\alpha}, b^{-m-\alpha}),$$
$$|\ln b - \ln a| \leq |b-a|^\alpha \max(a^{-\alpha}, b^{-\alpha})/\alpha.$$

If x, y and z are in $\mathbb{R}^n$, we can use the triangle inequality $||x-z| - |y-z|| \leq |x-y|$, as well as the fact that $\max(s,t) \leq s + t$, to conclude that

$$\left||x-z|^{-m} - |y-z|^{-m}\right| \leq m|x-y|^\alpha\{|x-z|^{-m-\alpha} + |y-z|^{-m-\alpha}\},$$
$$\left|\ln|x-z| - \ln|y-z|\right| \leq |x-y|^\alpha\{|x-z|^{-\alpha} + |y-z|^{-\alpha}\}/\alpha. \tag{3}$$

If we insert (3) into the definition of K_f, 10.1(2), we find for $n \geq 2$ that there is a universal constant C_n such that

$$|K_f(x) - K_f(y)| \leq C_n|x-y|^\alpha \sup_x \int_{\mathbb{R}^n} |x-y|^{2-n-\alpha}|f(y)|\,dy. \tag{4}$$

Using Hölder's inequality, we then have

$$|K_f(x) - K_f(y)|$$
$$\leq C_n|x-y|^\alpha \sup_x \left\{\int_{\text{supp}\{f\}} |x-y|^{(2-n-\alpha)p'}\,dy\right\}^{1/p'} \|f\|_p. \tag{5}$$

If $p > n/2$, then $p' < n/(n-2)$, so $|x|^{(2-n-\alpha)} \in L^{p'}_{\text{loc}}(\mathbb{R}^n)$ if $\alpha < 2 - n/p$. For such α, the integral in (5) is largest (given the volume of supp$\{f\}$) when supp$\{f\}$ is a ball and x is located at its center. The proof of this uses the simplest rearrangement inequality (see Theorem 3.4) and the fact that $|y|^{-1}$ is a symmetric-decreasing function. (5) proves (1).

The proof of (2) is essentially the same, except that we have to start with the representation 6.21(4) for the derivative $\partial_i K_f$. ∎

10.3 THEOREM (Higher differentiability of solutions of Poisson's equation)

Let f be in $C^{k,\alpha}(\mathbb{R}^n)$ with compact support, with $k \geq 0$ and $0 < \alpha < 1$, and let K_f be given by 10.1(2). Then
$$K_f \in C^{k+2,\alpha}(\mathbb{R}^n).$$

PROOF. Again we consider only $n \geq 2$ explicitly. It suffices to consider only $k = 0$ since 'differentiation commutes with Poisson's equation', i.e., $-\Delta u = f$ in $\mathcal{D}'(\mathbb{R}^n)$ implies that $-\Delta(\partial_i u) = \partial_i f$ in $\mathcal{D}'(\mathbb{R}^n)$. This follows directly from the fundamental definition of distributional derivative in terms of $C_c^\infty(\mathbb{R}^n)$ test functions. We assume $k = 0$ henceforth.

By Theorem 10.2 we know that $u \in C^{1,\alpha}(\mathbb{R}^n)$ with derivative given by 6.21(4). To show that $u \in C^2(\mathbb{R}^n)$ it suffices, by Theorem 6.10, to show that $\partial_i u$ has a distributional derivative that is a continuous function. We introduce a test function ϕ in order to compute this distributional derivative, i.e.,

$$-\int_{\mathbb{R}^n} (\partial_j \phi)(x)(\partial_i u)(x)\,\mathrm{d}x = \int_{\mathbb{R}^n} f(y) \int_{\mathbb{R}^n} (\partial_j \phi)(x)(\partial G_y/\partial x_i)(x)\,\mathrm{d}x\,\mathrm{d}y, \quad (1)$$

where Fubini's theorem has been used.

Note that we cannot integrate by parts once more, since $\partial_i \partial_j G_y(x)$ has a nonintegrable singularity. However, by dominated convergence the right side of (1) can be written as

$$\lim_{\varepsilon \to 0} \int_{\mathbb{R}^n} f(y) \int_{|x-y| \geq \varepsilon} (\partial_j \phi)(x)(\partial G_y/\partial x_i)(x)\,\mathrm{d}x\,\mathrm{d}y, \quad (2)$$

and it remains to compute the inner integral over x. Without loss of generality we can set $y = 0$. If we denote by e_j the vector with a one in position j and otherwise zeros, this inner integral is given by

$$\int_{|x| \geq \varepsilon} \mathrm{div}(e_j \phi)(x)(\partial G_0/\partial x_i)(x)\,\mathrm{d}x \quad (3)$$

which, by integration by parts and Gauss' theorem, is expressed as

$$-\varepsilon^{n-1}\int_{\mathbb{S}^{n-1}} \phi(\varepsilon\omega)(\partial G_0/\partial x_i)(\varepsilon\omega)\omega_j\,d\omega - \int_{|x|\geq\varepsilon} \phi(x)(\partial^2 G_0/\partial x_i\partial x_j)(x)\,dx, \tag{4}$$

where $\omega_j = x_j/|x|$.

To understand the second term one computes that

$$\int_{\mathbb{S}^{n-1}}(\partial^2 G_0/\partial x_i\partial x_j)(|x|\omega)\,d\omega = 0 \tag{5}$$

for all $|x|\neq 0$, since

$$(\partial^2 G_0/\partial x_i\partial x_j)(x) = \frac{1}{|\mathbb{S}^{n-1}|}|x|^{-n}(n\omega_i\omega_j - \delta_{ij}), \tag{6}$$

where $\delta_{ij} = 1$ if $i = j$ and $\delta_{ij} = 0$ otherwise. Thus, the second term in (4) can be replaced by

$$\int_{|x|\geq 1} \phi(x)(\partial^2 G_0/\partial x_i\partial x_j)(x)\,dx \\ + \int_{1\geq |x|\geq\varepsilon} (\phi(x)-\phi(0))(\partial^2 G_0/\partial x_i\partial x_j)(x)\,dx. \tag{7}$$

Inserting the first term of (4) in (2) and replacing 0 by y we obtain, by dominated convergence as $\varepsilon \to 0$,

$$\frac{1}{n}\delta_{ij}\int_{\mathbb{R}^n} \phi(y)f(y)\,dy. \tag{8}$$

Combining (7) with (2) yields

$$\int_{\mathbb{R}^n} \phi(x)\int_{|x-y|\geq 1} f(y)(\partial^2 G_y/\partial x_i\partial x_j)(x)\,dy\,dx \\ + \lim_{\varepsilon\to 0}\int_{\mathbb{R}^n}\phi(x)\int_{1\geq |x-y|\geq\varepsilon}(f(y)-f(x))(\partial^2 G_y/\partial x_i\partial x_j)(x)\,dy\,dx \tag{9}$$

by use of Fubini's theorem. Since $f \in C^{0,\alpha}(\mathbb{R}^n)$, the inner integral converges uniformly as $\varepsilon\to 0$ and hence, by interchanging this limit with the integral by Theorem 6.5 (functions are uniquely determined by distributions), we obtain the final formula

$$(\partial_i\partial_j u)(x) = \frac{1}{n}\delta_{ij}f(x) + \int_{|x-y|\geq 1} f(y)(\partial^2 G_y/\partial x_i\partial x_j)(x)\,dy \\ + \lim_{\varepsilon\to 0}\int_{1\geq |x-y|\geq\varepsilon}(f(x)-f(y))(\partial^2 G_y/\partial x_i\partial x_j)(x)\,dy \tag{10}$$

for almost every x in $\mathbb{R}^n$.

The first term on the right side of (10) is clearly Hölder continuous. So, too, is the second term, provided we recall that f has compact support. The third term is the interesting one. We can clearly take the limit $\varepsilon \to 0$ inside the integral by dominated convergence because $|f(y) - f(x)| < C|x - y|^\alpha$, and hence the integrand is in $L^1(\mathbb{R}^n)$. Let us call this third term $W_{ij}(x)$. It is defined for all x by the integral in (10) with $\varepsilon = 0$.

We want to show that
$$|W_{ij}(x) - W_{ij}(z)| \leq (\text{const.})|x - z|^\alpha.$$

If we change the integration variable in the integral for $W_{ij}(x)$ from y to $y + x$, and in $W_{ij}(z)$ from y to $y + z$, and then subtract the two integrals, we obtain
$$W_{ij}(x) - W_{ij}(z) = \int_{|y|<1} [f(x) - f(z) - f(y+x) + f(y+z)] H(y) \, dy, \quad (11)$$

with $H(y) := (\partial^2 G_0/\partial x_i \partial x_j)(y)$ given in (6). Note that $|H(y)| \leq C_1 |y|^{-n}$. Obviously, the factor [ ] in (11) is bounded above by $2C_2 |y|^\alpha$, where C_2 is the Hölder constant for f, i.e., $|f(x) - f(z)| \leq C_2 |x - z|^\alpha$.

By appealing to translation invariance, it suffices to assume $z = 0$, which we do henceforth for convenience. The integration domain $0 < |y| < 1$ in (11) can be written as the union of $A = \{y : 0 < |y| \leq 4|x|\}$ and $B = \{y : 4|x| < |y| < 1\}$. The second domain is empty if $|x| \geq 1/4$. For the first domain, A, we use our bound $|y|^\alpha$ to obtain the bound
$$2C_1 C_2 C_3 \int_0^{4|x|} r^{-n} r^\alpha r^{n-1} \, dr = C_4 |x|^\alpha$$

for the integral over A in (11), which is precisely our goal.

For the second domain, B, we observe that $\int_B [f(x) - f(0)] H(y) \, dy = 0$ since, by (5), the angular integral of H is zero. For the third term, $f(y+x)$, we change back to the original variables $y + x \to y$, and thus the third plus fourth terms in (11) become
$$I := -\int_D f(y) H(y - x) \, dy + \int_B f(y) H(y) \, dy, \quad (12)$$

where $D = \{y : 4|x| < |y - x| < 1\}$.

To calculate the first integral we can write $B = (B \cap D) \cup (B \sim D)$. For the second we can write $D = (B \cap D) \cup (D \sim B)$. On the common domain we have
$$I_1 = \int_{B \cap D} f(y)[H(y) - H(y - x)] \, dy.$$

Section 10.3

But $|H(y) - H(y-x)| \leq C_5|x||y|^{-n-1}$ when $y \in B \cap D$ and moreover, $B \cap D \subset \{y : 3|x| < |y| < 1 + |x|\}$. Thus

$$|I_1| \leq C_3 C_5 |x| \int_{3|x|}^{1+|x|} r^\alpha r^{-n-1} r^{n-1} \, dr \leq \frac{C_6}{1-\alpha} |x|^\alpha.$$

(Recall $|x| < 1/4$.) Here, *for the first time*, we require $\alpha < 1$ instead of merely $\alpha \leq 1$.

The domain $B \sim D$ essentially has two parts. We can write $B \sim D \subset E \cup G$ where $E = \{y : 4|x| < |y| < 5|x|\}$ and $G = \{y : 1-|x| < |y| < 1\}$. Then

$$\left| \int_E f(y) H(y) \, dy \right| \leq C_1 C_2 C_3 \int_{4|x|}^{5|x|} r^\alpha r^{-n} r^{n-1} \, dr \leq C_7 |x|^\alpha,$$

$$\left| \int_G f(y) H(y) \, dy \right| \leq C_1 C_2 C_3 \int_{1-|x|}^{1} r^\alpha r^{-n} r^{n-1} \, dr \leq C_8 |x|^\alpha.$$

A similar estimate holds for the $D \sim B$ contribution to the second integral in (12). Thus, the last term in (10) is Hölder continuous of order α. ∎

Chapter 11

Introduction to the Calculus of Variations

11.1 INTRODUCTION

As an illustration of the use of the mathematics developed in this book, we give three additional examples (beyond those of Chapter 4) of solving optimization problems. The first comes from quantum mechanics and is the problem of determining the energy of an atom—primarily the lowest one. The second is a classical type minimization problem—the Thomas–Fermi problem—that arises in chemistry. The third is a problem in electrostatics, namely the capacitor problem. In all cases the difficult part is showing the existence of a minimizer, and hence of a solution to a partial differential equation. Needless to say, the following considerations (known as **the direct method in the calculus of variations**) for establishing a solution to a differential equation are not limited to these elementary examples, but should be viewed as a general strategy to attack optimization problems.

Historically, and even today in many places, it is customary to dispense with the question of existence as a mere subtlety. By simply assuming that a minimizer or maximizer exists, however, and then trying to derive properties for it, one can be led to severe inconsistencies—as the following amusing example taken from [L. C. Young] and attributed to Perron shows: "Let N be the largest natural number. Since $N^2 \geq N$ and N is the largest natural number, $N^2 = N$ and hence $N = 1$." What this example tells us is that even if the 'variational equation', here $N^2 = N$, can be solved explicitly, the resulting solution need not have anything to do with the problem we started out to solve.

Let us continue this overview with some general remarks about minimization of functions. A general theorem in analysis says that a bounded continuous real function f defined on a bounded and closed set K in $\mathbb{R}^n$ attains its minimum value. To prove this, pick a sequence of points x^j such that
$$f(x^j) \to \lambda := \inf_{x \in K} f(x) \quad \text{as } j \to \infty.$$
Since K is bounded and closed, there exists a subsequence, again denoted by x^j, and a point $x \in K$ such that $x^j \to x$ as $j \to \infty$. Hence, since f is continuous,
$$\lambda := \lim_{j \to \infty} f(x^j) = f(x),$$
and the minimum value is attained at x.

Instead of $\mathbb{R}^n$, consider now $L^2(\Omega, \mathrm{d}\mu)$ and let $\mathcal{F}(\psi)$ be some functional defined on this space. In many examples $\mathcal{F}(\psi)$ is strongly continuous, i.e., $\mathcal{F}(\psi^j) \to \mathcal{F}(\psi)$ as $j \to \infty$ whenever $\|\psi^j - \psi\|_2 \to 0$ as $j \to \infty$. Suppose we wish to show that the infimum of $\mathcal{F}(\psi)$ is attained on $K := \{\psi \in L^2(\Omega, \mathrm{d}\mu) : \|\psi\|_2 \leq 1\}$. This set is certainly closed and bounded, but for a bounded sequence $\psi^j \in K$ there need not be a strongly convergent subsequence (see Sect. 2.9).

The idea now is to relax the strength of convergence. Indeed, if we use the notion of weak convergence instead of strong convergence, then, by Theorem 2.18 (bounded sequences have weak limits), every sequence in K has a weakly convergent subsequence. In this way, the set of convergent sequences has been enlarged—but a new problem arises. The functional $\mathcal{F}(\psi)$ need not be weakly continuous—and it rarely is. Thus, to summarize, the more sequences exist that have convergent subsequences the less likely it is that $\mathcal{F}(\psi)$ is continuous on these sequences. The way out of this apparent dilemma is that in many examples the functional turns out to be **weakly lower semicontinuous**, i.e.,
$$\liminf_{j \to \infty} \mathcal{F}(\psi^j) \geq \mathcal{F}(\psi) \quad \text{if } \psi^j \rightharpoonup \psi \text{ weakly.} \tag{1}$$
Thus, if ψ^j is a minimizing sequence, i.e., if
$$\mathcal{F}(\psi^j) \to \inf\{\mathcal{F}(\psi) : \psi \in C\} = \lambda,$$
then there exists a subsequence ψ^j such that $\psi^j \rightharpoonup \psi$ weakly, and hence
$$\lambda = \lim_{j \to \infty} \mathcal{F}(\psi^j) \geq \mathcal{F}(\psi) \geq \lambda.$$
Therefore, $\mathcal{F}(\psi) = \lambda$, and our goal is achieved!

11.2 SCHRÖDINGER'S EQUATION

The **time independent Schrödinger equation** [Schrödinger] for a particle in $\mathbb{R}^n$, interacting with a force field $F(x) = -\nabla V(x)$, is

$$-\Delta \psi(x) + V(x)\psi(x) = E\psi(x). \tag{1}$$

The function $V : \mathbb{R}^n \to \mathbb{R}$ is called a **potential** (not to be confused with the potentials in Chapter 9). The 'wave function' ψ is a complex-valued function in $L^2(\mathbb{R}^n)$ subject to the **normalization condition**

$$\|\psi\|_2 = 1. \tag{2}$$

The function $\rho_\psi(x) = |\psi(x)|^2$ is interpreted as the probability density for finding the particle at x. An $L^2(\mathbb{R}^n)$ solution to (1) may or may not exist for any E; often it does not. The special real numbers E for which such solutions exist are called **eigenvalues** and the solution, ψ, is called an **eigenfunction**.

Associated with (1) is a variational problem. Consider the following functional defined for a suitable class of functions in $L^2(\mathbb{R}^n)$ (to be specified later):

$$\mathcal{E}(\psi) = T_\psi + V_\psi, \tag{3}$$

with

$$T_\psi = \int_{\mathbb{R}^n} |\nabla \psi(x)|^2 \, dx \quad \text{and} \quad V_\psi = \int_{\mathbb{R}^n} V(x) |\psi(x)|^2 \, dx. \tag{4}$$

Physically, T_ψ is called the **kinetic energy** of ψ, V_ψ is its **potential energy** and $\mathcal{E}(\psi)$ is the **total energy** of ψ.

The variational problem we shall consider is to minimize $\mathcal{E}(\psi)$ subject to the constraint $\|\psi\|_2 = 1$.

As we shall show in Sect. 11.5, a minimizing function ψ_0, if one exists, will satisfy equation (1) with $E = E_0$, where

$$E_0 := \inf\{\mathcal{E}(\psi) : \int |\psi|^2 = 1\}.$$

Such a function ψ_0 will be called a **ground state**. E_0 is called the **ground state energy**.[1]

Thus the variational problem determines not only ψ_0 but also a corresponding eigenvalue E_0, which is the smallest eigenvalue of (1).

---

[1] Physically, the ground state energy is the lowest possible energy the particle can attain. It is a physical fact that the particle will settle eventually into its ground state, by emitting energy, usually in the form of light.

Our route to finding a solution to (1) takes us to the main problem: Show, under suitable assumptions on V, that a minimizer exists, i.e., show that there exists a ψ_0 satisfying (2) and such that

$$\mathcal{E}(\psi_0) = \inf\{\mathcal{E}(\psi) : \|\psi\|_2 = 1\}.$$

There are examples where a minimizer does *not* exist, e.g., take V to be identically zero.

In Sect. 11.5 we shall prove, under suitable assumptions on V, the existence of a minimizer for $\mathcal{E}(\psi)$. We shall also solve the corresponding relativistic problem, in which the kinetic energy is given by $(\psi, |p|\psi)$ instead, as defined in Sect. 7.11. In the nonrelativistic case ((1), (4)) it will be shown that the minimizers satisfy (1) in the sense of distributions. Higher eigenvalues will be explained in Sect. 11.6. The content of Sect. 11.7 is an application of the results of Chapter 10 to show that under suitable additional assumptions on V, the distributional solutions of (1) are sufficiently regular to yield classical solutions, i.e., solutions that are twice continuously differentiable.

A final question concerns uniqueness of the minimizer. In our Schrödinger example, $\mathcal{E}(\psi)$, uniqueness means that the ground state solution to (1) is unique, apart from an 'overall phase', i.e., $\psi_0(x) \to e^{i\theta}\psi_0(x)$ for some $\theta \in \mathbb{R}$. That uniqueness of the minimizer implies uniqueness of the solution to (1) with $E = E_0$ is not totally obvious; it is proved in Theorem 11.8. The tool that will enable us to prove uniqueness of the minimizer is the strict convexity of the map $\rho \to \mathcal{E}(\sqrt{\rho})$ for *strictly* positive functions $\rho : \mathbb{R}^n \to \mathbb{R}^+$. (See Theorem 7.8 (convexity inequality for gradients).) The hard part is to establish the strict positivity of a minimizer. Theorem 9.10 (lower bounds on Schrödinger wave functions) will be crucial here.

11.3 DOMINATION OF THE POTENTIAL ENERGY BY THE KINETIC ENERGY

Recall that the functional to consider is

$$\mathcal{E}(\psi) = \int_{\mathbb{R}^n} |\nabla \psi(x)|^2 \, dx + \int_{\mathbb{R}^n} V(x) |\psi(x)|^2 \, dx,$$

and the ground state energy E_0 is

$$E_0 = \inf\{\mathcal{E}(\psi) : \|\psi\|_2 = 1\}. \tag{1}$$

The kinetic energy is defined for any function in $H^1(\mathbb{R}^n)$ and the second term is defined at least for $\psi \in C_c^\infty(\mathbb{R}^n)$ if we assume that $V \in L^1_{\text{loc}}(\mathbb{R}^n)$. The first necessary condition for a minimizer to exist is that $\mathcal{E}(\psi)$ is bounded below by some constant independent of ψ (when $\|\psi\|_2 \leq 1$). The reader can imagine that when, e.g., $V(x) = -|x|^{-3}$, then $\mathcal{E}(\psi)$ is no longer bounded below. Indeed, for any $\psi \in C_c^\infty(\mathbb{R}^n)$ with $\|\psi\|_2 = 1$ and $\int V(x)|\psi(x)|^2\, dx < \infty$, define $\psi_\lambda(x) = \lambda^{n/2}\psi(\lambda x)$ and observe that $\|\psi_\lambda\|_2 = 1$. One easily computes

$$\mathcal{E}(\psi_\lambda) = \lambda^2 \int_{\mathbb{R}^n} |\nabla \psi(x)|^2\, dx - \lambda^3 \int_{\mathbb{R}^n} V(x)|\psi(x)|^2\, dx.$$

Clearly, $\mathcal{E}(\psi_\lambda) \to -\infty$ as $\lambda \to \infty$. One sees from this example that the assumptions on V must be such that V_ψ can be bounded below in terms of the kinetic energy T_ψ and the norm $\|\psi\|_2$.

Any inequality in which the kinetic energy T_ψ dominates some kind of integral of ψ (but not involving $\nabla \psi$) is called an **uncertainty principle**. The historical reason for this strange appellation is that such an inequality implies that one cannot make the potential energy very negative without also making the kinetic energy large, i.e., one cannot localize a particle simultaneously in both $\mathbb{R}^n$ and the Fourier transform copy of $\mathbb{R}^n$. The most famous uncertainty principle, historically, is Heisenberg's: In $\mathbb{R}^n$

$$(\psi, p^2\psi) \geq \frac{n^2}{4}(\psi, x^2\psi)^{-1} \tag{2}$$

for $\psi \in H^1(\mathbb{R}^n)$ and $\|\psi\|_2 = 1$. The proof of this inequality (which uses the fact that $\nabla \cdot x - x \cdot \nabla = n$) can be found in many textbooks and we shall not give it here because (2) is not actually very useful. Knowledge of $(\psi, x^2\psi)$ tells us little about T_ψ. The reason for this is that any ψ can easily be modified in an arbitrarily small way (in the $H^1(\mathbb{R}^n)$-norm) so that ψ concentrates somewhere, i.e., $(\psi, p^2\psi)$ is not small, but $(\psi, x^2\psi)$ is huge. To see this, take any fixed function ψ and then replace it by $\psi_y(x) = \sqrt{1-\varepsilon^2}\psi(x) + \varepsilon \psi(x-y)$ with $\varepsilon \ll 1$ and $|y| \gg 1$. To a very good approximation, $\psi_y = \psi$ but, as $|y| \to \infty$, $\|\psi_y\|_2 \to 1$ and $(\psi_y, x^2\psi_y) \to \infty$. Thus, the right side of (2) goes to zero as $|y| \to \infty$ while $T_{\psi_y} \approx T_\psi$ does not go to zero.

Sobolev's inequality (see Sects. 8.3 and 8.5) is much more useful in this respect. Recall that for functions that vanish at infinity on $\mathbb{R}^n$, with $n \geq 3$, there are constants S_n such that

$$T_\psi \geq S_n \left\{ \int_{\mathbb{R}^n} |\psi(x)|^{2n/(n-2)}\, dx \right\}^{(n-2)/n} = S_n \|\rho_\psi\|_{n/(n-2)}$$
$$= \frac{3}{4}(4\pi^2)^{2/3}\|\rho_\psi\|_3 \quad \text{for } n = 3. \tag{3}$$

For $n = 1$ and $n = 2$, on the other hand, we have

$$T_\psi + \|\psi\|_2^2 \geq S_{n,p}\|\rho_\psi\|_p \quad \text{for all } 2 \leq p < \infty, \quad n = 2, \tag{4}$$

$$T_\psi + \|\psi\|_2^2 \geq S_1 \|\rho_\psi\|_\infty, \qquad n = 1. \tag{5}$$

Moreover, when $n = 1$ and $\psi \in H^1(\mathbb{R}^1)$, ψ is not only bounded, it is also continuous.

An application of Hölder's inequality to (3) yields, for any potential $V \in L^{n/2}(\mathbb{R}^n)$, $n \geq 3$,

$$T_\psi \geq S_n \|\rho_\psi\|_{n/(n-2)} \geq S_n(\psi, V\psi) \|V\|_{n/2}^{-1}. \tag{6}$$

An immediate application of (6) is that

$$T_\psi + V_\psi \geq 0 \tag{7}$$

whenever $\|V\|_{n/2} \leq S_n$.

A simple extension of (6) leads to a lower bound on the ground state energy for $V \in L^{n/2}(\mathbb{R}^n) + L^\infty(\mathbb{R}^n), n \geq 3$, i.e., for V's that satisfy

$$V(x) = v(x) + w(x) \tag{8}$$

for some $v \in L^{n/2}(\mathbb{R}^n)$ and $w \in L^\infty(\mathbb{R}^n)$. There is then some constant λ such that $h(x) := -(v(x) - \lambda)_- = \min(v(x) - \lambda, 0) \leq 0$ satisfies $\|h\|_{n/2} \leq \frac{1}{2} S_n$ (exercise for the reader). In particular, by (6), $h_\psi \geq -\frac{1}{2} T_\psi$. Then we have

$$\begin{aligned} \mathcal{E}(\psi) &= T_\psi + V_\psi = T_\psi + (v - \lambda)_\psi + \lambda + w_\psi \\ &\geq T_\psi + h_\psi + \lambda + w_\psi \geq \frac{1}{2} T_\psi + \lambda - \|w\|_\infty \end{aligned} \tag{9}$$

and we see that $\lambda - \|w\|_\infty$ is a lower bound to E_0. Furthermore (9) implies that *the total energy effectively bounds the kinetic energy*, i.e., we have that

$$T_\psi \leq 2(\mathcal{E}(\psi) - \lambda + \|w\|_\infty). \tag{10}$$

When $n = 2$, the preceding argument, together with (4), gives a finite E_0 whenever $V \in L^p(\mathbb{R}^2) + L^\infty(\mathbb{R}^2)$ for any $p > 1$. Likewise, when $n = 1$ we can conclude that E_0 is finite whenever $V \in L^1(\mathbb{R}^1) + L^\infty(\mathbb{R}^1)$. In fact, a bit more can be deduced when $n = 1$. Since $\psi \in H^1(\mathbb{R}^1)$ implies that ψ is continuous, it makes sense to define $\int \psi(x) \mu(\mathrm{d}x)$ when $\mu = \mu_1 - \mu_2$ and when μ_1 and μ_2 are any bounded, positive Borel measures on $\mathbb{R}^1$. ('Bounded' means that $\int \mu_i(\mathrm{d}x) < \infty$.) A well-known example in the physics literature is $\mu(\mathrm{d}x) = c\, \delta(x)\, \mathrm{d}x$ where $\delta(x)$ is Dirac's 'delta function'. More precisely, $\int \psi(x) \mu(\mathrm{d}x) = c\, \psi(0)$. Then we can define

$$\mathcal{E}(\psi) = T_\psi + \int_{\mathbb{R}^n} |\psi(x)|^2 \mu(\mathrm{d}x) \tag{11}$$

and then (5) *et seq.* imply that E_0, defined as before, is finite. In short, in one dimension a 'potential' can be a bounded measure plus an $L^\infty(\mathbb{R})$-function.

So far we have considered the nonrelativistic kinetic energy $T_\psi = (\psi, p^2\psi)$. Similar inequalities hold for the relativistic case $T_\psi = (\psi, |p|\psi)$. The relativistic analogues of (3)–(5) are (12) and (13) below (see Sects. 8.4 and 8.5). There are constants S'_n for $n \geq 2$ and $S'_{1,p}$ for $2 \leq p < \infty$ such that

$$T_\psi \geq S'_n \|\rho_\psi\|_{n/(n-1)}, \qquad n \geq 2, \tag{12}$$

and $S'_3 = 2^{1/3}\pi^{2/3}$. When $n = 1$,

$$T_\psi + \|\psi\|_2^2 \geq S'_{1,p}\|\rho_\psi\|_p \quad \text{for all } 2 \leq p < \infty, \quad n = 1. \tag{13}$$

The results of this section can be summarized in the following statement. *In all dimensions $n \geq 1$, the hypothesis that V is in the space*

$$\text{nonrelativistic} \quad \begin{cases} L^{n/2}(\mathbb{R}^n) + L^\infty(\mathbb{R}^n), & n \geq 3, \\ L^{1+\varepsilon}(\mathbb{R}^2) + L^\infty(\mathbb{R}^2), & n = 2, \\ L^1(\mathbb{R}^1) + L^\infty(\mathbb{R}^1), & n = 1, \end{cases} \tag{14}$$

$$\text{relativistic} \quad \begin{cases} L^n(\mathbb{R}^n) + L^\infty(\mathbb{R}^n), & n \geq 2, \\ L^{1+\varepsilon}(\mathbb{R}^1) + L^\infty(\mathbb{R}^1), & n = 1, \end{cases} \tag{15}$$

leads to the following two conclusions:

$$E_0 \text{ is finite}, \tag{16}$$

$$T_\psi \leq C\mathcal{E}(\psi) + D\|\psi\|_2^2 \tag{17}$$

when $\psi \in H^1(\mathbb{R}^n)$ (nonrelativistic), or $\psi \in H^{1/2}(\mathbb{R}^n)$ (relativistic), for suitable constants C and D. Furthermore, in the nonrelativistic case in one-dimension, V can be generalized to be a bounded Borel measure.

The existence of minimum energy—or ground state—functions will be proved for the one-body problem under fairly weak assumptions. The principal ingredients are the Sobolev inequality (Theorems 8.3–8.5), and the Rellich–Kondrashov theorem (Theorems 8.7, 8.9). The following definition is convenient:

$$H^\#(\mathbb{R}^n) \text{ denotes } \begin{cases} H^1(\mathbb{R}^n) \text{ in the nonrelativistic case}, \\ H^{1/2}(\mathbb{R}^n) \text{ in the relativistic case.} \end{cases}$$

The main technical result is the following theorem.

11.4 THEOREM (Weak continuity of the potential energy)

Let $V(x)$ be a function on $\mathbb{R}^n$ that satisfies the condition given in 11.3(14) (nonrelativistic case) or 11.3(15) (relativistic case). Assume, in addition, that $V(x)$ vanishes at infinity, i.e.,

$$|\{x : |V(x)| > a\}| < \infty \quad \text{for all } a > 0.$$

If $n = 1$ in the nonrelativistic case, V can be the sum of a bounded Borel measure and an $L^\infty(\mathbb{R})$-function ω that vanishes at infinity. Then V_ψ, defined in 11.2(4), is weakly continuous in $H^\#(\mathbb{R}^n)$, i.e., if $\psi^j \rightharpoonup \psi$ as $j \to \infty$, weakly in $H^\#(\mathbb{R}^n)$, then $V_{\psi^j} \to V_\psi$ as $j \to \infty$.

PROOF. Note that by Theorem 2.12 (uniform boundedness principle) $\|\psi^j\|_{H^\#}$ is uniformly bounded. First, assume that V is a function.

Define V^δ (when V is a function) by

$$V^\delta(x) = \begin{cases} V(x) & \text{if } |V(x)| \leq 1/\delta, \\ 0 & \text{if } |V(x)| \geq 1/\delta, \end{cases}$$

and note that $V - V^\delta$ tends to zero as $\delta \to 0$ (by dominated convergence) in the appropriate $L^p(\mathbb{R}^n)$ norm of 11.3(14), resp. 11.3(15). Since $\|\psi^j\|_{H^\#} \leq t$, Theorems 8.3–8.5 (Sobolev's inequality) imply that

$$\int (V - V^\delta)|\psi^j|^2 < C_\delta,$$

with C_δ independent of j and, moreover, $C_\delta \to 0$ as $\delta \to 0$. Thus, our goal of showing that $V_{\psi^j} \to V_\psi$ as $j \to \infty$ will be achieved if we can prove that $V^\delta_{\psi^j} \to V^\delta_\psi$ as $j \to \infty$ for each $\delta > 0$. If $n = 1$ and V is a measure, then V^δ is simply taken to be V itself.

The problem in showing that $V^\delta_{\psi^j} \to V^\delta_\psi$ as $j \to \infty$ comes from the fact that V^δ is known to vanish at infinity only in the weak sense. Fix δ and define the set

$$A_\varepsilon = \{x : |V^\delta(x)| > \varepsilon\}$$

for $\varepsilon > 0$. By assumption, $|A_\varepsilon| < \infty$. Then

$$V^\delta_{\psi^j} = \int_{A_\varepsilon} V^\delta |\psi^j|^2 + \int_{A^c_\varepsilon} V^\delta |\psi^j|^2. \tag{1}$$

The last term is not greater than $\varepsilon \int |\psi^j|^2 = \varepsilon$ (independent of j), and hence (since ε is arbitrary) it suffices to show that the first term in (1) converges, for a subsequence of ψ^j's, to $\int_{A_\varepsilon} V^\delta |\psi|^2$.

This is accomplished as follows. By Theorem 8.6 (weak convergence implies strong convergence on small sets), on any set of finite measure (that we take to be A_ε) there is a subsequence (which we continue to denote by ψ^j) such that $\psi^j \to \psi$ strongly in $L^r(A_\varepsilon)$. Here $2 \leq r < p$. The reader is invited to check (by writing

$$\left| |\psi^j|^2 - |\psi|^2 \right| \leq |\psi^j - \psi||\psi^j + \psi|)$$

that $|\psi^j|^2 \to |\psi|^2$ strongly in $L^{r/2}(A_\varepsilon)$. Since $V^\delta \in L^\infty(\mathbb{R}^n)$, we have that $V^\delta \in L^s(A_\varepsilon)$ for all $1 \leq s \leq \infty$. Thus, by taking $1/s + 2/r = 1$, our claim is proved. When $n = 1$ we leave it to the reader to check that $\psi^j(x) \to \psi(x)$ uniformly
on bounded intervals in $\mathbb{R}^1$, and hence that the same proof goes through in the nonrelativistic case when V is a bounded measure plus an $L^\infty(\mathbb{R}^1)$-function. ∎

11.5 THEOREM (Existence of a minimizer for E_0)

Let $V(x)$ be a function on $\mathbb{R}^n$ that satisfies the condition given in 11.3(14) (nonrelativistic case) or 11.3(15) (relativistic case). Assume that $V(x)$ vanishes at infinity, i.e.,

$$|\{x : |V(x)| > a\}| < \infty \quad \text{for all } a > 0.$$

When $n = 1$ in the nonrelativistic case V can be the sum of a bounded measure and a function $w \in L^\infty(\mathbb{R})$ that vanishes at infinity. Let $\mathcal{E}(\psi) = T_\psi + V_\psi$ as before and assume that

$$E_0 = \inf\{\mathcal{E}(\psi) : \psi \in H^\#(\mathbb{R}^n), \|\psi\|_2 = 1\} < 0.$$

By 11.3(16), $\mathcal{E}(\psi)$ is bounded from below when $\|\psi\|_2 = 1$.

Our conclusion is that there is a function ψ_0 in $H^\#(\mathbb{R}^n)$ such that $\|\psi_0\|_2 = 1$ and

$$\mathcal{E}(\psi_0) = E_0. \tag{1}$$

(We shall see in Sect. 11.8 that ψ_0 is unique up to a factor and can be chosen to be positive.) Furthermore, any minimizer ψ_0 satisfies the Schrödinger equation in the sense of distributions:

$$H_0\,\psi_0 + V\psi_0 = E_0\psi_0, \qquad (2)$$

where $H_0 = -\Delta$ (nonrelativistic) and $H_0 = (-\Delta + m^2)^{1/2} - m$ (relativistic). Note that (2) implies that the function $V\psi_0$ is also a distribution; this implies that $V\psi_0 \in L^1_{\text{loc}}(\mathbb{R}^n)$.

REMARKS. (1) From (2) we see that the distribution $(H_0+V)\psi_0$ is always a function (namely $E_0\psi_0$). This is true in the nonrelativistic case when $n=1$, even when V is a measure!

(2) Theorem 11.5 states that a minimizer satisfies the Schrödinger equation (2). Suppose, on the other hand, that ψ is some function in $H^\#(\mathbb{R}^n)$ that satisfies (2) in $\mathcal{D}'$, but with E_0 replaced by some real number E. Can we conclude that $E \geq E_0$ and, moreover, that $E = E_0$ if and only if ψ is a minimizer? The answer is *yes* and we invite the reader to prove this by taking a sequence $\phi^j \in C_0^\infty(\mathbb{R}^n)$ that converges to ψ as $j \to \infty$ and testing (2) with this sequence. By taking the limit $j \to \infty$, one can easily justify the equality $\mathcal{E}(\psi) = E\|\psi\|_2^2$. The stated conclusion follows immediately.

PROOF. Let ψ^j be a minimizing sequence, i.e., $\mathcal{E}(\psi^j) \to E_0$ as $j \to \infty$ and $\|\psi^j\|_2 = 1$. First we note that by 11.3(17) T_{ψ^j} is bounded by a constant independent of j. Since $\|\psi^j\|_2 = 1$, the sequence ψ^j is bounded in $H^\#(\mathbb{R}^n)$. Since bounded sets in $H^{1/2}(\mathbb{R}^n)$ and $H^1(\mathbb{R}^n)$ are weakly sequentially compact (see Sect. 7.18), we can therefore find a function ψ_0 in $H^\#(\mathbb{R}^n)$ and a subsequence (which we continue to denote by ψ^j) such that $\psi^j \rightharpoonup \psi_0$ weakly in $H^\#(\mathbb{R}^n)$. The weak convergence of ψ^j to ψ_0 implies that $\|\psi_0\|_2 \leq 1$. This function ψ_0 will be our minimizer as we shall show. Note that, since the kinetic energy T_ψ is weakly lower semicontinuous (see the end of Sect. 8.2), and since, by Theorem 11.4, V_ψ is weakly continuous in $H^\#(\mathbb{R}^n)$, we have that $\mathcal{E}(\psi)$ is weakly lower semicontinuous on $H^\#(\mathbb{R}^n)$. Hence

$$E_0 = \lim_{j \to \infty} \mathcal{E}(\psi^j) \geq \mathcal{E}(\psi_0)$$

and ψ_0 is a minimizer *provided* we know that $\|\psi_0\|_2 = 1$. By assumption however,

$$0 > E_0 \geq \mathcal{E}(\psi_0) \geq E_0\|\psi_0\|_2^2.$$

The last inequality holds by the definition of E_0 and, since $E_0 < 0$, it follows that $\|\psi_0\|_2 = 1$. This shows the existence of a minimizer.

To prove that ψ_0 satisfies the Schrödinger equation (2) we take any function $f \in C_c^\infty(\mathbb{R}^n)$ and we set $\psi^\varepsilon := \psi_0 + \varepsilon f$ for $\varepsilon \in \mathbb{R}$. The quotient $\mathcal{R}(\varepsilon) = \mathcal{E}(\psi^\varepsilon)/(\psi^\varepsilon, \psi^\varepsilon)$ is clearly the ratio of two second degree polynomials in ε and hence differentiable for small ε. Since its minimum, E_0, occurs (by assumption) at $\varepsilon = 0$, $\mathrm{d}\mathcal{R}(\varepsilon)/\mathrm{d}\varepsilon = 0$ at $\varepsilon = 0$. This yields

$$\frac{\mathrm{d}\mathcal{E}(\psi^\varepsilon)}{\mathrm{d}\varepsilon}\bigg|_{\varepsilon=0} = E_0 \frac{\mathrm{d}(\psi^\varepsilon, \psi^\varepsilon)}{\mathrm{d}\varepsilon}\bigg|_{\varepsilon=0}, \qquad (3)$$

which implies that

$$((H_0 + V)f, \psi_0) = E_0(f, \psi_0) \qquad (4)$$

for $f \in C_c^\infty(\mathbb{R}^n)$ and hence, by the definition of distributions and their derivatives in Chapter 6, equation (2) above is correct. ∎

● The next theorem is an extension of Theorem 11.5 to higher eigenvalues and eigenfunctions. The ground state energy E_0 is the first eigenvalue with ψ_0 as the first eigenfunction. Since $\mathcal{E}(\psi)$ is a quadratic form, we can try to minimize it over ψ in $H^1(\mathbb{R}^n)$ (resp. $H^{1/2}(\mathbb{R}^n)$ in the relativistic case) under the two constraints that ψ is normalized and ψ is orthogonal to ψ_0, i.e.,

$$(\psi, \psi_0) = \int_{\mathbb{R}^n} \overline{\psi(x)} \psi_0(x) \, \mathrm{d}x = 0.$$

This infimum we call E_1, **the second eigenvalue**, and, if it is attained, we call the corresponding minimizer, ψ_1, the **first excited state** or **second eigenfunction**. In a similar fashion we can define the $(k+1)^{th}$ eigenvalue recursively (under the assumption that the first k eigenfunctions $\psi_0, \ldots, \psi_{k-1}$ exist)

$$E_k := \inf\{\mathcal{E}(\psi) : \psi \in H^1(\mathbb{R}^n), \|\psi\|_2 = 1 \text{ and } (\psi, \psi_i) = 0, \; i = 0, \ldots, k-1\}.$$

$H^1(\mathbb{R}^n)$ has to be replaced by $H^{1/2}(\mathbb{R}^n)$ in the relativistic case.

In the physical context these eigenvalues have an important meaning in that their differences determine the possible frequencies of light emitted by a quantum-mechanical system. Indeed, it was the highly accurate experimental verification of this fact for the case of the hydrogen atom (see Sect. 11.10) that overcame most of the opposition to the radical idea of the quantum theory.

11.6 THEOREM (Higher eigenvalues and eigenfunctions)

Let V be as in Theorem 11.5 and assume that the $(k+1)^{th}$ eigenvalue E_k given above is negative. (This includes the assumption that the first k eigenfunctions exist.) Then the $(k+1)^{th}$ eigenfunction also exists and satisfies the Schrödinger equation

$$(H_0 + V)\psi_k = E_k \psi_k \qquad (1)$$

in the sense of distributions. In other words, the recursion mentioned at the end of the previous section does not stop until energy zero is reached. Furthermore each E_k can have only finite multiplicity, i.e., each number $E_k < 0$ occurs only finitely many times in the list of eigenvalues.

REMARK. There is no general theorem about the existence of a minimizer if $E_k = 0$.

PROOF. The proof of existence of a minimizer ψ_k is basically the same as the one of Theorem 11.5. Take a minimizing sequence $\psi_k^j, j = 1, 2, \ldots$, each of which is orthogonal to the functions $\psi_0, \ldots, \psi_{k-1}$. By passing to a subsequence we can find a weak limit in $H^1(\mathbb{R}^n)$ (resp. $H^{1/2}(\mathbb{R}^n)$ in the relativistic case) which we call ψ_k. As in Theorem 11.4, $\mathcal{E}(\psi_k) = E_k$ and $\|\psi_k\|_2 = 1$. The only thing we have to check is that ψ_k is orthogonal to $\psi_0, \ldots, \psi_{k-1}$. This, however, is a direct consequence of the definition of the weak limit.

The proof of (1) requires a few steps. First, as in the proof of Theorem 11.5, we conclude that the distribution $D := (H_0 + V - E_k)\psi_k$ is a distribution that satisfies $D(f) = 0$ for every $f \in C_c^\infty(\mathbb{R}^n)$ with the property that $(f, \psi_i) = 0$ for all $i = 0, \ldots, k-1$. By Theorem 6.14 (linear dependence of distributions), this implies that

$$D = \sum_{i=0}^{k-1} c_i \psi_i \qquad (2)$$

for some numbers $c_0, \ldots, c_{k-1}$. Our goal is to show that $c_i = 0$ for all i. Formally, this is proved by multiplying (2) by some ψ_j with $j \leq k-1$ and partially integrating to obtain (using the assumed orthogonality)

$$\int_{\mathbb{R}^n} \overline{\nabla \psi_j} \cdot \nabla \psi_k + \int_{\mathbb{R}^n} V \overline{\psi_j} \psi_k = c_j \ . \qquad (3)$$

On the other hand, taking the complex conjugate of (1) for ψ_j and multiplying it by ψ_k yields

$$\int_{\mathbb{R}^n} \overline{\nabla \psi_j} \cdot \nabla \psi_k + \int_{\mathbb{R}^n} V \overline{\psi_j} \psi_k = 0 \ . \qquad (4)$$

The justification of this formal manipulation is left as Exercise 3.

To prove that E_k has finite multiplicity, assume the contrary. This means that $E_k = E_{k+1} = E_{k+2} = \cdots$. By the foregoing there is then an orthonormal sequence $\psi_1, \psi_2, \ldots$ satisfying (1). By 11.3(10) the kinetic energies T_{ψ_j} remain bounded, i.e., $T_{\psi_j} < C$ for some $C > 0$. Since the ψ_j's are orthogonal, they converge weakly to zero in $L^2(\mathbb{R}^n)$, and hence in $H^1(\mathbb{R}^n)$ as well, as $j \to \infty$. But in Theorem 11.4 it was shown that $V_{\psi_j} \to 0$ as $j \to \infty$ and hence $E_k = \lim_{j \to \infty} T_{\psi_j} + V_{\psi_j} \geq 0$, which is a contradiction. ∎

11.7 THEOREM (Regularity of solutions)

Let $\mathcal{B}_1 \subset \mathbb{R}^n$ be an open ball and let u and V be functions in $L^1(\mathcal{B}_1)$ that satisfy

$$-\Delta u + Vu = 0 \quad \text{in } \mathcal{D}'(\mathcal{B}_1). \tag{1}$$

Then the following hold for any ball $\mathcal{B}$ concentric with $\mathcal{B}_1$ and with strictly smaller radius:

(i) $n = 1$: Without any further assumption on V, u is continuously differentiable.

(ii) $n = 2$: Without any further assumptions on V, $u \in L^q(\mathcal{B})$ for all $q < \infty$.

(iii) $n \geq 3$: Without any further assumptions on V, $u \in L^q(\mathcal{B})$ with $q < n/(n-2)$.

(iv) $n \geq 2$: If $V \in L^p(\mathcal{B}_1)$ for $n \geq p > n/2$, then for all $\alpha < 2 - n/p$,

$$|u(x) - u(y)| \leq C|x - y|^\alpha$$

for some constant C and all $x, y \in \mathcal{B}$.

(v) $n \geq 1$: If $V \in L^p(\mathcal{B}_1)$ for $p > n$, then u is continuously differentiable and its first derivatives $\partial_i u$ satisfy

$$|\partial_i u(x) - \partial_i u(y)| \leq C|x - y|^\alpha$$

for all $\alpha < 1 - n/p$, all $x, y \in \mathcal{B}$ and some constant C.

(vi) Let $V \in C^{k,\alpha}(\mathcal{B}_1)$ for some $k \geq 0$ and $0 < \alpha < 1$ (see Remark 2 in Sect. 10.1). Then $u \in C^{k+2,\alpha}(\mathcal{B})$.

PROOF. The assumption (1) implies that $Vu \in L^1_{\text{loc}}(\mathcal{B}_1)$. As explained in Sect. 10.1 regularity questions are purely local. Thus, applying Theorem 10.2(i), statements (i), (ii) and (iii) are readily obtained. To prove (iv) we use the 'bootstrap' argument. If $n = 2$ we know by (ii) that $u \in L^q(\mathcal{B}_2)$ for any $q < \infty$, and hence $Vu \in L^r(\mathcal{B}_2)$ for some $r > n/2$. Here $\mathcal{B} \subset \mathcal{B}_2 \subset \mathcal{B}_1$ and $\mathcal{B}_2$ is concentric with $\mathcal{B}_1$. Then Theorem 10.2(ii) implies that u is Hölder continuous, which shows that in fact $Vu \in L^p(\mathcal{B}_3)$. Again $\mathcal{B} \subset \mathcal{B}_3 \subset \mathcal{B}_2$ and $\mathcal{B}_3$ is concentric with $\mathcal{B}_2$. One more application of Theorem 10.2(ii) yields the result for $n = 2$, since the radii of the balls decrease by an arbitrarily small amount.

If $n \geq 3$, we proceed as follows. Suppose that $Vu \in L^{s_1}(\mathcal{B}_2)$ for some $1 < s_1 < n/2$ and some ball $\mathcal{B}_2$ concentric with $\mathcal{B}_1$ but of smaller radius. By Theorem 10.2(i), $u \in L^t(\mathcal{B}_3)$ for any $t < ns_1/(n - 2s_1)$ and $\mathcal{B}_3$ concentric with $\mathcal{B}_2$ with a smaller radius than that of $\mathcal{B}_2$, but as close as we please. Since $V \in L^p(\mathcal{B}_1)$ for $n/2 < p \leq n$, we can set $1/p = 2/n - \varepsilon$ with $0 < \varepsilon \leq 1/n$. By Hölder's inequality $Vu \in L^{s_2}(\mathcal{B}_3)$ for any $s_2 < s_1/(1 - \varepsilon s_1)$ and thus, in particular, for any $s_2 < s_1/(1 - \varepsilon)$. Iterating this estimate we arrive at the situation where, for some *finite* k, $Vu \in L^{s_k}(\mathcal{B}_{k+1})$, $s_k > n/2$. Then, by Theorem 10.2(ii), u is Hölder continuous. Now $Vu \in L^p(\mathcal{B})$ for some ball concentric with $\mathcal{B}_1$ but of smaller radius, and Theorem 10.2(ii) applied once more yields the result.

In the same fashion, by using Theorem 10.3 in addition, the reader can easily prove (v) and (vi). ∎

11.8 THEOREM (Uniqueness of minimizers)

Assume that $\psi_0 \in H^1(\mathbb{R}^n)$ is a minimizer for $\mathcal{E}$, i.e., $\mathcal{E}(\psi_0) = E_0 > -\infty$ and $\|\psi_0\|_2 = 1$. The only assumptions we make are that $V \in L^1_{\mathrm{loc}}(\mathbb{R}^n)$ and V is locally bounded from above (not necessarily from below) and, of course, $V|\psi_0|^2$ is summable. Then ψ_0 satisfies the Schrödinger equation 11.2(1) with $E = E_0$. Moreover ψ_0 can be chosen to be a strictly positive function and, most importantly, ψ_0 is the unique minimizer up to a phase.

In the relativistic case the same is true for an $H^{1/2}(\mathbb{R}^n)$ minimizer, but this time we need only assume that V is in $L^1_{\mathrm{loc}}(\mathbb{R}^n)$.

PROOF. Since

$$E_0 = \mathcal{E}(\psi_0) = \int_{\mathbb{R}^n} |\nabla \psi_0|^2 + \int_{\mathbb{R}^n} V(x)|\psi_0(x)|^2$$

and $\psi_0 \in H^1(\mathbb{R}^n)$, we must have that both

$$\int_{\mathbb{R}^n} [V(x)]_+ |\psi_0(x)|^2 \, \mathrm{d}x \quad \text{and} \quad \int_{\mathbb{R}^n} [V(x)]_- |\psi_0(x)|^2 \, \mathrm{d}x$$

are finite. Thus, in particular, $\int_{\mathbb{R}^n} V(x)u_0(x)\phi(x)\,dx$ is finite for every $\phi \in C_c^\infty(\mathbb{R}^n)$. Next, we compute for any $\phi \in C_c^\infty(\mathbb{R}^n)$

$$0 \leq \mathcal{E}(\psi_0 + \varepsilon\phi) - E_0 \|\psi_0 + \varepsilon\phi\|_2^2$$
$$= \mathcal{E}(\psi_0) - E_0 + 2\varepsilon \operatorname{Re} \int [\nabla \psi_0 \overline{\nabla \phi} + (V - E_0)\psi_0 \overline{\phi}]$$
$$+ \varepsilon^2 \int [|\nabla \phi|^2 + (V - E)|\phi|^2].$$

Every term is finite and, since $\mathcal{E}(\psi_0) = E_0$, the last two terms add up to something nonnegative. Since ε is arbitrary and can have any sign, this implies that

$$-\Delta \psi_0 + W\psi_0 = 0 \quad \text{in } \mathcal{D}'(\mathbb{R}^n), \tag{1}$$

where $W := V - E_0$.

Next we note that with $\psi_0 = f + ig$, f and g separately are minimizers. Since, by Theorem 6.17 (derivative of the absolute value), $\mathcal{E}(f) = \mathcal{E}(|f|)$ and $\mathcal{E}(g) = \mathcal{E}(|g|)$, we also have that $\phi_0 = |f| + i|g|$ is a minimizer. By Theorem 7.8 (convexity inequality for gradients) $\mathcal{E}(|\phi_0|) \leq \mathcal{E}(\phi_0)$, and hence there must be equality. The same Theorem 7.8 states that there is equality if and only if $|f| = c|g|$ for some constant c *provided* that either $|f(x)|$ or $|g(x)|$ is strictly positive for all $x \in \mathbb{R}^n$.

Since these functions are minimizers, they satisfy the Schrödinger equation (1) and, since V is locally bounded, so is W. By Theorem 9.10 (lower bounds on Schrödinger 'wave' functions) $|f(x)|$ and $|g(x)|$ are equivalent to strictly positive lower semicontinuous functions $\widetilde{f}$ and $\widetilde{g}$. Thus, up to a fixed sign, $f = \widetilde{f}$ and $g = \widetilde{g}$, and thus $f = cg$ for some constant c, i.e., $\psi_0 = (1 + ic)f$.

The proof for the relativistic case is similar except that the convexity inequality, Theorem 7.13, for the relativistic kinetic energy does not require strict positivity of the function involved. ∎

11.9 COROLLARY (Uniqueness of positive solutions)

Suppose that V is in $L^1_{\text{loc}}(\mathbb{R}^n)$, V is bounded above (uniformly and not just locally) and that $E_0 > -\infty$. Let $\psi \neq 0$ be any nonnegative function with $\|\psi\|_2 = 1$ that is in $H^1(\mathbb{R}^n)$ and satisfies the nonrelativistic Schrödinger equation 11.2(1) in $\mathcal{D}'(\mathbb{R}^n)$ or is in $H^{1/2}(\mathbb{R}^n)$ and satisfies the relativistic Schrödinger equation

$$\left[\sqrt{p^2 + m^2} - m\right]\psi + V\psi = E\psi \quad \text{in } \mathcal{D}'(\mathbb{R}^n). \tag{1}$$

Then $E = E_0$ and ψ is the unique minimizer ψ_0.

PROOF. The main step is to prove that $E = E_0$. The rest will then follow simply from Remark (2) in Sect. 11.5 (existence of a minimizer) and from Theorem 11.8 (uniqueness of minimizers). To prove $E = E_0$, we prove that $E \neq E_0$ implies the orthogonality relation $\int \psi \psi_0 = 0$. (We know that $E \geq E_0$ by Remark (2) in 11.5.) Since ψ_0 is strictly positive and ψ is nonnegative, this orthogonality is impossible.

To prove orthogonality when $E \neq E_0$ in the nonrelativistic case we take the Schrödinger equation for ψ_0, multiply it by ψ, integrate over $\mathbb{R}^n$ and obtain (formally)

$$\int_{\mathbb{R}^n} \nabla \psi \cdot \nabla \psi_0 + \int_{\mathbb{R}^n} (V - E_0) \psi \psi_0 = 0. \tag{2}$$

To justify this we note, from 11.2(1), that the distribution $\Delta \psi$ is a function and hence is in $L^1_{\text{loc}}(\mathbb{R}^n)$. Moreover, since ψ is nonnegative and V is bounded above, $\Delta \psi = f + g$ for some nonpositive functions $f \in L^1_{\text{loc}}(\mathbb{R}^n)$ and $g \in L^2(\mathbb{R}^n)$. Thus (2) follows from Theorem 7.7.

If we interchange ψ and ψ_0, we obtain (2) with E_0 replaced by E. If $E \neq E_0$, this is a contradiction unless $\int \psi \psi_0 = 0$.

The proof in the relativistic case is identical, except for the substitution of 7.15(3) in place of 7.7(2). ■

11.10 EXAMPLE (The hydrogen atom)

The potential V for the hydrogen atom located at the origin in $\mathbb{R}^3$ is

$$V(x) = -|x|^{-1}. \tag{1}$$

A solution to the Schrödinger equation 11.2(1) is found by inspection to be

$$\psi_0(x) = \exp(-\tfrac{1}{2}|x|), \qquad E_0 = -\tfrac{1}{4}. \tag{2}$$

Since ψ_0 is positive, it is the ground state, i.e., the unique minimizer of

$$\mathcal{E}(\psi) = \int_{\mathbb{R}^3} |\nabla \psi|^2 - \int_{\mathbb{R}^3} \frac{1}{|x|} |\psi(x)|^2 \, dx.$$

This fact follows from Corollary 11.9 (uniqueness of positive solutions). It is not obvious and is usually not mentioned in the standard texts on quantum mechanics.

We can note several facts about ψ_0 that are in accord with our previous theorems.

(i) Since V is infinitely differentiable in the complement of the origin, $x = 0$, the solution ψ_0 is also infinitely differentiable in that same region. This result can be seen directly from Theorem 11.7 (regularity of solutions). As a matter of fact, V is real analytic in this region (meaning that it can be expanded in a power series with some nonzero radius of convergence about every point of the region). It is a general fact, borne out by our example, that in this case ψ_0 is also real analytic in this region; this result is due to Morrey and can be found in [Morrey].

(ii) Since V is in $L^p_{\mathrm{loc}}(\mathbb{R}^n)$ for $3 > p > 3/2$, we also conclude from Theorem 11.7 that ψ_0 must be Hölder continuous at the origin, namely

$$|\psi_0(x) - \psi_0(0)| < c|x - y|^\alpha$$

for all exponents $1 > \alpha > 0$. In our example, ψ_0 is slightly better; it is Lipschitz continuous, i.e., we can take $\alpha = 1$.

• We turn now to our second main example of a variational problem—the Thomas–Fermi (TF) problem. See [Lieb–Simon] and [Lieb, 1981]. It goes back to the idea of L. H. Thomas and E. Fermi in 1926 that a large atom, with many electrons, can be approximately modeled by a simple nonlinear problem for a 'charge density' $\rho(x)$. We shall not attempt to derive this approximation from the Schrödinger equation but will content ourselves with stating the mathematical problem.

The potential function $Z/|x|$ that appears in the following can easily be replaced by

$$V(x) := \sum_{j=1}^{K} Z_j |x - R_j|^{-1}$$

with $Z_j > 0$ and $R_j \in \mathbb{R}^3$, but we refrain from doing so in the interest of simplicity.

Unlike our previous tour through the Schrödinger equation, this time we shall leave many steps as an exercise for the reader (who should realize that knowledge does not come without a certain amount of perspiration).

11.11 THE THOMAS–FERMI PROBLEM

TF theory is defined by an energy functional $\mathcal{E}$ on a certain class of *nonnegative* functions ρ on $\mathbb{R}^3$:

$$\mathcal{E}(\rho) := \frac{3}{5}\int_{\mathbb{R}^3} \rho(x)^{5/3}\,\mathrm{d}x - \int_{\mathbb{R}^3} \frac{Z}{|x|}\rho(x)\,\mathrm{d}x + D(\rho,\rho)\,,\tag{1}$$

where $Z > 0$ is a fixed parameter (the charge of the atom's nucleus) and

$$D(\rho,\rho) := \frac{1}{2}\int_{\mathbb{R}^3}\int_{\mathbb{R}^3}\rho(x)\rho(y)|x-y|^{-1}\,\mathrm{d}x\,\mathrm{d}y\tag{2}$$

is the Coulomb energy of a charge density, as given by 9.1(2). The class of admissible functions is

$$\mathcal{C} := \left\{\rho\,:\,\rho\geq 0,\,\int_{\mathbb{R}^3}\rho < \infty,\,\rho\in L^{5/3}(\mathbb{R}^3)\right\}.\tag{3}$$

We leave it as an exercise to show that each term in (1) is well defined and finite when ρ is in the class $\mathcal{C}$.

Our problem is to minimize $\mathcal{E}(\rho)$ under the condition that $\int\rho = N$, where N is any fixed positive number (identified as the 'number' of electrons in the atom). The case $N = Z$ is special and is called the neutral case. We define two subsets of $\mathcal{C}$:

$$\mathcal{C}_N := \mathcal{C}\cap\left\{\rho:\int_{\mathbb{R}^3}\rho = N\right\}\subset \mathcal{C}_{\leq N} := \mathcal{C}\cap\left\{\rho:\int_{\mathbb{R}^3}\rho \leq N\right\}.$$

Corresponding to these two sets are two energies: The 'constrained' energy

$$E(N) = \inf\{\mathcal{E}(\rho)\,:\,\rho\in\mathcal{C}_N\}\,,\tag{4}$$

and the 'unconstrained' energy

$$E_{\leq}(N) = \inf\{\mathcal{E}(\rho)\,:\,\rho\in\mathcal{C}_{\leq N}\}\,.\tag{5}$$

Obviously, $E_{\leq}(N) \leq E(N)$.

The reason for introducing the unconstrained problem will become clear later. A minimizer will not exist for the constrained problem (4) when $N > Z$ (atoms cannot be negatively charged in TF theory!). But a minimizer will always exist for the unconstrained problem. It is often advantageous, in variational problems, to relax a problem in order to get at a minimizer; in fact, we already used this device in the study of the Schrödinger equation. When a minimizer for the constrained problem does exist it will later be seen to be the ρ that is a minimizer for the unconstrained problem.

11.12 THEOREM (Existence of an unconstrained Thomas–Fermi minimizer)

For each $N > 0$ there is a unique minimizing ρ_N for the unconstrained TF problem (5), i.e., $\mathcal{E}(\rho_N) = E_\leq(N)$. The constrained energy $E(N)$ and the unconstrained energy $E_\leq(N)$ are equal. Moreover, $E(N)$ is a convex and nonincreasing function of N.

REMARK. The last sentence of the theorem holds only because our problem is defined on all of $\mathbb{R}^3$. If $\mathbb{R}^3$ were replaced by a bounded subset of $\mathbb{R}^3$, then $E(N)$ would not be a nonincreasing function.

PROOF. It is an exercise to show that $\mathcal{E}(\rho)$ is bounded below on the set $\mathcal{C}_{\leq N}$, so that $E_\leq(N) > -\infty$. Let ρ^1, ρ^2, ... be a minimizing sequence, i.e., $\mathcal{E}(\rho^j) \to E_\leq(N)$. It is a further exercise to show that $\|\rho^j\|_{5/3}$ is also a bounded sequence of numbers. Therefore, by passing to a subsequence we can assume that $\rho^j \rightharpoonup \rho_N$ weakly in $L^{5/3}(\mathbb{R}^3)$ for some $\rho_N \in L^{5/3}(\mathbb{R}^3)$, by Theorem 2.18 (bounded sequences have weak limits). Since ρ_N is the weak limit of the ρ^j, we can infer that $\int \rho_N \leq N$, and hence that $\rho_N \in \mathcal{C}_{\leq N}$. (Reason: If $\int \rho_N > N$, then $\int_B \rho_N > N$ for some sufficiently large ball, B, but this is a contradiction since $\chi_B \in L^{5/2}(\mathbb{R}^3)$.) The first term in $\mathcal{E}(\rho)$ is weakly lower semicontinuous (by Theorem 2.11 (lower semicontinuity of norms)). We also claim that the $D(\rho,\rho)$ term is lower semicontinuous, for the following reason. Since the sequence ρ^j is bounded in $L^1(\mathbb{R}^n)$ as well, the sequence is bounded in $L^{6/5}(\mathbb{R}^n)$, by Hölder's inequality. By passing to a further subsequence we can demand weak convergence in $L^{6/5}(\mathbb{R}^n)$ as well (to the same ρ_N, of course). Using the weak Young inequality of Sect. 4.3 and Theorem 9.8 (positivity properties of the Coulomb energy) it is an exercise to show that $D(\rho, \rho)$ is also weakly lower semicontinuous.

We want to show that the whole functional is weakly lower semicontinuous. We will then have that ρ_N is a minimizer because

$$E_\leq(N) = \lim_{j \to \infty} \mathcal{E}(\rho^j) \geq \mathcal{E}(\rho_N) \geq E_\leq(N) \ .$$

Since the negative term, $-Z \int_{\mathbb{R}^3} |x|^{-1} \rho(x) \, dx$, is obviously upper semicontinuous (because of the minus sign), we have to show that this term is in fact continuous. This is easy to do (compare Theorem 11.4).

To prove that ρ_N is unique we note that the functional $\mathcal{E}(\rho)$ is a *strictly* convex functional of ρ on the convex set $\mathcal{C}_{\leq N}$. (Why?) If there were two different minimizers, ρ^1 and ρ^2, in $\mathcal{C}_{\leq N}$, then $\rho = (\rho^1 + \rho^2)/2$, which is also in $\mathcal{C}_{\leq N}$, has strictly lower energy than $E_\leq(N)$, which is a contradiction.

This reasoning also shows that $E_\leq(N)$ is a convex function. That $E_\leq(N)$ is nonincreasing is a simple consequence of its definition.

As we said above, $E(N) \geq E_\leq(N)$, by definition. To prove the reverse inequality, we can suppose that $\int \rho_N = M < N$, for otherwise the desired conclusion is immediate. Take any nonnegative function $g \in L^{5/3}(\mathbb{R}^3) \cap L^1(\mathbb{R}^3)$ with $\int g = N - M$ and consider, for each $\lambda > 0$, the function $\rho^\lambda(x) := \rho_N(x) + \lambda^3 g(\lambda x)$. As $\lambda \to 0$, $\rho^\lambda \to \rho_N$ strongly in every $L^p(\mathbb{R}^3)$ with $1 < p \leq 5/3$. Therefore, $\mathcal{E}(\rho^\lambda) \to \mathcal{E}(\rho_N)$. On the other hand, $\mathcal{E}(\rho^\lambda) \geq E(N)$, and hence $E(N) \leq E_\leq(N)$. (It is here that we use the fact that our domain is the whole of $\mathbb{R}^3$.) ∎

11.13 THEOREM (Thomas–Fermi equation)

The minimizer of the unconstrained problem, ρ_N, is not the zero function and it satisfies the following equation, in which $\mu \geq 0$ is some constant that depends on N:

$$\rho_N(x)^{2/3} = Z/|x| - \big[|x|^{-1} * \rho_N\big](x) - \mu \quad \text{if } \rho_N(x) > 0 \tag{1a}$$

$$0 \geq Z/|x| - \big[|x|^{-1} * \rho_N\big](x) - \mu \quad \text{if } \rho_N(x) = 0 \,. \tag{1b}$$

REMARK. An equivalent way to write (1) is

$$\rho_N(x)^{2/3} = \bigg[Z/|x| - \big[|x|^{-1} * \rho_N\big](x) - \mu\bigg]_+ \,. \tag{2}$$

PROOF. Clearly, $E_\leq(N)$ is strictly negative because we can easily construct some small ρ for which $\mathcal{E}(\rho) < 0$. This implies that $\rho_N \not\equiv 0$.

For any function $g \in L^{5/3}(\mathbb{R}^3) \cap L^1(\mathbb{R}^3)$ and all $0 \leq t \leq 1$ consider the family of functions

$$\rho_t(x) := \rho_N(x) + t\bigg(g(x) - \bigg[\int g \bigg/ \int \rho_N\bigg]\rho_N(x)\bigg),$$

which are defined since $\rho_N \not\equiv 0$. Clearly, $\int \rho_t = \int \rho_N$, and it is easy to check that $\rho_t(x) \geq 0$ for all $0 \leq t \leq 1$ provided that g satisfies the two conditions: $g(x) \geq -\rho_N(x)/2$ and $\int g \leq \int \rho_N/2$. Define the function $F(t) := \mathcal{E}(\rho_t)$, which certainly has the property that $F(t) \geq E_\leq(N)$ for $0 \leq t \leq 1$. Hence, the derivative, $F'(t)$, if it exists, satisfies $F'(0) \geq 0$. Indeed, the $\int \rho^{5/3}$ term in 11.11(1) is differentiable, by Theorem 2.6 (differentiability of norms). The

Sections 11.13–11.14

second and third terms in 11.11(1) are trivially differentiable, since they are polynomials. Thus, if we define the function

$$W(x) := \rho_N^{2/3}(x) - Z|x|^{-1} + \left[|x|^{-1} * \rho_N\right](x) , \tag{3}$$

and set

$$\mu := -\int_{\mathbb{R}^3} \rho_N(x) W(x) \, \mathrm{d}x \bigg/ \int_{\mathbb{R}^3} \rho_N(x) \, \mathrm{d}x , \tag{4}$$

the condition that $F'(0) \geq 0$ is

$$\int_{\mathbb{R}^3} g(x)[W(x) + \mu] \, \mathrm{d}x \geq 0 \tag{5}$$

for all functions g with the properties stated above.

In particular, (5) holds for all nonnegative functions g with

$$\int_{\mathbb{R}^3} g \leq \frac{1}{2} \int_{\mathbb{R}^3} \rho_N,$$

and hence (5) holds for all nonnegative functions in $L^{5/3}(\mathbb{R}^3) \cap L^1(\mathbb{R}^3)$. From this it follows that $W(x) + \mu \geq 0$ a.e., which yields (1b). From (4) we see that $-\mu$ is the average of W with respect to the measure $\rho_N(x)\,\mathrm{d}x$, and hence the condition $W(x) + \mu \geq 0$ forces us to conclude that $W(x) + \mu = 0$ wherever $\rho_N(x) > 0$; this proves (1a).

The last task is to prove that $\mu \geq 0$. If $\mu < 0$, then (1a) implies that for $|x| > -\mu/Z$, $\rho_N(x)^{2/3}$ equals an $L^6(\mathbb{R}^3)$-function plus a constant function, i.e., $-\mu$. If ρ_N had this property, it could not be in $L^1(\mathbb{R}^3)$. ∎

• The Thomas–Fermi equation 11.13(2) reveals many interesting properties of ρ_N and we refer the reader to [Lieb–Simon] and [Lieb, 1981] for this theory. Here we shall give but one example—using the potential theory of Chapter 9—which demonstrates the relation between ρ_N and the solution of the constrained problem as stated in Sect. 11.11.

11.14 THEOREM (The Thomas–Fermi minimizer)

As before, let ρ_N be the minimizer for the unconstrained problem. Then

$$\int_{\mathbb{R}^3} \rho_N(x) \, \mathrm{d}x = N \quad \text{if } 0 < N \leq Z, \tag{1}$$

$$\rho_N = \rho_Z \quad \text{if } N \geq Z. \tag{2}$$

In particular, (1) implies that ρ_N is the minimizer for the constrained problem when $N \leq Z$. If $N > Z$, there is no minimizer for the constrained problem.

The number μ is 0 if and only if $N \geq Z$ and in this case $\rho_N(x) \equiv \rho_Z(x) > 0$ for all $x \in \mathbb{R}^3$.

*The **Thomas–Fermi potential** defined by*

$$\Phi_N(x) := Z/|x| - \bigl[|x|^{-1} * \rho_N\bigr](x) \tag{3}$$

satisfies $\Phi_N(x) > 0$ for all $x \in \mathbb{R}^3$. Hence, when $\mu = 0$, corresponding to $N = Z$, the TF equation becomes

$$\rho_Z(x)^{2/3} = \Phi_Z(x) . \tag{4}$$

PROOF. We shall start by proving that there is a minimizer for the constrained problem if and only if $\int \rho_N = N$, in which case the minimizer is then obviously ρ_N. If $\int \rho_N = N$, then ρ_N is a minimizer for $E(N)$. If the $E(N)$ problem has a minimizer (call it ρ^N), then $\int \rho^N = N$ and, by the monotonicity statement in Theorem 11.12, ρ^N is a minimizer for the unconstrained problem. Since this minimizer is unique, $\rho^N = \rho_N$.

Now suppose that there is some $M > 0$ for which $M > \int \rho_M =: N_c$ (we shall soon see that $N_c = Z$). By uniqueness, we have that $E(M) = E(N_c)$. Then two statements are true:

a) $\int \rho_N = N_c$ and $\rho_N = \rho_{N_c}$ for all $N \geq N_c$, and

b) $\int \rho_N = N$ for all $N \leq N_c$.

To prove a) suppose that $N \geq N_c$. We shall show that $E(N) = E(N_c)$ (recall that $E(N) = E_{\leq}(N)$), and hence that $\rho_N = \rho_{N_c}$ by uniqueness. Clearly, $E(N) \leq E(N_c)$. If $E(N) < E(N_c)$ and if $N < M$, we have a contradiction with the monotonicity of the function E. If $E(N) < E(N_c)$ and if $N > M$, we have a contradiction with the convexity of the function E. Thus, $E(N) = E(N_c)$ and statement a) is proved. Statement b) follows from a), for suppose that $\int \rho_N =: P < N$. Then the conclusion of a) holds with N_c replaced by P and M replaced by N. Thus, by a), $\int \rho_Q = P$ for all $Q \geq P$. By choosing

$$Q = N_c \geq N > P,$$

we find that $N_c = \int \rho_{N_c} = P$, which is a contradiction.

We have to show that $N_c = Z$, and this will be done in conjunction with showing the nonnegativity of the TF potential.

Let $A = \{x \in \mathbb{R}^3 : \Phi_N(x) < 0\}$. By Lemma 2.20 (convolutions of functions in dual $L^p(\mathbb{R}^n)$-spaces are continuous), Φ_N is continuous away

second and third terms in 11.11(1) are trivially differentiable, since they are polynomials. Thus, if we define the function

$$W(x) := \rho_N^{2/3}(x) - Z|x|^{-1} + \bigl[|x|^{-1} * \rho_N\bigr](x) , \tag{3}$$

and set

$$\mu := -\int_{\mathbb{R}^3} \rho_N(x) W(x) \, \mathrm{d}x \Big/ \int_{\mathbb{R}^3} \rho_N(x) \, \mathrm{d}x , \tag{4}$$

the condition that $F'(0) \geq 0$ is

$$\int_{\mathbb{R}^3} g(x)[W(x) + \mu] \, \mathrm{d}x \geq 0 \tag{5}$$

for all functions g with the properties stated above.

In particular, (5) holds for all nonnegative functions g with

$$\int_{\mathbb{R}^3} g \leq \frac{1}{2} \int_{\mathbb{R}^3} \rho_N ,$$

and hence (5) holds for all nonnegative functions in $L^{5/3}(\mathbb{R}^3) \cap L^1(\mathbb{R}^3)$. From this it follows that $W(x) + \mu \geq 0$ a.e., which yields (1b). From (4) we see that $-\mu$ is the average of W with respect to the measure $\rho_N(x) \, \mathrm{d}x$, and hence the condition $W(x) + \mu \geq 0$ forces us to conclude that $W(x) + \mu = 0$ wherever $\rho_N(x) > 0$; this proves (1a).

The last task is to prove that $\mu \geq 0$. If $\mu < 0$, then (1a) implies that for $|x| > -\mu/Z$, $\rho_N(x)^{2/3}$ equals an $L^6(\mathbb{R}^3)$-function plus a constant function, i.e., $-\mu$. If ρ_N had this property, it could not be in $L^1(\mathbb{R}^3)$. ∎

● The Thomas–Fermi equation 11.13(2) reveals many interesting properties of ρ_N and we refer the reader to [Lieb–Simon] and [Lieb, 1981] for this theory. Here we shall give but one example—using the potential theory of Chapter 9—which demonstrates the relation between ρ_N and the solution of the constrained problem as stated in Sect. 11.11.

11.14 THEOREM (The Thomas–Fermi minimizer)

As before, let ρ_N be the minimizer for the unconstrained problem. Then

$$\int_{\mathbb{R}^3} \rho_N(x) \, \mathrm{d}x = N \quad \text{if } 0 < N \leq Z, \tag{1}$$

$$\rho_N = \rho_Z \quad \text{if } N \geq Z. \tag{2}$$

In particular, (1) implies that ρ_N is the minimizer for the constrained problem when $N \leq Z$. If $N > Z$, there is no minimizer for the constrained problem.

The number μ is 0 if and only if $N \geq Z$ and in this case $\rho_N(x) \equiv \rho_Z(x) > 0$ for all $x \in \mathbb{R}^3$.

The **Thomas–Fermi potential** *defined by*

$$\Phi_N(x) := Z/|x| - \left[|x|^{-1} * \rho_N\right](x) \tag{3}$$

satisfies $\Phi_N(x) > 0$ for all $x \in \mathbb{R}^3$. Hence, when $\mu = 0$, corresponding to $N = Z$, the TF equation becomes

$$\rho_Z(x)^{2/3} = \Phi_Z(x) \ . \tag{4}$$

PROOF. We shall start by proving that there is a minimizer for the constrained problem if and only if $\int \rho_N = N$, in which case the minimizer is then obviously ρ_N. If $\int \rho_N = N$, then ρ_N is a minimizer for $E(N)$. If the $E(N)$ problem has a minimizer (call it ρ^N), then $\int \rho^N = N$ and, by the monotonicity statement in Theorem 11.12, ρ^N is a minimizer for the unconstrained problem. Since this minimizer is unique, $\rho^N = \rho_N$.

Now suppose that there is some $M > 0$ for which $M > \int \rho_M =: N_c$ (we shall soon see that $N_c = Z$). By uniqueness, we have that $E(M) = E(N_c)$. Then two statements are true:

a) $\int \rho_N = N_c$ and $\rho_N = \rho_{N_c}$ for all $N \geq N_c$, and

b) $\int \rho_N = N$ for all $N \leq N_c$.

To prove a) suppose that $N \geq N_c$. We shall show that $E(N) = E(N_c)$ (recall that $E(N) = E_{\leq}(N)$), and hence that $\rho_N = \rho_{N_c}$ by uniqueness. Clearly, $E(N) \leq E(N_c)$. If $E(N) < E(N_c)$ and if $N < M$, we have a contradiction with the monotonicity of the function E. If $E(N) < E(N_c)$ and if $N > M$, we have a contradiction with the convexity of the function E. Thus, $E(N) = E(N_c)$ and statement a) is proved. Statement b) follows from a), for suppose that $\int \rho_N =: P < N$. Then the conclusion of a) holds with N_c replaced by P and M replaced by N. Thus, by a), $\int \rho_Q = P$ for all $Q \geq P$. By choosing

$$Q = N_c \geq N > P,$$

we find that $N_c = \int \rho_{N_c} = P$, which is a contradiction.

We have to show that $N_c = Z$, and this will be done in conjunction with showing the nonnegativity of the TF potential.

Let $A = \{x \in \mathbb{R}^3 : \Phi_N(x) < 0\}$. By Lemma 2.20 (convolutions of functions in dual $L^p(\mathbb{R}^n)$-spaces are continuous), Φ_N is continuous away from $x = 0$ and vanishes uniformly as $|x| \to \infty$. (Why?) Hence A is an open set. In some small neighborhood of $x = 0$ $\Phi_N(x)$ is clearly positive (again using Lemma 2.20), so $0 \notin A$. From the TF equation (with $\mu \geq 0$), we see that $\rho_N(x) = 0$ for $x \in A$. But

$$\Delta \Phi_N = 4\pi \rho_N = 0 \quad \text{in} \quad A,$$

and Theorem 9.3 tells us that Φ_N is harmonic in A. Since Φ_N is continuous, Φ_N vanishes on the boundary of A. Since Φ_N also vanishes uniformly at ∞, the strong maximum principle, Theorem 9.4, states that $\Phi_N(x) \equiv 0$ for $x \in A$. Thus, A is empty, as claimed. We leave the proof that Φ_N is strictly positive as an exercise.

Let $N > Z$ and consider the unconstrained optimizer ρ_N. We claim that $\int \rho_N \leq Z$. By the fact that ρ_N is a radial function we get from equation 9.7(5) (Newton's theorem), that

$$\left[|x|^{-1} * \rho_N\right](x) = |x|^{-1} \int_{|y| \leq |x|} \rho_N(x)\, \mathrm{d}x + \int_{|y| > |x|} |y|^{-1} \rho_N(x)\, \mathrm{d}x \ .$$

From this and the definition of Φ_N it follows easily that $\lim_{|x| \to \infty} |x| \Phi_N(x) = Z - \int \rho_N$. Hence $\int \rho_N \leq Z$, for otherwise it would contradict the positivity of Φ_N. Thus, for $N > Z$ the constrained TF problem does not have a minimizer and we conclude that $N_c \leq Z$.

Because $E(N_c)$ is the absolute minimum of $\mathcal{E}(\rho)$ on $\mathcal{C}$, and because ρ_{N_c} is the absolute minimizer, a proof analogous to that of Theorem 11.13 (indeed, an even simpler proof), shows that this ρ_{N_c} satisfies the TF equation with $\mu = 0$. Since Φ_N is nonnegative, this is equation (4) with ρ_Z replaced by ρ_{N_c}. We have seen that $\Phi_N(x)$ behaves like $(Z - N_c)/|x|$ for large $|x|$. If $N_c < Z$, then, from (4), $\rho_{N_c} \notin L^{5/3}(\mathbb{R}^3)$, which is a contradiction. ∎

11.15 THE CAPACITOR PROBLEM

The following two problems further illustrate some of the ideas developed in this book. The first (Sect. 11.16) has its roots in antiquity while the second (Sect. 11.17) goes back to [Pólya–Szegő], which can be consulted for several other problems of this genre.

The proper definition of the (electrostatic) **capacity**, Cap(A), of a bounded set $A \subset \mathbb{R}^n$ with $n \geq 3$ is a subtle matter, so let us begin with a heuristic discussion. There are several approaches, and four will be discussed here. The fourth will serve as our final definition, in terms of which a theorem will be formulated in Sect. 11.16. The definition of capacity for sets in 1- and 2-dimensions poses additional problems with which we prefer not to deal.

The first formulation begins by asking the question: How can we spread a unit amount of electric charge over A in such a way as to minimize its Coulomb energy, as given in 9.1(2)? This minimum energy is defined to be $\frac{1}{2}\operatorname{Cap}(A)^{-1}$. Thus,

$$\frac{1}{2\ \operatorname{Cap}(A)} := \inf\left\{\mathcal{E}(\rho)\ :\ \int_A \rho = 1\right\} \tag{1}$$

where

$$\mathcal{E}(\rho) := \frac{1}{2}\int_A\int_A \rho(x)\rho(y)|x-y|^{2-n}\,\mathrm{d}x\,\mathrm{d}y. \tag{2}$$

Thus, a large set has larger capacity than a small one because the charge can be spread out more. It is true, although not obvious, that one can restrict ρ to be nonnegative in (1). In other words, allowing both signs of charge (with unit total charge) can only increase the energy $\mathcal{E}$. It is perfectly correct to take (1) as the definition of capacity, but it has a drawback. A minimizing ρ can be shown to exist if A is a closed set, but it will be a measure, not a function. This measure will be concentrated on the 'surface' of A, and for this reason we cannot expect a minimizer to exist, even as a measure supported in A, if A is not closed. For instance, if A is a ball or a sphere of radius R, then the optimum distribution for the charge will be a 'delta function' of the radius, $|x|$, i.e.,

$$\rho(x) = |\mathbb{S}^{n-1}|^{-1}R^{1-n}\delta(|x|-R)$$

and $\operatorname{Cap}(A) = R^{2-n}$. Thus, in order to prove that a minimizer exists for (1) we will have to extend the class of functions to measures and then take limits in this class. That is, we will have to extend (2) to measures, μ, by defining

$$\mathcal{E}(\mu) := \frac{1}{2}\int_A\int_A |x-y|^{2-n}\mu(\mathrm{d}x)\mu(\mathrm{d}y) \tag{3}$$

with the side condition $\mu(A) = 1$. While this can certainly always be done, and a minimizing measure for (1) can be shown to exist if A is closed, we prefer not to follow this route here because we wish to exploit the machinery we have so far developed for functions, and not yet developed for measures, and because we do not wish to restrict ourselves to closed sets.

The second approach is to define the capacity as the largest charge that can be placed on A so that the potential is at most 1 everywhere on A. (This explains the etymology of the word 'capacity'.) The potential generated by a measure μ is

$$\phi(x) = \int_A |x-y|^{2-n}\mu(\mathrm{d}y), \tag{4}$$

and
$$\operatorname{Cap}(A) = \sup \{\mu(A) : \phi(x) \leq 1, \text{ for all } x \in A\}. \tag{5}$$

The μ that minimizes (1) and the μ that maximizes (5) are the same, in fact. The reason, heuristically at least, is that a minimizer for (1) satisfies an equation similar to the Thomas–Fermi equation 11.13(2) (and for a similar reason), namely

$$[\,|x|^{-1} * \mu](x) = \phi(x) = \lambda \quad \text{for all } x \in A, \tag{6}$$

where λ is some constant. Integration of (6) against $\mu(\mathrm{d}x)$ shows that $\mathcal{E}(\mu) = \lambda/2$. The important point is that a minimizer for the first problem yields a potential that is automatically constant on A, and this potential must be a minimizer for the second problem (because there can be only one solution of (6) with $\lambda = 1$, at least if A has a nonempty connected interior).

The third formulation tries to deal directly with the potential, ϕ, by expressing the energy, $\mathcal{E}(\rho)$, in terms of ϕ. That is, from 6.19(2) and 6.21, $-\Delta\phi = (n-2)|\mathbb{S}^{n-1}|\rho$, and hence

$$\mathcal{E}(\rho) = \frac{1}{2}[\,(n-2)|\mathbb{S}^{n-1}|\,]^{-1} \int_{\mathbb{R}^n} |\nabla\phi(x)|^2 \,\mathrm{d}x.$$

We can then set

$$\operatorname{Cap}(A) = \inf\Big\{[\,(n-2)|\mathbb{S}^{n-1}|\,]^{-1} \int_{\mathbb{R}^n} |\nabla\phi(x)|^2 \,\mathrm{d}x : \\ \phi \in D^1(\mathbb{R}^n) \cap C^0(\mathbb{R}^n) \text{ and } \phi(x) \geq 1 \text{ for all } x \in A\Big\}. \tag{7}$$

This might look a bit odd, at first. Instead of $1/\operatorname{Cap}(A)$ as in (1) we have $\operatorname{Cap}(A)$ here. We also have $\phi(x) \geq 1$ here instead of $\phi(x) \leq 1$, as in (3). The difference arises, of course, from the fact that in one case the total charge is fixed, whereas in the other the potential is fixed. The reader is urged to work these relations through.

The condition in (7) that ϕ must be continuous is crucial in many cases. For example, without continuity definition (7) would give zero capacity for a set of zero measure (because a $D^1(\mathbb{R}^n)$-function can be set equal to zero on a set of Lebesgue measure zero without changing the function in the $D^1(\mathbb{R}^n)$ sense), but this is certainly *not* in accordance with the notion of capacity in (1). Indeed, it is a simple exercise to prove that a ball and a sphere of equal radii have the same capacity. Another easy exercise leads to the conclusion that *a set of zero capacity always has zero measure.*

On the other hand, if we include the requirement of continuity, as in (7), then we see that the capacity of a set A and its closure $\overline{A}$ are the same.

This 'conclusion' does not agree with the capacities obtained with the first formulation, (1), which we regard as the most physical and fundamental. An amusing and easy exercise is to construct a set in which $\text{Cap}(A) \neq \text{Cap}(\overline{A})$ in the sense of (1). Therefore, while (7) looks reasonable, it is really inadequate.

Our fourth and, for the purposes of this book, *actual definition* of $\text{Cap}(A)$ combines the first three in some way, but it always agrees in the end with the first definition (1). We shall first give it and then explain what it has to do with (1).

Definition of capacity:

$$\text{Cap}(A) := \inf \left\{ C_n \int_{\mathbb{R}^n} f^2 \ : \ f \in L^2(\mathbb{R}^n) \right.$$
$$\left. \text{and } [\ |x|^{1-n} * f](x) \geq 1 \text{ for all } x \in A \right\}, \tag{8}$$

where
$$C_n := \pi^{n/2+1} \Gamma((n-2)/2) / \Gamma((n-1)/2)^2.$$

Note that with this definition it is not necessary to assume that A be measurable. Note, also, that $|x|^{1-n} * f \in L^{2n/(n-2)}(\mathbb{R}^n)$ by the HLS inequality, 4.3.

In some sense, (8) is a halfway house between the first and third formulations. To understand it, think of the charge density ρ as being known and think of f as equal to $C_n^{-1} |x|^{1-n} * \rho$. From formulas 5.10(3) and 5.9(1) we have that
$$C_n |x|^{2-n} = |x|^{1-n} * |x|^{1-n}.$$
Thus,
$$2\mathcal{E}(\rho) = C_n \int_{\mathbb{R}^n} f^2, \qquad \phi = |x|^{1-n} * f,$$
and the condition in (8) is the same as the condition in (7). The inverse relation is $f = (\text{const.}) \sqrt{-\Delta} \, \phi$, and not $f = (\text{const.}) |\nabla \phi|$.

The significant difference between (7) and (8) is that it is unnecessary in (8) to specify any continuity. The function $\phi := |x|^{1-n} * f$ cannot be changed in an arbitrary way on a set of measure zero (although f can be changed arbitrarily on such a set). Indeed, a certain amount of continuity will be inherent in ϕ. To see this, note first that the 'inf' in (8) can be taken over nonnegative f without loss of generality because replacing f by $|f|$ does not change $f(x)^2$ but it can only increase $[\ |x|^{1-n} * f](x)$. If f is positive, then ϕ is automatically *lower semicontinuous*, a fact that follows from Fatou's lemma, i.e., if $x_j \mapsto x \in \mathbb{R}^n$, then $|x_j - y|^{1-n} \mapsto |x - y|^{1-n}$ pointwise everywhere.

Lower semicontinuity can actually occur. The minimizer, ϕ, found in Theorem 11.16, is continuous in 'decent' cases, but it can sometimes be only lower semicontinuous. An example of this occurs at the tip of what is known as 'Lebesgue's needle'.

Although (7) is not generally correct as it stands, it can be made correct by demanding that ϕ only be lower semicontinuous rather than $\phi \in C^0(\mathbb{R}^n)$, as in (7). It is an exercise to prove that then (7) will agree with (8) and (1). However, the imposition of lower semicontinuity rather than continuity in (7) might be seen as somewhat artificial.

We wish to address the question of the existence of a minimizing f for (8). Note the obvious fact that the definition of the capacity of a set is independent of the existence of a minimizer. In 'decent' cases there will be a minimizer, but exceptions can occur. As an example, a single point x_0 has zero capacity, cf. Exercise 12, but there is no f with $\int f^2 = 0$ and $[\,|x|^{1-n} * f\,](x_0) \geq 1$. What is true is that there always exists an f that minimizes $\int f^2$ but satisfies the slightly weaker condition that $\phi(x) = [\,|x|^{1-n} * f\,](x) \geq 1$ everywhere on A *except* for a set of zero *capacity* (which necessarily has zero measure). In the case of the single point, the zero function is the minimizer in the foregoing sense.

With these preparations behind us, we are now ready to state our main result precisely.

11.16 THEOREM (Solution of the capacitor problem)

For any bounded set $A \subset \mathbb{R}^n, n \geq 3$, there exists a unique $f \in L^2(\mathbb{R}^n)$ that satisfies the following two conditions:
  a) $\mathrm{Cap}(A) = \int_{\mathbb{R}^n} f^2$.
  b) $\phi := |x|^{1-n} * f$ *satisfies* $\phi(x) \geq 1$ *for all* $x \in A \sim B$, *where B is some (possibly empty) subset of A with* $\mathrm{Cap}(B) = 0$.

This function satisfies $0 \leq \phi(x) \leq 1$ everywhere (in particular, $\phi(x) = 1$ on $A \sim B$) and has the following additional properties:
  c) ϕ *is superharmonic on $\mathbb{R}^n$, i.e., $\Delta\phi \leq 0$.*
  d) ϕ *is harmonic outside of $\overline{A}$, the closure of A, i.e., $\Delta\phi = 0$ in $\overline{A}^c$.*
  e) $\mathrm{Cap}(A) = [\,(n-2)|\mathbb{S}^{n-1}|\,]^{-1} \int_{\mathbb{R}^n} |\nabla\phi(x)|^2\,\mathrm{d}x$.

REMARK. As stated in Sect. 11.15, f is nonnegative, ϕ is lower semicontinuous and ϕ is in $L^{2n/(n-2)}(\mathbb{R}^n)$. This, together with e) above, says that $\phi \in D^1(\mathbb{R}^n)$.

PROOF. The first goal is to find an f satisfying a) and b). The uniqueness of this f follows immediately from the strict convexity of the map $f \mapsto \int f^2$.

The proof is a bit subtle and it illustrates the usefulness of Mazur's Theorem 2.13 (strongly convergent convex combinations). In order to bring out the force of that theorem we shall begin by trying to follow the method used in the previous examples in this chapter, i.e., taking weak limits and using lower semicontinuity of $\int f^2$. At a certain point we shall reach an impasse from which Theorem 2.13 will rescue us.

We start with a minimizing sequence f^j, $j = 1, 2, 3, \ldots$, i.e.,

$$C_n \int_{\mathbb{R}^n} (f^j)^2 \to \mathrm{Cap}(A)$$

and $\phi^j := |x|^{1-n} * f^j$ satisfies $\phi^j(x) \geq 1$ for all $x \in A$. [Note that there actually exist functions in $L^2(\mathbb{R}^n)$ for which $|x|^{1-n} * f \geq 1$ on A because A is a bounded set.] Since this sequence is bounded in $L^2(\mathbb{R}^n)$, there is an f such that $f^j \rightharpoonup f$ weakly. By lower semicontinuity, $\mathrm{Cap}(A) \geq C_n \int_{\mathbb{R}^n} f^2$, and thus f would be a good candidate for a minimizer *provided* $\phi := |x|^{1-n} * f \geq 1$ on A. This need not be true; indeed it will not be true in cases such as Lebesgue's needle. The problem is that the function $|x|^{1-n}$ is not in $L^2(\mathbb{R}^n)$ and so the weak $L^2(\mathbb{R}^n)$ convergence of f^j to f is insufficient for deducing pointwise properties of ϕ.

Now we introduce Theorem 2.13. Since f^j converges weakly to f, there are convex combinations of the f^j's, which we shall denote by F^j, such that F^j converges *strongly* to f in $L^2(\mathbb{R}^n)$. Thus,

$$\mathrm{Cap}(A) \geq C_n \int_{\mathbb{R}^n} f^2 = C_n \lim_{j \to \infty} \int_{\mathbb{R}^n} (F^j)^2.$$

On the other hand, $C_n \lim_{j \to \infty} \int_{\mathbb{R}^n} (F^j)^2 \geq \mathrm{Cap}(A)$ because each F^j is an admissible function. Therefore,

$$\mathrm{Cap}(A) = C_n \int_{\mathbb{R}^n} f^2. \tag{1}$$

What is needed now is a proof that $\phi = 1$ on A, except for a set of zero capacity. For each $\varepsilon > 0$ define the sets

$$B_\varepsilon = \left\{ x \in A \ : \ \phi(x) \leq 1 - \varepsilon \right\},$$
$$V_\varepsilon^j = \left\{ x \ : \ \left| [\, |x|^{1-n} * F^j \,](x) - [\, |x|^{1-n} * f \,](x) \right| \geq \varepsilon \right\},$$
$$T_\varepsilon^j = \left\{ x \ : \ [\, |x|^{1-n} * \frac{|F^j - f|}{\varepsilon} \,](x) \geq 1 \right\}.$$

Clearly, $B_\varepsilon \subset V_\varepsilon^j \subset T_\varepsilon^j$ for all j, and hence, by the obvious monotonicity of capacity,
$$\operatorname{Cap}(B_\varepsilon) \leq \operatorname{Cap}(V_\varepsilon^j) \leq \operatorname{Cap}(T_\varepsilon^j).$$

However, by definition,
$$\operatorname{Cap}(T_\varepsilon^j) \leq \varepsilon^{-2} \|F^j - f\|_2^2,$$

and this converges to zero as $j \to \infty$. Therefore, $\operatorname{Cap}(B_\varepsilon) = 0$.

If we now define
$$B = \{x \in A \ : \ \phi(x) < 1\},$$

we have that $B \subset \bigcup_{k=1}^\infty B_{1/k}$. But, it is easy to see directly from 11.15(8) that capacity is countably subadditive (cf. Exercise 11), and therefore
$$\operatorname{Cap}(B) \leq \sum_{k=1}^\infty \operatorname{Cap}(B_{1/k}) = 0,$$

as required.

Our next goal is to deduce properties c)–e) of ϕ, as well as $\phi \leq 1$ on A. Item c) is proved as follows. Let η be any *nonnegative* function in $C_c^\infty(\mathbb{R}^n)$, in which case $\Delta \eta$ is also in $C_c^\infty(\mathbb{R}^n)$. For $\varepsilon > 0$ let $f_\varepsilon := f - \varepsilon g$ with $g = |x|^{1-n} * \Delta \eta$. Correspondingly,
$$\phi_\varepsilon := |x|^{1-n} * f_\varepsilon = \phi - \varepsilon(|x|^{1-n} * |x|^{1-n}) * \Delta \eta.$$

(We are using Fubini's theorem here to exchange the order of integration in the repeated convolution.) By Theorem 6.21 (solution of Poisson's equation) and the fact that $|x|^{1-n} * |x|^{1-n} = C_n |x|^{2-n}$ we have that
$$-(|x|^{1-n} * |x|^{1-n}) * \Delta \eta = C_n' \eta,$$

with $C_n' > 0$. Therefore, f_ε is an admissible function for every $\varepsilon > 0$, because $\phi_\varepsilon \geq \phi$. Since f was a minimizer, we can conclude that
$$0 \leq -2\varepsilon \int_{\mathbb{R}^n} fg + \varepsilon^2 \int_{\mathbb{R}^n} g^2.$$

This holds for all $\varepsilon > 0$, so $\int_{\mathbb{R}^n} fg \leq 0$. In other words,
$$0 \geq \int_{\mathbb{R}^n} f|x|^{1-n} * \Delta \eta = \int_{\mathbb{R}^n} \phi \Delta \eta$$

for every nonnegative $\eta \in C_c^\infty(\mathbb{R}^n)$. (Fubini's theorem has been used again.) This means, by definition, that $\Delta \phi \leq 0$ in the distributional sense, and c) is proved.

A similar argument, but now without imposing the condition that $\eta \geq 0$, proves d).

Item e) is left to the reader as an exercise with Fourier transforms.

The proof that $\phi \leq 1$ is a bit involved. Since ϕ is superharmonic, and since ϕ vanishes at infinity, Theorem 9.6 (subharmonic functions are potentials) shows that $\phi = |x|^{2-n} * d\mu$, where μ is a positive measure. Therefore, by Fubini's theorem,

$$|x|^{2-n} * |x|^{1-n} * d\mu = |x|^{1-n} * \phi = C_n |x|^{2-n} * f.$$

Taking the Laplacian of both sides we conclude, by Theorem 6.21, that $f = |x|^{1-n} * d\mu$ as distributions, and hence as functions by Theorem 6.5 (functions are uniquely determined by distributions). We conclude, therefore, that

$$\mathrm{Cap}(A) = C_n \int_{\mathbb{R}^n} f^2 = 2\mathcal{E}(\mu) = C_n \int_{\mathbb{R}^n} \phi \, d\mu.$$

Now, let $\phi_0(x) := \min\{1, \phi(x)\}$, which is also superharmonic. (Why?) Again, by Theorem 9.6, $\phi_0 = |x|^{2-n} * d\mu_0$. Then

$$\int_{\mathbb{R}^n} \phi \, d\mu \geq \int_{\mathbb{R}^n} \phi_0 \, d\mu = \int_{\mathbb{R}^n} \phi \, d\mu_0 \ [\text{by Fubini}] \geq \int_{\mathbb{R}^n} \phi_0 \, d\mu_0.$$

Thus, if we define $f_0 = |x|^{1-n} * d\mu_0$, we see that f_0 satisfies the correct conditions and gives us a lower value for $\mathrm{Cap}(A)$, which is a contradiction unless $\phi = \phi_0$. ∎

• As an application of rearrangements we shall solve the following problem: *Which set has minimal capacity among all bounded sets of fixed measure?* The answer is given in the following theorem.

11.17 THEOREM (Balls have smallest capacity)

Let $A \subset \mathbb{R}^n, n \geq 3$, be a bounded set with Lebesgue measure $|A|$ and let B_A be the ball in $\mathbb{R}^n$ with the same measure. Then

$$\mathrm{Cap}(B_A) \leq \mathrm{Cap}(A).$$

PROOF. Let ϕ be the minimizing potential for Cap(A). Since ϕ is nonnegative and $\phi \in D^1(\mathbb{R}^n)$, the rearrangement inequality for the gradient (Lemma 7.17) yields that $\int_{\mathbb{R}^n} |\nabla \phi^*|^2 \leq \int_{\mathbb{R}^n} |\nabla \phi|^2$, where ϕ^* is the symmetric-decreasing rearrangement of ϕ (see Sect. 3.3). By the equimeasurability of the rearrangement, $\phi^* = 1$ on B_A.

Let ϕ_b denote the potential for the ball problem, B_A. We claim that $\int_{\mathbb{R}^n} |\nabla \phi^*|^2 \geq \int_{\mathbb{R}^n} |\nabla \phi_b|^2$, which will prove the theorem. Both ϕ^* and ϕ_b are radial and decreasing functions. Outside of B_A, we have $\phi_b(r) = (R/r)^{2-n}$, where R is the radius of B_A. (Why?) Now

$$\int_{\mathbb{R}^n} |\nabla \phi^*|^2 = \int_{|x|>R} |\nabla \phi^*|^2 \geq \left\{ \int_{|x|>R} \nabla \phi^* \cdot \nabla \phi_b \right\}^2 \Big/ \int_{|x|>R} |\nabla \phi_b|^2$$

by Schwarz's inequality and the fact that $\phi^*(x) = 1$ for $x \in B_A$. However, with the aid of polar coordinates, we see that $\int_{|x|>R} \nabla \phi^* \cdot \nabla \phi_b$ is proportional to $\int_{r>R} (d\phi^*/dr) \, dr$, which is proportional to $\phi^*(0) = 1$ by the fundamental theorem of calculus for distributional derivatives, Theorem 6.9. (Why is ϕ^* continuous?) In other words, $\int_{\mathbb{R}^n} |\nabla \phi^*|^2$ is bounded below by a quantity that depends only on $\phi^*(0)$, and which is therefore identical to the same quantity with ϕ^* replaced by ϕ_b. ∎

Exercises for Chapter 11

1. Compute the capacity of a ball of radius 1 in $\mathbb{R}^n$ by verifying that $\phi_b(x) = |x|^{2-n}$ as stated in Sect. 11.17. Use c) and d) of Theorem 11.16.

2. Prove that the right side of 11.15(7) is zero in dimensions 1 and 2.

3. Justify the formal manipulation in the proof of Theorem 11.6 by first approximating ψ_j, and then ψ_k, by a sequence of $C_c^\infty(\mathbb{R}^n)$-functions.

4. Referring to Sect. 11.11, prove that all terms in the Thomas–Fermi energy are well defined when $\rho \in \mathcal{C}$.

5. Prove that $\mathcal{E}(\rho)$ is bounded below on the set $\mathcal{C}_{\leq N}$, as stated in the proof of Theorem 11.12.

6. Use the various inequalities in this book to show that $\|\rho^j\|_{5/3}$ is a bounded sequence when ρ^j is a minimizing sequence on $\mathcal{C}_{\leq N}$, as claimed in the proof of Theorem 11.12.

7. Show that $D(\rho,\rho)$ is weakly lower semicontinuous on $L^{6/5}(\mathbb{R}^3)$ as asserted in the proof of Theorem 11.12. That proof seemed to imply that it is necessary to pass to a subsequence of the ρ^j sequence in order to get the $L^{6/5}(\mathbb{R}^3)$ weak limit; this is not so. (Why?)

8. Show that $-\int_{\mathbb{R}^3} Z|x|^{-1} \rho^j(x)\,\mathrm{d}x$ converges to $-\int_{\mathbb{R}^3} Z|x|^{-1} \rho_N(x)\,\mathrm{d}x$ in the proof of Theorem 11.12.

9. Prove that the capacity of a ball and a sphere in $\mathbb{R}^n$ of the same radius have the same capacity.

10. If $\mathrm{Cap}(A) = 0$, then $\mathcal{L}^n(A) = 0$.

11. Prove the **countable subadditivity of capacity**. That is, let $A_1, A_2, \ldots$ be a sequence of bounded subsets of $\mathbb{R}^n$ and assume that

$$\sum_{i=0}^{\infty} \mathrm{Cap}(A_i) < \infty.$$

Set $A := \bigcup_{i=0}^{\infty} A_i$, which is also assumed to be a bounded set. Then

$$\mathrm{Cap}(A) \leq \sum_{i=0}^{\infty} \mathrm{Cap}(A_i).$$

Do not assume here that the A_j are disjoint. Construct a proof that does not use the existence of a minimizing f for 11.15(7).

12. Show that a single point has zero capacity. Hence, by Exercise 11, the capacity of countably many points is zero.

13. Construct a set A for which $\mathrm{Cap}(A) \neq \mathrm{Cap}(\overline{A})$.

14. Complete the proof of item e) in Theorem 11.16 (solution of the capacitor problem).

15. Prove that if we replace the condition $\phi \in C^0(\mathbb{R}^n)$ in (7) by the weaker condition that ϕ need only be lower semicontinuous, then the minimum in (7) will, indeed, be the same as that in (8).

    ▶ *Hints.* Show that there is a minimizer for (7), in the sense of 'up to a set of zero capacity' and that it is superharmonic. An important point will be the verification of countable subadditivity. Verify that this superharmonic function is the one in Theorem 11.16.

List of Symbols

References

Index

List of Symbols

$\mathcal{B}$	Borel sigma-algebra	4
$\mathbb{C}$	Complex numbers	2
$C^k(\Omega)$	k-times differentiable functions on $\Omega \subset \mathbb{R}^n$	3
$C^{k,\alpha}_{\text{loc}}(\Omega)$	functions whose k^{th} derivative is Hölder continuous of order α	222
$C^\infty(\Omega)$	Infinitely differentiable functions on $\Omega \subset \mathbb{R}^n$	3
$C_c(\Omega)$	continuous functions of compact support in $\Omega \subset \mathbb{R}^n$	151
$C_c^\infty(\Omega)$	Infinitely differentiable functions of compact support in $\Omega \subset \mathbb{R}^n$	3
$\mathcal{D}(\Omega)$	Test function space	128
$\mathcal{D}'(\Omega)$	Distributions	128
$D(f,g)$	Coulomb energy	201
$D^1(\mathbb{R}^n)$	Functions that vanish at infinity with gradient in L^2	185
$D^{1/2}(\mathbb{R}^n)$	Functions that vanish at infinity with 1/2-derivative	
D^α	Multi-derivative	131
$e^{t\Delta}(x,y)$	Heat kernel	166
ess supp$\{f\}$	Support of a measurable function	13
$G_y(x)$	Green's function for the Laplacian with source at y	147
$\mathcal{H}$	Hilbert space	65
$H^1(\Omega)$	Sobolev space for 'one derivative'	157

$H_0^1(\Omega)$	Completion of $C_c^\infty(\Omega)$ in the H^1-norm	160
$H^{1/2}(\mathbb{R}^n)$	Sobolev space for 'half of a derivative'	167
$H_A^1(\mathbb{R}^n)$	Sobolev space with magnetic fields	178
$\operatorname{Im} z$	Imaginary part of $z \in \mathbb{C}$	12
j_ε	A typical mollifier	58
$\mathcal{L}^n$	Lebesgue measure on $\mathbb{R}^n$	6
$L^p(\Omega)$	Space of pth power summable functions	35
$L^p(\Omega)^*$	Dual of $L^p(\Omega)$	48
$L_{\text{loc}}^p(\Omega)$	Space of locally p^{th} power summable functions	129
$O(n)$	Orthogonal group	102
p^2	Physicist's notation for $-\Delta$	168
$\mathbb{R}$	Real numbers	2
$\mathbb{R}^+$	Nonnegative real numbers	14
$\mathbb{R}^n$	n-dimensional Euclidean space	2
$\operatorname{Re} z$	Real part of $z \in \mathbb{C}$	12
$S_f(t)$	Level set of a function f	12
$\mathbb{S}^{n-1}$	Unit sphere in $\mathbb{R}^n$	6
$|\mathbb{S}^{n-1}|$	area of $\mathbb{S}^{n-1}$	6
$\operatorname{supp}\{f\}$	Support of a continuous function f	3
$W^{m,p}(\Omega)$	Sobolev spaces	133
$W_{\text{loc}}^{m,p}(\Omega)$	Sobolev spaces (local)	133
$W_0^{1,p}(\Omega)$	Sobolev spaces	196
δ_y	Delta-measure, or 'function', centered at $y \in \mathbb{R}^n$	5, 130
Δ	Laplacian	147
∇	Gradient	131
μ	Measure	5
μ a.e.	Almost everywhere with respect to μ	7
$\partial\Omega$	Boundary of Ω	160
Σ	Sigma-algebra	4
χ_A	Characteristic function of a set A	3
$\chi_{\{f>t\}}$	Characteristic function of the level set $S_f(t)$	15
(Ω, Σ, μ)	Measure space	5
$\varnothing$	Empty set	4
$f_\pm$	Positive (negative) part of the function f	15
f^*	Symmetric decreasing rearrangement of a function	72

List of Symbols

$\langle f \rangle$	Average of a function f	38
$\langle f \rangle_{x,R}$	Average of a function f in a ball	202
$[f]_{x,R}$	Average of a function f over a sphere	202
$\widehat{f}$	Fourier transform of f	115
$f^\vee$	Inverse Fourier transform	120
$\|f\|_{H^1(\Omega)}$	H^1-norm of f	158
$\|f\|_{H^{1/2}(\mathbb{R}^n)}$	$H^{1/2}$-norm of f	167
$\|\cdot\|_p, \|\cdot\|_{L^p}$	L^p-norm	36
$\overline{z}$	Complex conjugate of z	2
$|x|$	Euclidean length of $x \in \mathbb{R}^n$	2
$|\alpha|$	Multi-index magnitude	131
$\overline{A}$	Closure of the set A	3
A^c	Complement of A	4
$M^\perp$	Orthogonal complement of M	66
A^*	Symmetric rearrangement of a set, $A \subset \mathbb{R}^n$	72
$|A|$	Volume (Lebesgue measure) of a set A	6
$\cap$	Intersection	4
$\cup$	Union	4
$\oplus$	Orthogonal sum	66
$A \times B$	Cartesian product, $\{(a,b) : a \in A, b \in B\}$	7
$f * g$	Convolution of f and g	57
$B \sim A$	Complement of A in B	4
(x, y)	Inner product of x and y	65
(a, b)	Open interval in $\mathbb{R}$	3
$[a, b]$	Closed interval in $\mathbb{R}$	3
$\{a : b\}$	Things of type a for which property b holds	3
$\in$	Member of	3
$a := b$	a is defined by b (also $b =: a$)	2
$\subset$	Subset of	4
$\rightharpoonup$	Weak convergence	49
$x \mapsto f(x)$	x is mapped to $f(x)$	3

References

Adams, R. A., *Sobolev spaces*, Academic Press, New York, 1975.

Aizenman, M. and Simon, B., *Brownian motion and Harnack's inequality for Schrödinger operators*, Comm. Pure Appl. Math. **35** (1982), 209–271.

Almgren, F. J. and Lieb, E. H., *Symmetric rearrangement is sometimes continuous*, J. Amer. Math. Soc. **2** (1989), 683–773.

Ball, K., Carlen, E. and Lieb, E. H., *Sharp uniform convexity and smoothness inequalities for trace norms*, Invent. Math. **115** (1994), 463–482.

Babenko, K. I., *An inequality in the theory of Fourier integrals*, Izv. Akad. Nauk SSR, Ser. Mat. **25** (1961), 531–542; English transl. in Amer. Math. Soc. Transl. **44** (1965), no. 2, 115–128.

Beckner, W., *Inequalities in Fourier analysis*, Ann. of Math. **102** (1975), 159–182.

Bliss, G. A., *An integral inequality*, J. London Math. Soc. **5** (1930), 404–406.

Brascamp, H. J. and Lieb, E. H., *Best constants in Young's inequality, its converse, and its generalization to more than three functions*, Adv. in Math. **20** (1976), 151–173.

Brascamp, H. J., Lieb, E. H., and Luttinger, J. M., *A general rearrangement inequality for multiple integrals*, J. Funct. Anal. **17** (1974), 227–237.

Brézis, H., *Analyse fonctionelle: Théorie et applications*, Masson, Paris, 1983.

Brézis, H. and Lieb, E. H., *A relation between pointwise convergence of functions and convergence of functionals*, Proc. Amer. Math. Soc. **88** (1983), 486–490.

Brothers, J. and Ziemer, W. P., *Minimal rearrangements of Sobolev functions*, J. Reine Angew. Math. **384** (1988), 153–179.

Burchard, A., *Cases of equality in the Riesz rearrangement inequality*, Annals of Math. **143** (1996), 499–527.

Carlen, E. and Loss, M., *Extremals of functionals with competing symmetries*, J. Funct. Anal. **88** (1990), 437–456.

Chiarenza, F., Fabes, E., and Garofalo, N., *Harnack's inequality for Schrödinger operators and the continuity of solutions*, Proc. Amer. Math. Soc. **98** (1986), 415–425.

Chiti, G., *Rearrangement of functions and convergence in Orlicz spaces*, Appl. Anal. **9** (1979), 23–27.

Crandall, M. G. and Tartar, L., *Some relations between nonexpansive and order preserving mappings*, Proc. Amer. Math. Soc. **78** (1980), 385–390.

Dubrovin, A., Fomenko, A. T., and Novikov, S. P., *Modern geometry—Methods and applications*, vol. 1, Springer–Verlag, Heidelberg, 1984.

Earnshaw, S., *On the nature of the molecular forces which regulate the constitution of the luminiferous ether*, Trans. Cambridge Philos. Soc. **7** (1842), 97–112.

Erdelyi, A., Magnus, W., Oberhettinger, F., and Tricomi, F. G., *Tables of integral transforms*, vol. 1, McGraw Hill, New York, 1954. See 2.4(5).

Fabes, E. B. and Stroock, D. W., *The L^p integrability of Green's functions and fundamental solutions for elliptic and parabolic equations*, Duke Math. J. **51** (1984), 997–1016.

Gilbarg, D. and Trudinger, N. S., *Elliptic partial differential equations of second order*, Second Edition, Springer–Verlag, Heidelberg, 1983.

Hanner, O., *On the uniform convexity of L^p and l^p*, Ark. Math. **3** (1956), 239–244.

Hardy, G. H. and Littlewood, J. E., *On certain inequalities connected with the calculus of variations*, J. London Math. Soc. **5** (1930), 34–39.

_____, *Some properties of fractional integrals* (1), Math. Z. **27** (1928), 565–606.

Hardy, G. H., Littlewood, J. E., and Pólya, G., *Inequalities*, Cambridge University Press, 1959.

Hausdorff, F., *Eine Ausdehnung des Parsevalschen Satzes über Fourierreihen*, Math. Z. **16** (1923), 163–169.

Hilden, K., *Symmetrization of functions in Sobolev spaces and the isoperimetric inequality*, Manuscripta Math. **18** (1976), 215–235.

Hinz, A. and Kalf, H., *Subsolution estimates and Harnack's inequality for Schrödinger operators*, J. Reine Angew. Math. **404** (1990), 118–134.

Hörmander, L., *The analysis of linear partial differential operators*, second edition, Springer–Verlag, Heidelberg, 1990.

Kato, T., *Schrödinger operators with singular potentials*, Israel J. Math. **13** (1972), 133–148.

Leinfelder, H. and Simader, C. G., *Schrödinger operators with singular magnetic vector potentials*, Math. Z. **176** (1981), 1–19.

Lieb, E. H., *Gaussian kernels have only Gaussian maximizers*, Invent. Math. **102** (1990), 179–208.

_____, *Sharp constants in the Hardy–Littlewood–Sobolev and related inequalities*, Ann. of Math. **18** (1983), 349–374.

_____, *Thomas–Fermi and related theories of atoms and molecules*, Rev. Modern Phys. **53** (1981), 603–641. Errata, **54**, (1982), 311.

Lieb, E. H. and Simon, B., *Thomas–Fermi theory of atoms, molecules and solids*, Adv. in Math. **23** (1977), 22–116.

Mazur, S., *Über konvexe Mengen in linearen normierten Räumen*, Studia Math. **4** (1933), 70–84.

Meyers, N. and Serrin, J., *$H = W$*, Proc. Nat. Acad. Sci. U.S.A. **51** (1964), 1055–1056.

Morrey, C., *Multiple integrals in the calculus of variations*, Springer–Verlag, Heidelberg, 1966.

Newton, I., *Philosphia Naturalis Principia Mathematica* (1687), Book 1, Propositions 71, 76, Transl. A. Motte, revised by F. Cajori, University of California Press, Berkeley, 1934, pp. 193.

Pólya, G. and Szegő, G., *Isoperimetric inequalities in mathematical physics*, Princeton University Press, Princeton, 1951.

Reed, M. and Simon, N., *Methods of Modern Mathematical Physics*, Vols. I and II, Academic Press, New York, 1975.

Riesz, F., *Sur une inégalité intégrale*, J. London Math. Soc. **5** (1930), 162–168.

Rudin, W., *Functional analysis*, second edition, McGraw Hill, New York, 1991.

———, *Real and complex analysis*, third edition, McGraw Hill, New York, 1987.

Schrödinger, E., *Quantisierung als Eigenwertproblem*, Annalen Phys. **79** (1926), 361–376. See also ibid **79** (1926), 489–527, **80** (1926), 437–490, **81** (1926), 109–139.

Schwartz, L., *Théorie des distributions*, Hermann, Paris, 1966.

Simon, B., *Maximal and minimal Schrödinger forms*, J. Operator Theory **1** (1979), 37–47.

Sobolev, S. L., *On a theorem of functional analysis*, Mat. Sb. (N.S.) **4** (1938), 471–479, English transl., A.M.S. Transl. Ser. 2, vol. 34, 1963, pp. 39–68.

Sperner, E., Jr., *Symmetrisierung für Funktionen mehrerer reeller Variablen*, Manuscripta Math. **11** (1974), 159–170.

Stein, E. M., *Singular integrals and differentiability properties of functions*, Princeton University Press, Princeton, 1970.

Stein, E. M. and Weiss, G., *Introduction to Fourier analysis on Euclidean spaces*, Princeton University Press, Princeton, 1971.

Talenti, G., *Best constant in Sobolev inequality*, Ann. Mat. Pura Appl. **110** (1976), 353–372.

Thomson, W., *Demonstration of a fundamental proposition in the mechanical theory of electricity*, Cambridge Math. J. **4** (1845), 223–226.

Titchmarsh, E. C., *A contribution to the theory of Fourier transforms*, Proc. London Math. Soc. (2) **23** (1924), 279–289.

Young, L. C., *Lectures on the calculus of variations and optimal control theory*, Saunders, Philadelphia, 1969.

Young, W. H., *On the determination of the summability of a function by means of its Fourier constants*, Proc. London Math. Soc. (2) **12** (1913), 71–88.

Ziemer, W. P., *Weakly differentiable functions*, Springer–Verlag, Heidelberg, 1989.

Index

A

absolute value, derivative of, 144
additivity, 5
algebra of sets, 9
almost everywhere, 7
   with respect to μ, 7
anticonformal, 103
approximation of L^p-functions by
      C^∞-functions, 58
   of L^p-functions by C_c^∞-functions, 63
   of distributions by C^∞-functions, 139
area of a sphere, 6

B

Banach-Alaoglu theorem, 62
bathtub principle, 28
Bessel inequality, 67
bootstrap process, 223
Borel measure, 5
   sets, 4
boundary, 160
bounded linear functional, 48
   sequences have weak limits, 62
bounded variation, 27

C

$C_c^\infty(\mathbb{R}^n)$ is dense in $H_A^1(\mathbb{R}^n)$, 180
Cantor diagonal argument, 62
capacitor problem, 253
   solution of, 257
   minimal, 260
Carathéodory criterion, 29
Cauchy sequence, 45

chain rule, 142
characteristic functions, 3
closed interval, 3
   sets, 3
closure, 3
compact sets, 3
competing symmetries, 109
complement, 4
completeness of $H^1(\Omega)$, 158
   of L^p, 45
completion, 6
concave, 38
cone property, 197
conformal group, 103
   transformations, 102
connections, 177
continuous linear functional, 48
convergence in $\mathcal{D}$, 128
   in $\mathcal{D}'$, 128
   convex function, 38
   set, 38
convex sets, projection on, 47
convexity inequality for gradients, 163
   for the relativistic kinetic energy, 171
convexity of the norm, 36
convolution, 57
   and distributions, 134
   and Fourier transform, 122
   of a distribution with a C_c^∞-function, 139
   of distributions by C^∞-functions, 138

273

of functions and continuity, 64
Coulomb energy, 201
   positivity properties of, 214
Coulomb potential, 201
countable additivity, 5
counting measure, 34
countable subadditivity of capacity, 262
covariant derivative, 178

D

$D^1(\mathbb{R}^n)$, definition of, 185
$D^{1/2}(\mathbb{R}^n)$, definition of, 185
delta-measure, 5
delta-'function', 130
density of $C^\infty(\Omega)$ in $H^1(\Omega)$, 160
   of $C^\infty(\Omega)$ in $W^{1,p}_{\text{loc}}(\Omega)$, 141
   of $C_c^\infty(\mathbb{R}^n)$ in $H^{1/2}(\mathbb{R}^n)$, 171
derivative of the absolute value, 144
   of distributions, 131
   of distributions and classical derivative, 136
   left and right, 38
diamagnetic inequality, 179
dimension of a Hilbert-space, 68
direct method in the calculus of variations, 231
directional derivative, 45
distribution, definition of, 128
distributional derivative, 131
   gradient, 131
   Laplacian, 147
distributions and convolution, 134
distributions and derivative, 131
   and the fundamental theorem of calculus, 135
   approximation by C^∞-functions, 139
   convergence of, 128
   determined by functions, 130
   linear dependence of, 140
   positivity and measures, 151
dominated convergence theorem, 19
dual of L^p, 48, 55
   index, 37
   space, 128

E

Earnshaw theorem, 208
eigenfunction, 233

eigenfunctions, higher, 242
eigenvalues, 233
   higher, 242
elliptic regularity theory, 222
equimeasurability, 73
essential support, 13
   supremum, 36
Euclidean distance, 2
   space n-dimensional, 2
Euclidean group, 102

F

Fatou lemma, 18
   missing term in, 21
finite cone, 197
first excited state, 241
Fourier characterization of $H^1(\mathbb{R}^n)$, 165
Fourier series, 67
   coefficients, 68
Fourier transform and convolutions, 122
   definition of, 115
   in L^2, 119
   in L^p, 120
   of $|x|^{\alpha-n}$, 122
   of a Gaussian function, 117
   inversion formula, 120
Fubini theorem, 16, 25
function as a distribution, 130
   vanishing at infinity, 72
fundamental theorem of calculus for distributions, 135

G

Gateaux derivative, 45
gauge invariance, 177
Gaussian function, 90, 113
   and Fourier transform, 117
general rearrangement inequality, 85
gradients, convexity inequality for, 163
   vanishing on the inverse of small sets, 146
Gram–Schmidt procedure, 67
Green's function of the Laplacian, 147
   and Poisson equation in $\mathbb{R}^n$, 147
ground state, 233
   energy, 233

Index

H

$H^1(\Omega)$ completeness of, 158
  definition of, 157
  density of $C^\infty(\Omega)$ in, 160
  multiplication by functions in $C^\infty(\Omega)$, 159
$H^1(\mathbb{R}^n)$, Fourier characterization of, 165
H^1 and $W^{1,2}$, 160
$H_0^1(\Omega)$, 160
$H_A^1(\mathbb{R}^n)$, density of $C_c^\infty(\mathbb{R}^n)$ in, 180
H_A^1, 177
  definition of, 178
$H^{1/2}(\mathbb{R}^n)$, density of $C_c^\infty(\mathbb{R}^n)$ in, 171
$H^{1/2}$, definition of, 167
Hahn–Banach theorem, 50
Hanner inequality, 43
Hardy–Littlewood–Sobolev inequality, 98
  conformal invariance of, 106
  sharp version of, 98
harmonic functions, 202
Harnack inequality, 209
Hausdorff–Young inequality, 121
heat kernel, 166
Helly's selection principle, 81, 110
Hessian matrix, 204
Hilbert-space, 65, 159
  separable, 67
hydrogen atom, 246
hypoelliptic, 223
Hölder inequality, 39

I

inequality, Bessel's, 67
  convexity for gradients, 163
  convexity for relativistic kinetic energy, 171
  diamagnetic, 179
  fully generalized Young, 92
  general rearrangement inequality, 85
  Hanner, 43
  Hardy–Littlewood–Sobolev, 98
  Harnack, 209
  Hausdorff–Young, 121
  Hölder's, 39
  Jensen's, 38
  mean value inequality for $\Delta - \mu^2$, 215
  mean value inequality for Laplacian, 203
  Minkowski, 41
  nonexpansivity of rearrangement, 75
  rearrangement, simplest, 74
  Riesz rearrangement inequality, 79
  Riesz rearrangement inequality in one-dimension, 76
  Schwarz, 40
  Sobolev for $W^{m,p}(\Omega)$, 197
  Sobolev for $|p|$, 188
  Sobolev for gradients, 186
  Sobolev in 1 and 2 dimensions, 189
  strict rearrangement inequality, 85
  triangle, 2, 36, 41
  weak Young, 99
  Young, 90
infinitesimal generator of the heat kernel, 167
inner product, 65, 159
  space, 65
inner regularity, 7
integrable functions, 14
  locally, 129
integrals, 12
interior regularity, 222
inversion formula for the Fourier transform, 120
inversion on the unit sphere, 103
isometry, 104, 118

J

Jensen inequality, 38

K

kernel, 140
kinetic energy, 158, 178, 233
  relativistic, 168
  with magnetic field, 178

L

L^2 Fourier transform, 119
L^p-spaces and convolution, 64
  completeness of, 45
  definition of, 35
  dual of, 55
  local, 129
  separability of, 61
L^p and Fourier transform, 120
Laplacian, 147
  Green's function of, 147
  infinitesimal generator of the heat kernel, 167
layer cake representation, 26
Lebesgue measure, 6
level set, 12
linear dependence of distributions, 140
linear functionals, 48

separation property, 50
Lipschitz continuity, 156, 222
locally Hölder continuous, 222
    summable functions, 129
    p^{th}-power summable functions, 129
lower semicontinuity of norms, 51
lower semicontinuous 12, 203

M

magnetic fields, 177
maximum of $W^{1,p}$-functions, 145
maximizers, 90
mean value inequality for $\Delta - \mu^2$, 215
    for Laplacian, 203
measure, 5
    Borel, 5
    counting, 34
    outer, 29
    restriction, 5
    space, 5
    theory, 4
    and distributions, 151
measurable, 4
    functions, 12
    sets, 5, 29
minimum of $W^{1,p}$-functions, 145
minimizers, 90
    existence of, 239
Minkowski inequality, 41
monotone class, 9
    theorem, 9
monotone convergence theorem, 17

N

negative part, 15
Newton theorem, 213
nonexpansivity of rearrangement, 75
norm, 36
    differentiability of, 44
norm closed set, 47
normalization condition, 233
normal vector, 66
null-space, 140

O

open balls, 4
    interval, 3
    sets, 3
optimizer, 90
order preserving, 73
orthogonal, 65
    complement, 66
    group, 102
    sum, 66
orthonormal basis, 67
    set, 66
outer measure, 29, 152
outer regularity, 7

P

parallelogram identity, 43, 65
partial integration for functions in
    $H^1(\mathbb{R}^n)$, 161
Plancherel theorem, 118
points of a set, 4
Poisson kernel, 169
Poisson equation, continuity of solutions, 224
    first differentiability of solutions, 224
    higher differentiability of solutions, 226
    solution of, 149
polarization, 119
positive distributions, 151
positive measure, 5
positive part, 15
potential, 233
potential energy, 233
    domination of the by the kinetic energy, 234
    weak continuity of, 238
product measure, 7, 23
    associativity of, 24
    commutativity of, 24
product sigma-algebras, 7
product space, 7
projection on convex sets, 47
Pythagoras theorem, 65

R

rearrangement inequality, general, 85
    nonexpansivity, 75
    Riesz, 79
    Riesz in one dimension, 76
    simplest, 74
    strict, 85
rearrangement, decreasing of kinetic energy, 174
    nonexpansivity of, 75
    of functions, 72
    of sets, 72
rectangles, 7
relativistic kinetic energy, 168
    convexity inequality for, 171

Rellich-Kondrashov theorem, 198
restriction of a measure, 5
Riemann integrable, 14
Riemann integral, 14
Riemann–Lebesgue lemma, 116
Riesz–Markov representation theorem, 151
Riesz rearrangement inequality, 79
    in one-dimension, 76
Riesz representation theorem, 55

S

scalar product, 159
scaling symmetry, 103
Schrödinger equation,
    existence of minimizer, 239
    lower bound on the wave function, 217
    regularity of solutions, 243
    uniqueness of minimizers, 244
    uniqueness of positive solutions, 245
Schwarz symmetrization, 80
    inequality, 40
second eigenfunction, 241
    eigenvalue, 241
section property, 8
semicontinuous, 12
separability of L^p, 61
sigma-algebra, 4
    smallest, 4
sigma-finiteness, 7
signed measure, 27
Sobolev inequalities for $W^{m,p}(\Omega)$, 197
    inequalities in 1 and 2 dimensions, 189
    inequality for $|p|$, 188
    inequality for gradients, 186
    spaces, 133
spherical charge distributions, and point charges, 212
standard metric on $\mathbb{S}^n$, 105
    volume element on $\mathbb{S}^n$, 105
Steiner symmetrization, 79
stereographic coordinates, 104
    projection, 103
strict rearrangement inequality, 85
strictly convex, 38
strictly positive measurable function, 13
strictly symmetric-decreasing, 73
strong convergence, 45, 129, 133
strong maximum principle, 207

strongly convergent convex combinations, 54
subharmonic functions, 202
    and potentials, 210
summable function, 14
    locally, 129
subspace of a Hilbert-space, 66
    closed, 66
superharmonic functions, 202
support of a continuous function, 3
support plane, 38
symmetric-decreasing rearrangement, 72
    of a function, 72
    of a set, 72
symmetrization, Schwarz, 80
    Steiner, 79

T

tangent plane, 38
test functions (the space $\mathcal{D}(\Omega)$), 128
n Thomas–Fermi energy, 248
    TF equation, 250
    TF minimizer, 249, 251
    TF potential, 252
    TF problem, 248
time independent Schrödinger equation, 233
total energy, 233
translation invariance, 102
triangle inequality, 2, 36, 41

U

uncertainty principles, 184, 235
uniform boundedness principle, 52
uniform convexity, 42
unitary transformation, 119
upper semicontinuous 12, 203
Urysohn lemma, 4, 33

V

vanish at infinity, 72
vector potential, 177

W

$W^{1,2}$ and H^1, 160
$W^{1,p}(\Omega)$, definition of, 132
$W_0^{1,p}(\Omega)$, definition of, 196
$W_{\text{loc}}^{1,p}(\Omega)$, definition of, 132
    density of $C^\infty(\Omega)$ in, 141
weak L^q-space, 98
weak continuity of the potential energy, 234

weak convergence, 48, 129, 133
   implying a.e. convergence, 196
   implying strong convergence, 192
weak derivative, 131
weak limits, 176
   bounded sequences and, 62
weak Young inequality, 99

weakly lower semicontinuous, 51, 232
Weyl's lemma, 219

Y

Young inequality, 90
   fully generalized, 92